MULTIPHASE TECHNOLOGY

This volume consists of papers presented at the 1st North American Conference on Multiphase Technology held in Banff, Canada on 10–11 June 1998.

ACKNOWLEDGEMENTS

The valuable assistance of the Technical Advisory Committee and panel of referees is gratefully acknowledged.

TECHNICAL ADVISORY COMMITTEE

Prof J P Brill	TUFFP, The University of Tulsa, USA
Ms A Cutler	Technology Initiatives Limited
Dr G A Gregory	Neotechnology Consultants Limited, Canada
Mr J A R Miur	BHR Group Limited, UK
Dr M M Sarshar	CALTEC (BHR Group Limited), UK

INTERNATIONAL TECHNICAL ADVISORY COMMITTEE

Prof J P Brill	TUFFP, USA
Mr D Elgie	Canadian Hunter, Canada
Dr G Gregory	Neotechnology Consultants Limited, Canada
Dr E Caetano	Petrobras, Brazil
Dr E Guevara	INTEVEP SA, Venezuela
Prof P Jepson	Ohio University, USA
Dr G Kouba	Chevron Petroleum Technology Company, USA
Mr C Mothersele	Gulf Canada Resources Limited, Canada
Dr C Redus	Texaco, USA
Mr D Richardson	NeoCorr Engineering Limited, Canada
Dr M M Sarshar	CALTEC (BHR Group Limited), UK

CO-SPONSORS

Neotechnology Consultants Limited
The Tulsa University Fluid Flow Projects (TUFFP)

EVENT SPONSORS

Computer Modelling Group Limited
TransCanada PipeLines
VECO Engineering Limited
Neotechnology Consultants Limited

ORGANIZED AND SPONSORED BY

BHR Group Limited
Cranfield
Bedfordshire MK43 0AJ
UK
Tel: +44 (0) 1234 750422
Fax: +44 (0) 1234 750074
EMail: ccox@bhrgroup.co.uk
World Wide Web: http://www.bhrgroup.co.uk/confsite/index.html

1ST NORTH AMERICAN CONFERENCE ON

MULTIPHASE TECHNOLOGY

Technology from the Arctic to the Tropics

Edited by J P Brill and G A Gregory

BHR Group Conference Series
Publication No.31

Papers presented at the *1st North American Conference on Multiphase Technology*, organized and sponsored by BHR Group Limited.
Held in Banff, Canada on 10–11 June 1998.

Professional Engineering Publishing

Professional Engineering Publishing Limited, Bury St Edmunds and London, UK.

bHrGroup

Excellence in Fluid Engineering

ISBN 1 86058 139 0

Printed and bound in Great Britain by Antony Rowe Limited, Chippenham, Wiltshire, UK.

A CIP catalogue record for this book is available from the British Library.

Other Titles in the BHR Group Conference Series:

1 *Offshore Loss Prevention – A Systematic Approach*
2 *Fluid Power –The Future for Hydraulics*
3 *Effective Membrane Processes – New Perspectives*
4 *Multiphase Production*
5 *Bioreactor and Bioprocess Fluid Dynamics*
6 *Slurry Handling and Pipeline Transport – Hydrotransport 12*
7 *Pipe Protection*
8 *Advances in Water and Effluent Treatment*
9 *Fluid Sealing*
10 *Water Pipeline Systems*
11 *Hydraulic Modelling*
12 *Aerodynamics and Ventilation of Vehicle Tunnels*
13 *Jet Cutting Technology – Applications and Opportunities*
14 *Multiphase 95 - Where are We on the 'S' Curve?*
15 *Loss Prevention in the Oil and Gas Industry – Improving Safety in the Low-cost Environment*
16 *Pipeline Protection*
17 *Water and Waste Water Treatment – Towards Cleaner, Leaner Operation*
18 *Process Intensification for the Chemical Industry – Practical Applications of New Technology*
19 *Pressure Surges and Fluid Transients in Pipelines and Open Channels*
20 *Slurry Handling and Pipeline Transport – Hydrotransport 13*
21 *Jetting Technology – Applications and Opportunities*
22 *Profitable Condition Monitoring – Fluids and Machinery Performance Monitoring*
23 *Water Pipeline Systems – Leakage Management, Network Optimization, and Pipeline Rehabiliation Technology*
24 *Multiphase '97 – How Deep? How Far? How Soon?*
25 *Bioreactor & Bioprocess Fluid Dynamics*
26 *Fluid Sealing – Meeting Users' Needs*
27 *Aerodynamics and Ventilation of Tunnels – Developments for the 21st Century*
28 *Process Intensification in Practice – Applications and Opportunities*
29 *Pipeline Protection – New Techniques for the Management of the Pipeline Asset*
30 *Protecting the Environment – in Process Industry*

Related Titles

Title	Author/Editor	ISBN
Fluid Machinery for the Oil, Petrochemical, and Related Industries	IMechE Conference 1996–4	0 85298 994 6
Pumping Sludge and Slurry	IMechE Seminar 1998–1	1 86058 155 2
An Introductory Guide to Industrial Flow	Roger Baker	0 85298 983 0
An Introductory Guide to Flow Measurement	Roger Baker	0 85298 670 X
Subsea Control and Data Acquisition	L Adriaansen, R Phillips, C Rees and J Cattanach	0 85298 993 8
Hydrocyclones '96	D Claxton, L Svarovsky and M Thew	1 86058 032 7

For the full range of titles published by Professional Engineering Publishing contact:

Sales Department
Professional Engineering Publishing Limited
Northgate Avenue
Bury St Edmunds
Suffolk
IP32 6BW
UK
Tel: +44 (0) 1284 724384
Fax: +44 (0) 1284 718692

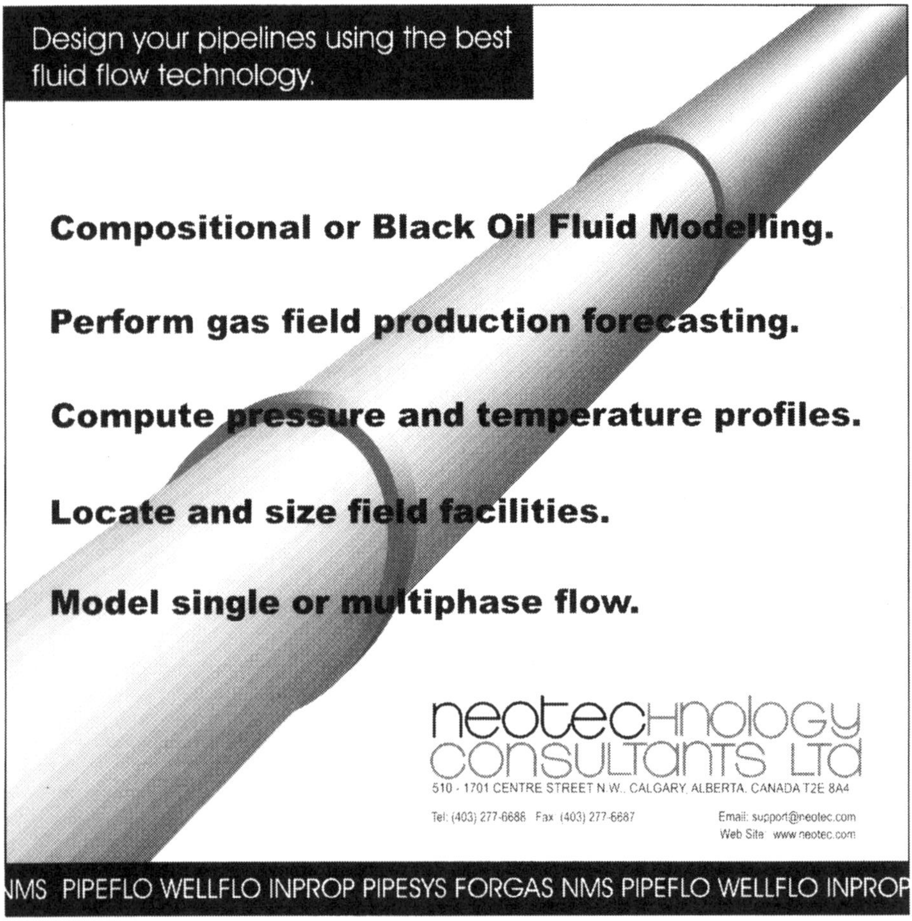
Design your pipelines using the best fluid flow technology.
Compositional or Black Oil Fluid Modelling.
Perform gas field production forecasting.
Compute pressure and temperature profiles.
Locate and size field facilities.
Model single or multiphase flow.
neotechnology consultants Ltd
510 - 1701 CENTRE STREET N.W., CALGARY, ALBERTA, CANADA T2E 8A4
Tel: (403) 277-6686 Fax: (403) 277-6687
Email: support@neotec.com
Web Site: www.neotec.com
NMS PIPEFLO WELLFLO INPROP PIPESYS FORGAS NMS PIPEFLO WELLFLO INPROP

TransCanada

TransCanada PipeLines, a global leader in energy transmission, is pleased to support continued development of our understanding of multiphase flow technology.

VECO Engineering Ltd.
VECO Engineering provides multi-discipline engineering, design, procurement, and construction management services to petroleum and resource development industries.

Contents

1st North American Conference on
MULTIPHASE TECHNOLOGY
Technology from the Arctic to the Tropics
Banff, Canada: 10–11 June 1998

Foreword

Foreword

In the 15 years since the BHR Group (then known as BHRA) sponsored its first (1983) multiphase flow international conference in Coventry, England, there have been many significant advances and improvements in the technology of multiphase systems. Upon reviewing papers contained in the Proceedings of the seven subsequent conferences, it is most interesting to note the steady progression of some areas (e.g. transient and mechanistic models, multiphase pumps) from inception and development stages, through early testing and trials, to case studies and general industry acceptance. The work in these areas continues, however, now with a focus on refinements and broadening of the original range of applications.

Some problems have proven to be very difficult to handle and have remained with us (well.... just how long can a slug be....?)

To the uninitiated, the topic of Multiphase Flow may appear to be rather narrow and restrictive. The Proceedings of the conferences held to date suggest otherwise. They show a steady growth trend in the scope of problems for which multiphase flow can be identified as an important factor. Strong and growing interest can be seen in the areas of corrosion analysis, wax deposition, underbalanced drilling in horizontal wells, metering and separation equipment, three phase flows, and the list goes on.

The BHR multiphase flow conferences have always been important events because they attract attendees from most of the world's active multiphase flow research centres, from the oil and gas producers and other companies that need the technology, and from universities and training centres in which the technology is passed along to the next generation of engineers. On the other hand, they have also all taken place in Europe. While much of the technology is independent of any international bounds, this has tended to give a distinctly European (i.e. North Sea) emphasis to the case studies that are presented.

It is with great pleasure, therefore, that we serve as co-sponsors of the first (and hopefully, not the last) North American BHR Multiphase Flow Conference, being held in beautiful Banff, Alberta. We hope that you find this conference to be challenging and informative, and consistent with the high standards set by those at previous venues. We urge you to participate in the discussions and question periods that follow the paper presentations and share your experiences with the rest of us. We hope also that you enjoy the social arrangements, have ample chance to meet with friends and colleagues, and time to enjoy some of the local spectacular scenery.

James P Brill
Tulsa University Fluid Flow Projects

Garry A Gregory
Neotechnology Consultants Limited

Field Measurements

Multiphase pipeline blowdown simulation and comparison to field data

G SHOUP and **J J XIAO**
Amoco Corporation, Houston, Texas, USA
J O ROMMA
Scandpower, USA

1.0 ABSTRACT

One design strategy for deepwater multiphase pipelines is to insulate so that the flowing fluid temperatures remain above the hydrate formation temperature under normal flow. Continuous hydrate inhibitor injection is then not required. Following an unscheduled shut-down, one method of preventing hydrate plugs upon cooling to ambient temperatures is to blowdown the line before hydrate formation conditions are reached. For pipelines connecting production wells to deepwater floating production systems, the ability to predict the liquid movement during blowdown is critical for determining the remaining hydrostatic head at the sea floor at the end of the blowdown event.

In 1997, Deepstar conducted blowdown field tests using an onshore gas condensate pipeline. Two different choke sizes were tested, generating both slow and rapid depressurization events. The transient multiphase flow simulator OLGA was used to model the blowdown tests. This paper describes the OLGA simulation model and compares the simulation predictions with the field measurements. Good agreement was obtained between field measurements and the simulation predictions.

This paper presents a valuable comparison between simulation and field-scale blowdown tests. The OLGA modeling is discussed in detail, and this discussion will be helpful to others modeling blowdown events.

2.0 INTRODUCTION

Deepwater production system design imposes new and challenging demands on transient simulation codes. Examples are the ability to simulate long risers, flows with solid wax or hydrate phases, thermal analysis of complex bundled pipes, multi-well manifolding of wells or reservoirs with different compositions, and pipeline blowdown. This paper addresses the ability of state-of-the-art transient simulators to accurately model blowdown events.

Accurate blowdown simulation can be important to some deepwater production system designs. One such deepwater system uses an insulated pipeline such that under normal flows the produced fluids arrive at the process facility above the hydrate formation temperature. For this flow situation continuous injection of hydrate inhibitor is not required. However, some method of preventing hydrates must be identified for the case of an unscheduled shut-in of the line. In this scenario the line is shut-in at both the production facility and the pipeline inlet due to process upset conditions. Some action by the operator must now occur before the pipeline cools to the hydrate formation temperature. One option is to blowdown or depressure the line so that the fluids remaining in the line do not form hydrates upon cooling to ambient conditions.

For this strategy to be successful, enough of the liquid phase must be "swept," or removed from the line by the blowdown event so that the resulting hydrostatic head of the remaining fluids does not cause hydrates. The long production risers present in deepwater systems can make removal of the liquid phase difficult. For typical hydrocarbon multiphase fluids the hydrostatic head must be less than 150-200 psia to prevent hydrate formation. Some increase in allowable head can be achieved by taking credit for the salinity of the produced water.

A blowdown event is a time-dependent transient event that requires a transient multiphase simulator for analysis. Several such simulators are available to the industry. The OLGA program marketed by SCANDPOWER is the simulation tool used for this study. Although OLGA predictions have been compared to both field data and experimental data for many years, the comparison to high pressure blowdown field data tests has not been done due to the high cost of such tests.

Because of the importance of accurate blowdown simulations to deepwater production system design, Deepstar decided to undertake blowdown field measurements in 1997. The purpose of the tests was to verify the ability of a state-of-the-art transient simulator to predict the field test results.

3.0 PRODUCTION SYSTEM CONFIGURATION FOR THE BLOWDOWN TESTS

The blowdown field tests were conducted in a 17,319 ft, 4-inch onshore gas-condensate pipeline. The pipeline connects a producing well to a larger diameter gathering line. At the wellhead the produced fluids are heated, separated, and metered, and then the gas and condensate are re-combined and put into the pipeline. All separated water is dumped to a tank. The pipeline is buried approximately five feet underground and is normally operated at line pressures between 800 and 1,000 psig.

For the tests the pipeline was instrumented to measure pressure and temperature at six sites. The pipeline profile and location of the measurement sites are shown in Figure 1. A separator, gas meter and flare line were installed at site 6. Liquids are measured at site 6 by determining levels in a storage tank at atmospheric pressure. A gas orifice meter was located in the flare line just downstream of the separator. The orifice pressure, temperature and differential pressure were used to perform real-time gas flow rate calculations. The pressure and temperature measurement uncertainties are calculated to be 11.25 psig and 1.5 °F.

Pigging was used to determine the liquid volume in the line before and after each blowdown event.

Prior to each blowdown test, the pipeline was operated at a steady-state flowrate of 1.0 MMSCFD. At this rate the flowing inlet and outlet pressures were measured as 747 psig and 700 psig. After steady-state flow is achieved, the line is shut-in for two hours to allow the liquids to drain into the low spots. The pipeline was then blown down to an average pressure of 100 psig. Two choke sizes were tested (0.5 inch and 1.0 inch).

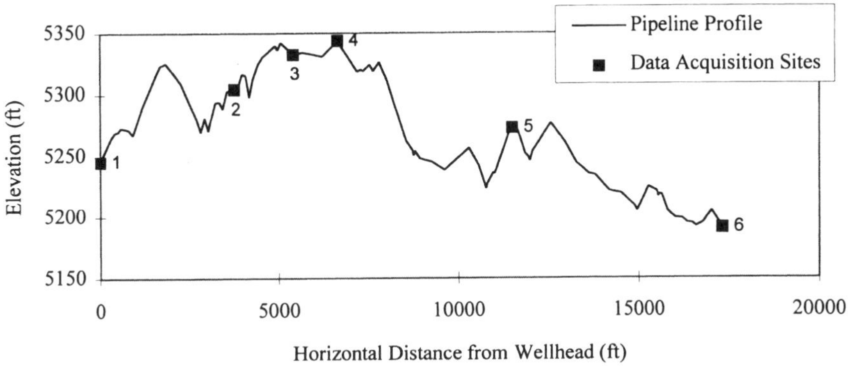

Figure 1 - Pipeline Profile and Data Acquisition Sites

4.0 PIPELINE TRANSIENT MODEL DESCRIPTION

The blowdown event was simulated with OLGA. OLGA is a dynamic, one dimensional, two-fluid model for two-phase hydrocarbon flow simulation in pipelines and pipeline networks. OLGA can also model separator vessels, valves and chokes. OLGA was used to model the coupled pipeline and process facilities at site 6.

OLGA was developed by the SINTEF/IFE two-phase flow project between 1984 - 1992. The theory and development of OLGA are reported by Bendiksen et al.[(1)] OLGA version 3.4 was used in this study.

4.1 Pipeline Data

The 4-inch nominal pipeline has an inner diameter of 3.688 inches. The total length of the pipeline is 17,345 ft and the total volume of the line is 229 BBL. An internal pipe roughness of 0.0018 inches was assumed.

The OLGA pipeline geometry model is as shown in Figure 1. The pipeline is divided into pipe elements of constant cross sectional area and inclination. Each pipe element is subdivided into computation cells using the grid-generator feature of Pre-OLGA. The pipeline model consists of 86 pipe elements and 209 cells.

4.2 Separator and Choke

The receiving facility at site 6 in the OLGA model is illustrated in Figure 2. The separator has an inner diameter of 3.0 ft and a length of 10.0 ft, with a total volume of 12.6 BBL. The assumed gas-liquid separation efficiency is 100% (no liquid carryover to the gas outlet). A separator liquid drain valve with a controller is used to regulate the liquid level in the separator. A 60 ft long, 2.9-inch ID pipe connects the main pipeline to the separator, and a 70 ft long 2.9-inch ID flare line connects the separator to the gas outlet.

A valve is placed just upstream of the separator and a choke is placed 20 ft downstream of the separator in the flare line. The choke is modeled with a standard orifice discharge coefficient of 0.84. The choke orifice has a diameter of 0.5 inch during the slow blowdown, and 1.0 inch during the rapid blowdown.

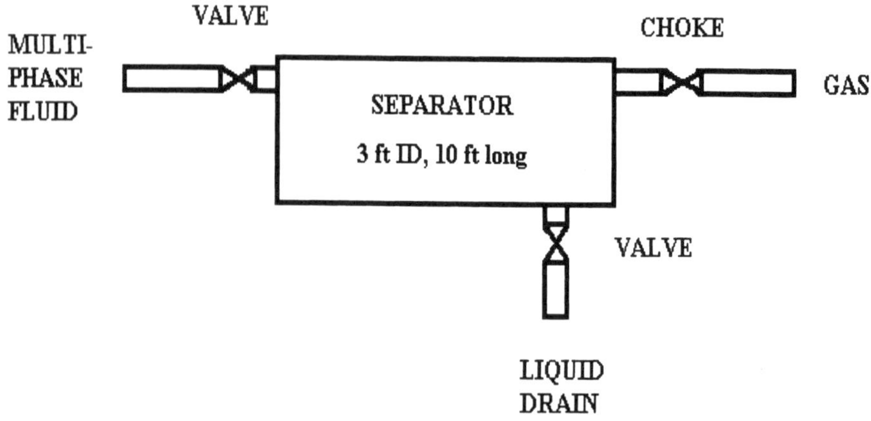

Figure 2 - Pipeline Receiving Facilities at Site 6 (OLGA model)

4.3 Temperature Calculation

The most rigorous heat transfer calculation option available in OLGA is used. The heat transfer calculation consists of the heat transfer between the fluid and the inner pipe wall, the heat conduction in the pipe and the soil, the heat storage in both the pipe and the soil.

The pipeline has a wall thickness of 0.156 inch. A 10.2 ft thick soil ring surrounding the pipe is used to model the 5.0 ft pipeline burial depth. The soil ring is divided into several sub-layers to avoid numerical dispersion in OLGA. The ambient earth temperature is 65 °F. The fluid temperature entering the line is 84 °F. The thermal properties used for the steel pipe and the soil are given in Table 1.

Table 1 - Pipe and Soil Thermal Properties

Material	Density (lb/ft^3)	Conductivity (Btu/hr-ft-°F)	Heat Capacity (Btu/lb-°F)
Steel Pipe	490	28.9	0.119
Soil	100	0.801	0.238

4.4 Fluid Property Data

The fluid flowing in the test line is a gas condensate. At standard conditions the condensate specific gravity is 0.746, the gas gravity is 0.80 relative to air, and the condensate to gas ratio is 11 STB/MMSCF.

A compositional fluid model is used for the OLGA simulations. The compositional PVT fluid data file is generated using PVTOL with the Soave-Redlich-Kwong equation of state. To improve the accuracy of the PVT data, the molecular weight of C25+ for the reservoir fluid is adjusted to match the experimentally measured saturation point (50 °F, 4,200 psia). The calculated saturation pressure at 50 °F is 3,950 psia, which is approximately 6% lower than the experimental value of 4,200 psia. The measured liquid density and viscosity are used to tune the correlations used for these physical properties.

The calculated liquid content (or loading) at wellhead conditions (85 °F and 1,055 psig) and at standard conditions have been determined and compared to both the field data and the experimental measurements in Table 2 and Table 3. The influence of reduced atmospheric pressure at 5000 ft elevation is ignored in this paper (12.2 psia vs. 14.7 psia). Flash calculations show the elevation pressure effect is less than 5% on liquid volume measurement and prediction.

Table 2 - Liquid Loading at Standard Conditions

Method	Liquid Loading at 60 °F &14.7 psia (STBO/MMSCF)
PVTOL Single Step Calculation: 60 °F & 14.7 psia	6.1
PVTOL Two Steps Calculation: 85 °F & 1,055 psig 60 °F & 14.7 psia	10.9
PVTOL Two Steps Calculation: 60 °F & 700 psig 60 °F & 14.7 psia	12.9
Field Average*	18.6

*Average of many wells.

Table 3 - Liquid Loading at Wellhead Separator Conditions

Method	Liquid Loading at 85 °F & 1,055 psig (BBL/MMSCF)
Single Step PVTOL Flash	18.3
Experimental PVT Lab Data	22.1

The field average reported in Table 2 is an average of many wells. This value is determined from a co-mingled flow stream at a central gas processing plant. Table 2 suggests the PVTOL predicted condensate loading from a two step flash is a reasonable approximation to the field average. The particular well producing into the test line has a lower condensate yield than the field average.

4.5 Boundary Conditions

To simulate steady state flow prior to the blowdown event, the pipeline inlet is modeled as a source with a specified total mass flow rate. The standard volumetric gas and liquid flow rates are converted to mass rates by multiplying by the phase densities at standard conditions.

During the blowdown simulations the pipeline inlet is isolated (closed) and atmospheric pressure is applied as an outlet boundary.

5.0 OLGA SIMULATION RESULTS

5.1 Steady State Simulation

The following steps were used to model the test events:

Steady state production at 1.0 MMSCFD was established at a separator pressure of 700 psig. The OLGA simulation was run until the pipeline reached equilibrium holdup and pressure drop. Both the valve upstream of the separator and the choke located in the flare line were fully open with an orifice equal to the 2.9 inches pipe inner diameter.

At 1.0 MMSCFD, OLGA predicts a pipeline inlet pressure of 734 psig and a total pressure drop of 34 psi across the pipeline. The simulations show the fluid temperature decreases from the inlet temperature of 84 °F to the ambient temperature of 65 °F in the first 5000 ft of pipe. The predicted average superficial gas velocity in the pipeline is 2.7 ft/s and the superficial liquid velocity is 0.02 ft/s.

A measured field volume of 35.5 BBL was obtained by pigging the line. This volume was determined by measuring the swept volume dumped into a tank at atmospheric pressure. The 35.5 BBL was adjusted by calculation to 40 - 55 BBL volume to take into account the increase in volume of the insitu condensate caused by the dissolved gas at 700 psig line pressure. The range in the adjusted total insitu liquid holdup is caused by uncertainty in the PVT properties of the condensate.

The liquid holdup measurement obtained by pigging a line can be in error on the low side because pigs can bypass some of the liquid phase. However, since a new, oversized pig was used, it is assumed all liquid was swept by the pig.

OLGA predicts a total liquid holdup of 38.3 BBL, which is lower than the field volume estimate of 40 - 55 BBL. Based on former comparisons against field data, OLGA version 3.4 often underpredicts the total liquid holdup for wet gas pipelines at low to medium gas velocities. It is unknown what percentage of the OLGA underprediction is due to hydrodynamic effects and what percentage is caused by PVT effects.

5.2 Two Hour Pipeline Shut-in

After pipeline equilibrium holdup was reached at 1.0 MMSCFD, the pipeline was shut-in using the following procedure:

- The valve upstream of the separator and the choke located in the flare line downstream of the separator were both closed. The flowrate into the line was set equal to zero at the same time.
- The pipeline was shut-in for 2 hours to allow the liquid to drain to the low spots by gravity. The separator was drained during the shut-in period.
- At the end of the shut-in, the separator liquid drain was closed and the valve upstream of the separator was opened such that the line pressure could be re-established in the separator.

5.3 Pipeline Blowdown

The pipeline was blown down through 0.5-inch and 1.0-inch choke openings. The choke at the flare line downstream of the separator was opened linearly in time over 10 seconds. The flare line outlet was modeled with a 14.7 psia pressure boundary. The inlet of the pipeline was kept closed. The separator liquid drain was closed such that all liquid could be collected in the separator.

In the field test, the pipeline was blown down to an average pipeline pressure of 100 psig. The time required to reach this pressure was (field data) 58 minutes for the 0.5-inch choke and 24 minutes for the 1.0-inch choke. The field test measured the quantity of liquid removed from the pipeline during the blowdown, and the quantity of liquid remaining in the pipeline. The gas flow rate downstream of the separator and the pressure and temperature at five of the six site locations along the pipeline were also measured as a function of time.

OLGA blowdown simulations were conducted and the predictions were compared to the field data. The estimated liquid volume in the pipeline prior to the blowdown, the measured liquid volume exiting the pipeline during the blowdown, and the remaining liquid volume in the line measured by pigging are compared to the OLGA predictions. The results are summarized in Table 4 for the slow blowdown and Table 5 for the rapid blowdown.

The gas flow rate versus time from the field test and the OLGA simulation are plotted in Figure 3 for the slow blowdown and in Figure 7 for the rapid blowdown. The predicted and measured pipeline pressure were compared at site 1 (pipeline inlet), site 3 (midpoint) and site 6 (pipeline outlet). Figures 4 through 6 show the predicted versus measured pressure at the three sites during the slow blowdown, and Figures 8 through 10 show the predicted versus measured pressure during the rapid blowdown.

Table 4 - Field Measurements versus OLGA Predictions, 0.5-Inch Choke Opening - 58 Minutes Blowdown Time

Parameter	Field Test Results	OLGA Simulation Results
Initial Volume of Liquid in the Pipeline	40 - 55 BBLS	38.3 BBLS
Liquid Volume Flowing Out During the Blowdown	6.6 STBO	6.9 STBO
Remaining Liquid Volume in the Pipeline	23.7 STBO	22.9 STBO
Peak Gas Flow Rate	3.3 MMSCFD	3.7 MMSCFD

Table 5 - Field Measurements versus OLGA Predictions, 1.0-Inch Choke Opening - 24 Minutes Blowdown Time

Parameter	Field Test Results	OLGA Simulation Results
Initial Volume of Liquid in the Pipeline	40 - 55 BBLS	38.3 BBLS
Liquid Volume Flowing Out During the Blowdown	9.7 STBO	11.3 STBO
Remaining Liquid Volume in the Pipeline	22 STBO	18 STBO
Peak Gas Flow Rate	10.8 MMSCFD	12.5 MMSCFD

6.0 DISCUSSION

6.1 Slow Blowdown (0.5-Inch Opening)

The OLGA predictions compare well to the field data. The predicted and measured liquid volume flowing out of the pipeline during the blowdown are similar. The predicted and measured remaining liquid volume in the pipeline also compare well. OLGA over predicts the peak gas flow rate at the beginning of the blowdown by 12%.

Figures 3 through 6 show that OLGA slightly under predicts the gas flow rate and the pressure for the main part of the blowdown. This may suggest that more liquid is being vaporized than what is being predicted by OLGA.

At site 5, the initial fluid temperature and lowest fluid temperature measured during the slow blowdown event were 68 °F and 62 °F. The corresponding OLGA predicted temperatures were 65 °F and 63 °F.

A steady state simulation at reduced flow rate was conducted to achieve a steady state initial liquid holdup of 55 BBL prior to the blowdown event. Only minor differences could be seen in the gas flow rate and pressure predictions.

To investigate the sensitivity to the pipe roughness, another OLGA simulation was conducted with an increased roughness from 0.0018 to 0.018 inches. This did not change the simulation results significantly.

6.2 Rapid Blowdown (1.0-Inch Opening)

In general, the OLGA predictions compare well to the field data. The predicted and measured liquid volume recovered from the blowdown compare well. As seen for the slow blowdown, OLGA underpredicts the gas flow rate at the outlet and the pressures at the selected sites. The predicted peak gas flow rate in the beginning of the blowdown is 1.7 MMSCFD or 16% higher than the field value.

At site 5, the initial fluid temperature and lowest fluid temperature measured during the rapid blowdown event were 65 °F and 60 °F. The corresponding OLGA predicted temperatures were 65 °F and 62 °F.

7.0 CONCLUSIONS

1. OLGA does an excellent job in predicting the blowdown events with a minimum of tuning.
2. PVT behavior of the gas condensate provides the most uncertainty in the study. No experimental multistage flash data is available for tuning the PVT model.
3. OLGA blowdown predictions are not sensitive to pipeline internal roughness.

The field data described in this paper adds confidence to OLGA blowdown predictions. Although deepwater pipelines operate at higher pressures and have much greater elevation changes than the field test pipeline, this comparison demonstrates that OLGA is correctly capturing the physics of the blowdown process.

8.0 ACKNOWLEDGMENTS

We wish to thank Deepstar for permission to publish this paper. We also wish to thank Greg Hatton and Andy Barajas of Southwest Research Institute for their efforts on the study.

9.0 REFERENCES

1. Bendiksen, K. H., Malnes, D., Moe, R., and Nuland, S.:" The Dynamic Two-Fluid Model OLGA: Theory and Application," SPE Prod. Eng. (May 1992) 171-180.

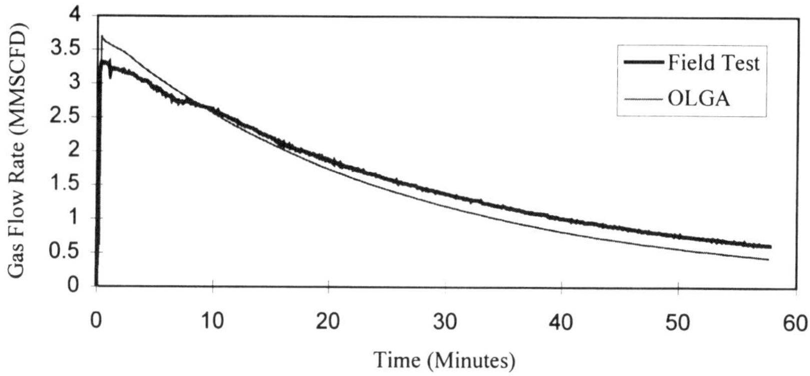

Figure 3 - Slow Blowdown - Gas Flow Rate at Site 6
OLGA Simulations versus Field Test

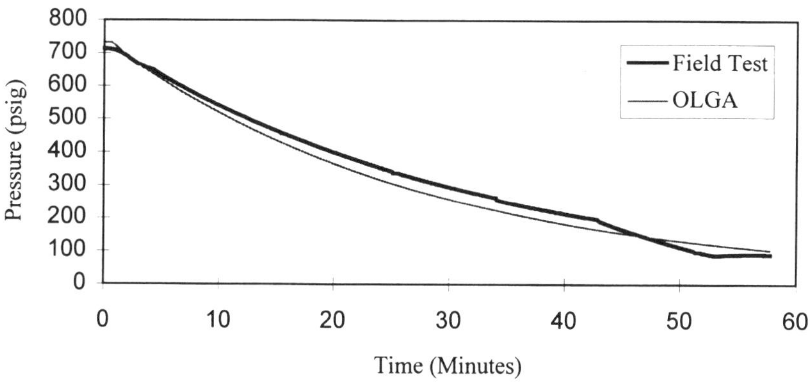

Figure 4 - Slow Blowdown - Pressure at Site 1
OLGA Simulations versus Field Test

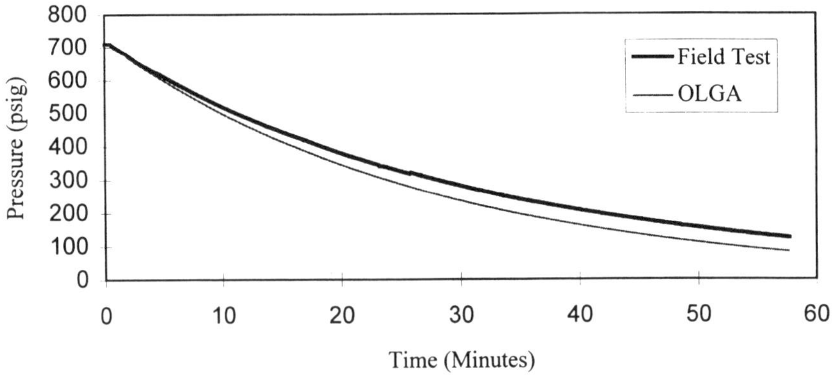

Figure 5 - Slow Blowdown - Pressure at Site 3
OLGA Simulations versus Field Test

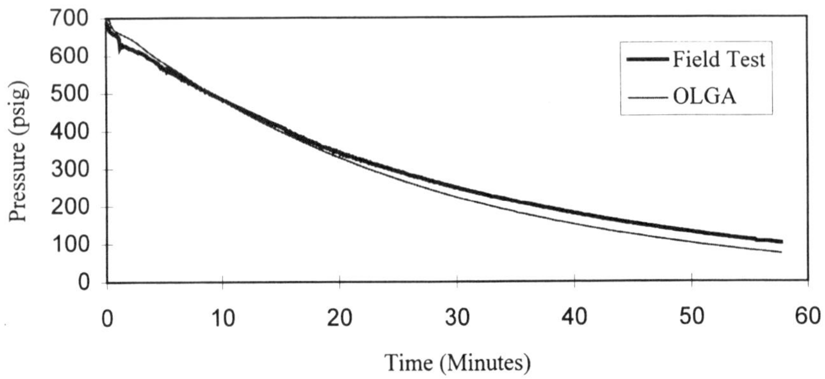

Figure 6 - Slow Blowdown - Pressure at Site 6
OLGA Simulations versus Field Test

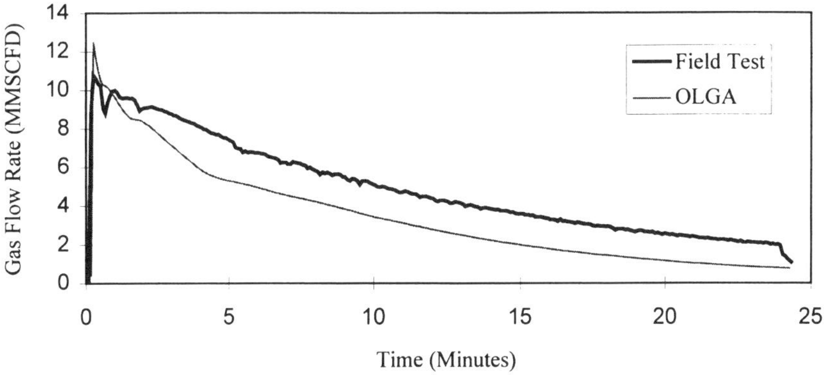

Figure 7 - Rapid Blowdown - Gas Flow Rate at Site 6
OLGA Simulations versus Field Test

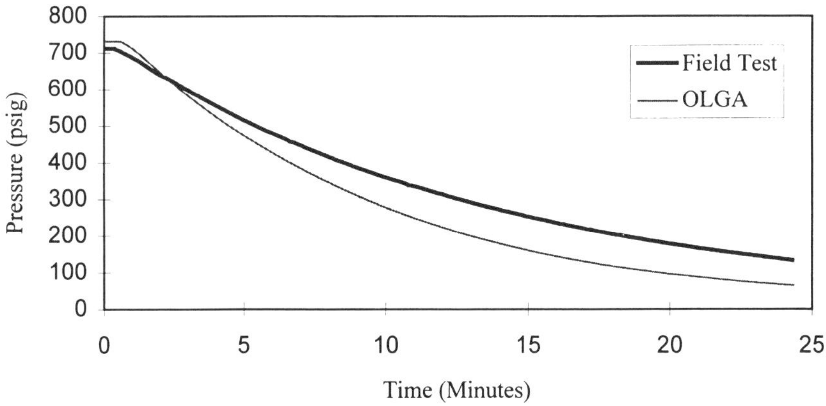

Figure 8 - Rapid Blowdown - Pressure at Site 1
OLGA Simulations versus Field Test

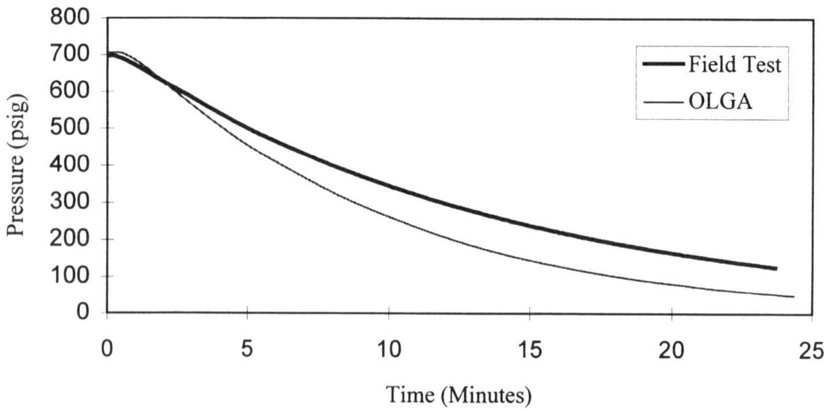

Figure 9 - Rapid Blowdown - Pressure at Site 3
OLGA Simulations versus Field Test

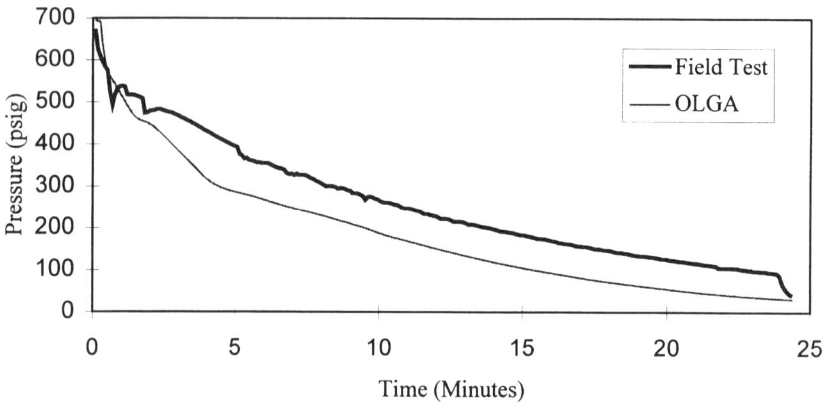

Figure 10 - Rapid Blowdown - Pressure at Site 6
OLGA Simulations versus Field Test

Measurements of gas-condensate field pressures to determine the phase split across a vertical impacting tee

M D TRICK
Neotechnology Consultants Limited, Calgary, Canada
A C MOGENSEN
ACM Consulting Limited, Canada

SYNOPSIS

Multiphase flow technology plays a vital role in providing the pressure loss information necessary for production optimization and forecasting. We need to know how "best" to model the split of phases at the flow junction between twinned pipelines. This paper presents field pressure measurements across a twinned 203 mm gas-condensate pipeline. Those pressure measurements are used to determine the phase and flow rate split across a 254 mm vertical impacting tee.

NOMENCLATURE

a	=	parameter used in Hong and Griston(5) model
A	=	area of pipe
b	=	distance in Hwang(3) model
b	=	parameter used in Hong and Griston(5) model
D	=	inside diameter of pipe
E	=	liquid mass entrainment fraction
F	=	parameter used in Hong and Griston(5) model
m	=	exponent in Hwang(3) model
n	=	exponent in Hwang(3) model
R	=	radius of curvature of streamline
S	=	slip ratio
u	=	velocity
V_L	=	liquid volume fraction
x	=	mass fraction of gas
β	=	angle between gas and liquid streamlines at point of impact

γ	=	angle of velocity triangle
δ	=	depth of zone of influence measured from pipe wall
Θ	=	angle between inlet and branch pipe
μ	=	viscosity
ρ	=	density
Φ	=	angle subtended at the center by the bounding chord of the zone of influence
ω	=	mass flow rate

Subscripts:

D	=	droplet
E	=	equal phase split
eq	=	equivalent
F	=	liquid film
G	=	gas
L	=	liquid
1	=	inlet pipe
2	=	branch pipe
3	=	other branch pipe

INTRODUCTION

Numerous authors(1,2,3,4,5) have observed that when multiphase flow encounters an impacting tee, the split of the phases is equal only if exactly half of the total flow exits each side. An impacting tee is orientated such that the inlet is at right angles to both outlet pipes and each outlet pipe is at 180 degrees to the other. Impacting tees are normally oriented such that both exits are in the same horizontal plane to eliminate the effect of gravity on the split of the fluid phases.

For practical applications involving "wet" steam injection distribution systems, it is important to predict the phase split across impacting tees to determine the steam quality and hence the enthalpy of the fluid injected into the reservoir through each well. In multiphase hydrocarbon transportation systems, it is not uncommon for lines to be "twinned" to increase the cross-sectional area available for flow to minimize pressure loss, or allow flow to take a different route to a terminal point. On uphill sections, twinning lines allows the flexibility to shut down one line when flow rates become lower. This helps to avoid increased hydrostatic pressure losses at lower flow rates. In all of these instances, the flow must be split at a tee to allow flow to enter the second line.

In "simple" looped pipelines where two or more lines of the same length are connected in parallel, the approach taken in most modeling software is to treat the system as a single line that is hydraulically equivalent to the looped system. If the length of that line is the same as the original lines, the *equivalent diameter* is given by equation (1).

$$D_{eq} = \left(D_2^{8/3} + D_3^{8/3}\right)^{3/8} \qquad (1)$$

For multiphase flow, a further assumption is made in that the gas/liquid ratio in each branch is assumed to be the same as in the inlet pipe.

This paper investigates the applicability of the work done to date on phase splitting at impacting tees for a gas-condensate pipeline where a 254 mm diameter pipeline splits vertically at an impacting tee into two, 203 mm horizontal lines. In this particular field, the pipeline is twinned to allow one 203 mm line to be shut off in the future when hydrostatic pressure losses become excessive for the steep uphill rise.

To forecast future gas-condensate production, the pressure loss across the twinned 203 mm lines must be determined as flow rates change. This is one of the major lines entering the Sexsmith gas plant and its pressure loss affects the flow rates of more than 14 upstream wells. This in turn affects the field's development strategy (size of compression, number of wells to drill, etc.). The forecast of surface pressure losses is done with FORGAS(6) while the reservoir is modelled using the three dimensional black oil reservoir simulator, ECLIPSE(7). These two programs communicate their computed bottom-hole flowing pressures and flow rates at each Newton iteration through the Open ECLIPSE(8) interface. For FORGAS/ECLIPSE modelling, an efficient method is required to determine the split of flow rate and phases at the impacting tee because of the large number of pressure loss calculations required over the 14 year forecast period. Figure 1 shows the location of the twinned 203 mm lines in the Sexsmith gathering system.

LITERATURE REVIEW

Various researchers(1,2,3,4,5) have measured the phase split at impacting tees. Figure 2 summarizes the range of conditions investigated. Note that 102 mm diameter lines were the largest sizes analyzed.

Hong(1) determined, using a single flow rate, that the split of phases would be equal when the downstream branch received between 15% and 85% of the total flow. He observed that outside of this range, the branch with the most flow preferentially received almost all of the liquids.

Work by Azzopardi et al(2) showed that the equal split of phases occurred only when exactly half of the flow came out of each branch. The arm with the lower total flow rate received more liquid. They suggested that the small diameter used by Hong(1) and its associated higher surface tension may have been the reason why Hong's and their measurements differed so drastically. Azzopardi, Purvis and Govan derived equations (see Appendix 1) to determine the split of phases at the tee as a function of two mechanisms:

a) gas drags the liquid film to an outlet

b) liquid drops impact at the stagnation surface and the liquid film is forced to one side by the pressure gradient.

Their method was deemed to accurately predict their measured phase splits, especially at low liquid flow rates. However, the measured values deviated from the predicted values at higher liquid rates.

Hwang et al(3) derived an analytical model based on the "dividing steamline" concept (a balance of centrifugal and volumetric drag forces) and the "zone of influence" (see Appendix 1). They experimentally determined that for mass extraction fractions (mass in one branch divided by inlet mass) up to 0.4, very little of the gas phase enters that particular branch. When the flow rate was equally split between the branches, the phase split was also equal. Once the mass extraction fraction in one branch exceeded 0.6, all of the gas went into that branch. These measurements were all done using low inlet gas fractions (0.002 to 0.004). Their model predicted the phase split within 25% for 95% of their data and of the Azzopardi et al(2) measurements.

Chien and Rubel(4) found that the extraction ratio (flow split between the inlet and the branch) was key to determining the phase split. The more the conditions deviated from an equal split of fluid phases across the impacting tee, the more "unequal" the split of the phases became. As the inlet fraction of gas increased, the amount of phase splitting was reduced. They measured less severe phase splitting than Hwang et al(3) but noted that this could be due to the very low inlet gas fraction which Hwang et al(3) were using. Chien and Rubel noted that the pressure and inlet velocity had only a minor effect on the phase split at an impacting tee. They derived an empirical relationship using the extraction ratio to predict the phase split (see Appendix 1). Their method matched their measured data to within 10%. When applied to the data published by Azzopardi et al(2), their method matched the data to within 10% in the high gas fraction range and to within 20% in the low gas fraction range.

Hong and Griston(5) measured phase splits for both air/water and steam systems. They noted that as the inlet air flow split between the arms became more unequal, so did the phase split. They observed that in the stratified flow regime for both air/water and steam systems, the arm with the lower total flow rate received more of the liquid, especially as the inlet liquid volume fraction and velocity increased. In contrast, in the annular-mist flow regime, they observed that the phases split more equally between the two outlet pipes. When using steam, they were able to obtain an equal split of phases under low velocities (less than 6 m/s). They derived an empirical correlation to determine the phase split (see Appendix 1) based on the gas velocity ratio between the arm and the inlet and the inlet liquid volume fraction. They found their method matched their measured air/water and steam data very well. They also used their correlation to predict the steam quality split measured by Chien and Rubel(4) and found good agreement except in the low vapour split and the low input steam quality cases.

All of these methods require some prior knowledge of the split of flow at the impacting tee. Azzopardi et al(2) requires the gas mass extraction factor. Hwang et al(3) needs a guess for the depth of the zone of influence for the gas. Chien and Rubel(4) requires an overall mass extraction ratio while Hong and Griston(5) needs the gas velocity in the arm and branch. Therefore, unfortunately, an iterative procedure is required to use any of these methods, which makes them very difficult to use in production forecasting software.

FIELD EXPERIMENT

Figure 3 shows the geometry of the phase splitting tee used in the Sexsmith, Alberta, gas-condensate field. A 1.8 m long vertical section (254 mm diameter, 4.95 mm wall thickness) ends in an impacting tee composed of two, 203 mm horizontal lines (each with wall thickness of 5.2 mm). The 203 mm lines are referred to as Leg A and Leg C. The 203 mm lines run horizontally for 1.8 m before going vertically downward for 1.8 m. After a 3 m horizontal section, both lines go below ground (approximately 1 m deep) into the same pipeline ditch and continue 8.8 km to the plant inlet. All of the pipe in the tee is above ground and insulated with approximately 25 mm of insulation.

The pipeline elevation profile over the 8.8 km twinned section is plotted in Figure 4. The 287 points came from photo surveys of the pipeline route which where digitized approximately every 30 m. Permanent SCADA pressure gauges are located at the impacting tee (7-22) and in close proximity to the well locations 8-28 and 6-32, whose locations are shown in this same figure.

The plant was shut down on December 3, 1997 and it was restarted on December 4, 1997. Pressure measurements were recorded during the period from December 3rd through 6th. The data recorded during the day of December 6 was used for the pressure loss calculations since the operation of the pipeline on this day was judged to most closely approximate steady state conditions.

Unfortunately, it was not possible to measure the flow rate at the impacting tee directly, and there was no direct way to determine the flow rates that went into each branch. The overall flow rate could be determined, however, from the sum of the fourteen individual wellhead flow rate measurements, which are made using Rosemont orifice plates and which have a claimed accuracy of 1% of the measured volume. Flow measurements from the wells are totalled daily, and on December 6, 1997, the total flow rate through the impacting tee was 1162.4 $10^3 m^3/d$. On that day, the wells upstream of the twinned line produced 23 hours.

There was no free water flowing in the pipelines as the line heaters installed at each wellhead and the insulated lines prevented the flowing temperature from declining below the water dewpoint. Condensate flow rates estimated from flash calculations based on the fluids analysis compared favorably with the wellhead separator condensate/gas ratios measured on December 6, 1997.

The recombined fluid analysis used was taken from well 14-13-74-07W6 on Nov. 4, 1996 and is shown in Figure 5. At a separator pressure of 4300 kPag and temperature of 44 °C, the measured condensate ratio was 16.97 $m^3/10^6 m^3$. The heptane plus fraction was characterized by 4 hypothetical components derived from the mole and mass fractions of the C_6 to C_{30} liquid analysis. The Peng-Robinson(9) equation of state in the fluid property program INPROP(10) predicted the condensate ratio at the separator pressure and temperature to be 15.6 $m^3/10^6 m^3$ after some adjustment of the critical pressure and critical temperature. This predicted value compared favorably with the measured ratio. This predicted condensate ratio was also very close to the measured condensate to gas ratios of the 14 wells (on average 17.9 $m^3/10^6 m^3$, as measured at their wellhead separators) which were producing through this pipeline on December 6, 1997.

A composition sample from each of the twinned lines would have helped determine the phase split across the impacting tee. Although personnel at the site were verbally instructed to take samples downstream of the impacting tee, no samples were, in fact, taken.

To make the required pressure measurements, a service company was hired to install and operate temporary pressure and temperature gauges over a 3 day period. Pressures were sampled every minute. Each gauge consisted of two transducers ("channels"), each with its own sampling point, which were located within 100 mm of each other. The transducers were electronic strain gauges with a claimed accuracy of 0.25% over their full scale. The gauges located at the impacting tee had a range of 35,000 kPa while the gauges at the other side of the twinned section had a full scale of 21,000 kPa. The resolution was 0.005% of full scale. Unfortunately, the service company had not been told the actual level of the pressures measured (3000 kPa), and thus supplied gauges with a range that was much too large. The measurement accuracy of the gauges could therefore only be expected to be within 87.5 kPa at the tee and 52.5 kPa at the end of the twinned section. At the impacting tee, a gauge was placed on each leg at 1 m beyond where the 203 mm line turns downward (see Figure 3). Just before the twinned lines rejoin, 8.8 km away, a gauge was placed on each line about 2 meters beyond where the pipeline goes above ground in a 'S' bend and 3 meters before the pipeline goes sharply upwards to join the other pipelines entering the plant. Because the gauges at the end of the twinned section did not start recording until 3 hours after the other gauges began, it was necessary to adjust all the time readings to correspond to a common start time. Figure 6 shows the measured pressures in Leg A at the tee and at the end of the twinned section. The same type of plot for Leg C is presented in Figure 7. Unfortunately the gauge at the tee on Leg A failed in the afternoon of December 5, 1997 due to a battery failure as a result of the cold temperatures (-14 °C).

Permanent pressure gauges (SCADA) were installed in the summer of 1995 at the impacting tee (7-22), at 8-28 (1911 m from 7-22) and at 6-32 (4989 m from 7-22) on both lines. These gauges were Rosemont transmitters bellows gauges with an claimed accuracy of 0.10% over their range of 7000 kPa. They have been calibrated twice a year and monitor the pressure continuously. Their measurements are averaged over each hour and are stored for a period of 21 days. The 7-22 pressure gauge is located exactly at the center of the impacting tee. The other two gauges are located above ground after 'S' bends. The pressures measured by these gauges are shown in Figure 8. Note that the 8-28 Leg C gauge appears to have malfunctioned, since it shows the same pressure throughout the period. Thus the only comparison of pressure between the legs is at 6-32 and the difference appears to be consistent and less than 30 kPa. The pressures measured by the SCADA gauge at the impacting tee and the temporary gauges located just past the impacting tee are compared in Figure 9. Note the large difference between the results of each gauge, and especially the difference between the different channels of the temporary gauges. For example, at 10 am on December 6, 1997 on Leg C, the difference between the channels is 150 kPa. Due to the large inaccuracy of the temporary gauges (87.5 kPa) compared to the SCADA gauge (7 kPa) and the large difference between the channels of the temporary gauges, it was decided to use the SCADA gauge data and ignore the readings from the temporary gauges.

The temperature at the impacting tee was assumed to be 35 °C since there was no thermal well located at that site to allow temperature measurement. The temperature at the downstream of the twinned section was measured by temporary temperature gauges (of claimed accuracy 0.25% of full scale) on December 6, 1997 to be, on average, 18 °C.

ANALYSIS OF RESULTS

Since the phase split could not be measured directly, an attempt was made to infer it from the measured pressure losses using the method proposed by Gregory and Fogarasi(11). The pressure loss was calculated from the measured upstream pressure (3016 kPaa at 10 am on December 6, 1997) to the end of the twinned section using an assumed flow rate and phase split in one of the 203 mm pipelines using the OLGAS(12) model in the steady state pipeline simulator, PIPEFLO(10). The OLGAS(12) correlation was chosen for these calculations because it is claimed that this correlation closely matches the actual hydraulic performance of gas transportation systems with low liquid loading. The temperature was specified at each end (upstream 35 °C and downstream 18 °C) and it was assumed to change linearly between the end points. The phase split at an impacting tee is not at thermodynamic equilibrium and the resulting composition in each branch cannot be predicted by an equation of state. Thus a gas gravity, liquid gravity, liquid viscosity (as looked-up from the fluid property table) and assumed liquid/gas ratio were entered directly into the software rather than using the fluid property table previously generated. This pressure loss calculation was repeated for several combinations of flow rate and phase splits. The calculated downstream pressure was plotted as a function of phase split (see Figure 10) for Leg A. The corresponding curve for the second 203 mm line can be constructed in the same way. For any given split, a possible operating point would be where the calculated downstream pressure of each pipeline was the same (for the correct total gas and liquid flow rates). Each different flow split had its own calculated downstream pressure and thus a locus of possible operating points could be drawn. It turned out, however, that for this pipeline, the pressure at 6-32 in almost totally insensitive to the phase split for a given flow rate. To avoid needlessly cluttering this diagram, only the pressures as computed for Leg A are plotted. From Figure 10 and the measured 6-32 pressure of 2787 kPaa (Leg A) and 2808 kPaa (Leg C), it was concluded that 50% of the total flow rate entered each arm. The calculated pressures show that between 0.0% and 100% of the liquid could be entering that arm. Most of the research to date indicates that an equal total flow rate split results in an equal phase split, so it would be reasonable to assume that the split of liquids is equal in this case.

To confirm the results predicted by the OLGAS (12) model, pressure loss calculations were done using other calculation methods for the equal flow rate and equal phase split condition. Figure 11 compares the calculated pressures from the OLGAS(12) method, Eaton et al(13) for liquid holdup and Oliemans(14) for frictional pressure loss, Eaton(13) with Oliemans(14) including the Flanigan(15) head factor, the Oliemans(16) mechanistic model and the Beggs and Brill(17) methods. The results are very similar, with the smallest pressure loss being calculated by the Beggs and Brill(17) method. The highest pressure loss comes from the combination of the Eaton(13), Oliemans(14) and Flanigan(15) head factor.

For production forecasting, we need to be able to accurately predict the pressure loss in the twinned lines. The pressure loss depends on both the flow rate and phase spit at the impacting tee. However, various researchers have shown that the phase split depends on the flow rate split at the tee, which is also an unknown and thus we have to make an assumption. In this case study in which each pipeline has the same inside diameter, follows the same route and neither has flow added or removed along its length, the logical choice is to assume an equal split of flow rate. As described earlier, an equal split of flow results in an equal split of phases. If the pipelines were of a different diameter, somehow the split of flow would have to be determined first before the phase split could be calculated using the methods described in Appendix 1. A split of flow rate could be assumed and then the split of phases calculated using one of the methods described in Appendix 1. The pressure loss to the end of the pipeline could then be computed. In the other pipeline, the rest of the flow rate and the corresponding amount of liquids could be used to compute the pressure to its end. If the pressures do not match, then new guesses for the flow split must be made until the computed downstream pressures agree. Unfortunately this would be a very time-consuming process, especially for production forecasting models which are required to do these calculations for several timesteps for each of many years.

CONCLUSIONS

The measured pressure loss across a twinned 203 mm gas-condensate pipeline was used to infer the flow rate and phase split at a 254 mm vertical impacting tee. The calculated pressures indicated that the total flow rate was split evenly at the tee. However, it was impossible to determine the split of phases at the tee because the calculated pressure loss was very insensitive to the amount of liquids. Therefore the accuracy of any of the methods for determining the phase split at an impacting tee could not be determined.

The major difficulty in applying the flow splitting technology to forecasting production rates is that the flow rate split between the branches is not known and it is the flow rate split that is the most important parameter in determining the phase split at an impacting tee. If the flow rate split is guessed and the resulting phase split is calculated from the methods described in Appendix 1, the model can compute the pressure losses to the end of the twinned pipeline. The assumed flow rate split can then be altered until the same pressure is computed at the end of the twinned pipeline for each leg. Unfortunately, this would be a very time-consuming process, which may make it unsuitable for use in production forecasting tools which are required to do multiple calculations over several timesteps per year over a period of many years.

This paper illustrates some of the problems involved in measuring and using field pressures. It is important that written instructions of all the parameters to be measured (e.g. composition of each leg) and how the data should be gathered should be sent in advance to those arranging for the equipment and those doing the actual measurements. These instructions should detail the expected pressure, flow rate and temperature values, such that gauges with the appropriate scale (i.e. *not* larger than required) be chosen. It would also useful to include why the measurements are required and how they will be used to let the field personnel know what is important. Finally, the person who will be using the measurements should be present at the start of the measurements to ensure that the measurements are done as requested.

More investigation needs to be conducted to identify the effect of impacting tees on the pressure loss downstream for cases in which the downstream lines do not have equivalent diameters or do not follow the same elevation traverse.

ACKNOWLEDGEMENTS

The authors would like to thank Greg Caswell of AEC Oil and Gas Partnership for gathering all of the data reported in this paper. Without his assistance, this paper could not have been written. Doug Matchett at NeoCorr Engineering Ltd. kindly provided the elevation profile information and PIPEFLO input data file used in this study. Dr. Garry Gregory of Neotechnology Consultants Ltd. provided invaluable guidance and encouragement (especially when everything went wrong).

REFERENCES

1 Hong, K.C.: "Two-Phase Flow Splitting at a Pipe Tee", *JPT* (Feb. 1978) p 290-296.

2 Azzopardi, B.J., Purvis, A., and Govan, A.H.: "Annular Two-Phase Flow Split at an Impacting T", *Int. J. Multiphase Flow*, Vol. **13**, No. 5 (1987) p 605-614.

3 Hwang, S.T., Soliman, H.M., Lahey, R.T. Jr.: "Phase Separation in Impacting Wyes and Tees", *Int. J. Multiphase Flow*, Vol. **15**, No. 6 (1989) p 965-975.

4 Chien, Sze-Foo, and Rubel, Mark T.: "Phase Splitting of Wet Steam in Annular Flow Through a Horizontal Impacting Tee", *SPE Production Engineering* (Nov. 1992) p 368-374.

5 Hong, K.C., and Griston, Suzanne: "Two-Phase Flow Splitting at an Impacting Tee", *SPE Production & Facilities* (Aug. 1995) p 184-190.

6 *FORGAS User Manual*, Neotechnology Consultants Ltd., Calgary, Canada (1997).

7 *ECLIPSE 100 Reference Manual*, Schlumberger GeoQuest (1997).

8 *Open ECLIPSE Developers Kit*, Schlumberger GeoQuest (1997).

9 Peng, D.Y., and Robinson, D.B.: "A New Two Constant Equation of State", *Ind. Eng. Chem. Fund.*, Vol. **15** (1976) p 59.

10 *NMS User Manual*, Neotechnology Consultants Ltd., Calgary, Canada (1997).

11 Gregory, G.A., and Fogarasi, M., "Estimation of pressure drop in two-phase oil-gas looped pipeline systems", *JCPT* (Mar.-Apr. 1982) p 75-81.

12 Bendiksen, K.H., Maines, D., Moe, R., and Nuland, S.: "The Dynamic Two-Fluid Model OLGA: Theory and Application", *SPE Production Engineering* (May 1991) p. 171.

13 Eaton, B.A., Andrews, D.E., Knowles, C.R., Silberberg, I.H., and Brown, K.E.: "The Prediction of Flow Patterns, Liquid Holdup and Pressure Losses Occurring

During Continuous Two-Phase Flow in Horizontal Pipelines", *JPT*, Vol. **19** (June 1967) p 815.

14 Oliemans, R.V.A.: "Two-Phase Flow in Gas-Transmission Pipelines", paper 76-Pet-25 presented at the Joint Petroleum Mechanical Engineering and Pressure Vessels and Piping Conference, Mexico City, Mexico, Sept. 19-24 (1976).

15 Flanigan, O.: "Effect of Uphill Flow on Pressure Drop in Design of Two-Phase Gathering Systems", *O&GJ*, Vol. 56, No. 10 (Mar. 10, 1958) p 132.

16 Oliemans, R.V.A.: "Modelling of Gas-Condensate Flow In Horizontal and Inclined Pipes", Proc. ASME Pipeline Eng. Symp., ETCE, Dallas, Texas (Feb. 1987) p 73.

17 Beggs, H.D., and Brill, J.P.: "A Study of Two-Phase Flow in Inclined Pipes", *JPT* (May 1973) p 607.

18 Hwang, S.T.: *A study on phase separation phenomena in branching conduits*, Ph.D. Thesis, Rensselaer Polytechnic Inst., Troy, N.Y. (1986).

APPENDIX 1 PUBLISHED METHODS TO PREDICT THE PHASE SPLITS AT IMPACTING TEES

Azzopardi, Purvis and Govan(2)

To solve for the liquid fraction resulting from the film that travels to the branch (ω_{3LF}/ω_{1L}), use the assumed gas mass ratio (ω_{3G}/ω_{1G}) between the arm and inlet and equation (2).

$$\frac{\omega_{3G}}{\omega_{1G}} = \frac{1}{2\pi}\left[\frac{2\pi\frac{\omega_{3LF}}{\omega_{1L}}}{1-E} - \sin\left(\frac{2\pi\frac{\omega_{3LF}}{\omega_{1L}}}{1-E}\right)\right] \qquad (2)$$

Calculate the liquid fraction carried by the droplets going into the branch (ω_{3LD}/ω_{1L}) by first computing the angle subtended at the center by the bounding chord of the zone of influence (Φ) using the assumed gas mass ratio (ω_{3G}/ω_{1G}) between the arm and inlet and equation (3). Then use that computed angle and the liquid mass entrainment fraction (E) and equation (4).

$$\frac{\omega_{3G}}{\omega_{1G}} = \frac{1}{2\pi}(\Phi - \sin\Phi) \qquad (3)$$

$$\frac{\omega_{3LD}}{\omega_{1L}} = \frac{E}{2}\left(1 - \cos\frac{\Phi}{2}\right) \tag{4}$$

The total liquid fraction in the arm is the sum of the liquid fraction from the film and from the droplets, as shown in equation (5).

$$\frac{\omega_{3L}}{\omega_{1L}} = \frac{\omega_{3LF}}{\omega_{1L}} + \frac{\omega_{3LD}}{\omega_{1L}} \tag{5}$$

where:

E	=	liquid mass entrainment fraction
Φ	=	angle subtended at the center by the bounding chord of the zone of influence
ω_{1G}	=	gas mass flow rate in inlet
ω_{3G}	=	gas mass flow rate in arm
ω_{1L}	=	liquid mass flow rate in inlet
ω_{3LD}	=	liquid mass flow rate in arm due to droplets
ω_{3LF}	=	liquid mass flow rate in arm due to liquid film

The split of the gas phase between the branches must be assumed first and then the liquid phase split can be calculated using the above equations. The pressure loss to the end of each pipeline can then be computed and compared. The solution is reached once the pressure loss across each pipeline is the same. The liquid entrainment fraction (fraction of the liquid which is carried in droplets) can be difficult to obtain since very few correlations calculate and report this parameter.

Hwang, Soliman and Lahey(3)

Calculate the slip ratio (S) using equation (6).

$$S = \frac{u_{1G}}{u_{1L}} \tag{6}$$

At equal phase split, calculate the corresponding mass ratio using equation (7), which assumes that the area of each outlet pipe is the same.

$$\left(\frac{\omega_3}{\omega_1}\right)_E = 0.5\left(1 - \frac{A_3}{A_1}\cos\theta\right) \tag{7}$$

Solve for the angle of the zone of influence (Φ) corresponding to an equal phase split using equation (8).

$$\left(\frac{\omega_3}{\omega_1}\right)_E = \frac{[\Phi - 0.5\sin(2\Phi)]}{\pi} \tag{8}$$

The depth of the zone of influence at equal phase split is then calculated from equation (9).

$$\frac{\delta_E}{D_1} = \frac{1 - \cos\Phi}{2} \tag{9}$$

Calculate the depth of zone of influence for liquid (δ_L/D_1) using an assumed liquid mass extraction ratio (ω_{3L}/ω_{1L}) and the mass distribution of liquid in the inlet, which is described in Hwang(18). The depth of the zone of influence for liquid can then be used in equations (10) and (11).

$$n_L = 5 + 20\exp\left[-53\frac{\delta_L}{D_1}\right] \tag{10}$$

$$\frac{b_L}{D_1} = \frac{\frac{\delta_L}{D_1} - \frac{\delta_E}{D_1}}{1 - \frac{\delta_E}{D_1}} \tag{11}$$

The minimum radius of curvature is calculated using equation (12).

Next the exponent m_L needs to be solved iteratively using equations (13) and (14).

$$\left(\frac{R_L}{D_3}\right)_{min} = \frac{\left[1 + \left(2\frac{D_1}{D_3}\sin\theta\right)^2\right]^{1.5}}{2\frac{D_1}{D_3}\sin^2\theta} \tag{12}$$

$$\frac{R_L}{D_3} = \frac{\left[1 + \left(m_L\frac{D_1}{D_3}\frac{b_L}{D_1}\sin\theta\right)^2\right]^{1.5}}{m_L\,(m_L - 1)\,\frac{D_1}{D_3}\frac{b_L}{D_1}\sin^2\theta} \tag{13}$$

$$\frac{R_L}{D_3} = \frac{\left(\frac{R_L}{D_3}\right)_{min}}{\left(\frac{b_L}{D_1}\right)^{n_L}} \tag{14}$$

Assume the depth of the zone of influence for the gas (δ_G/D_1) and solve for the distance (b_G/D_1), as shown in equation (15).

$$\frac{b_G}{D_1} = \frac{b_L}{D_1} + \left(\frac{\delta_G}{D_1} - \frac{\delta_L}{D_1}\right) \tag{15}$$

The gas exponent (n_G) can then be calculated using equation (16).

$$n_G = 5 + 20\exp\left[-53\,\frac{\delta_G}{D_1}\right] \tag{16}$$

The minimum radius of curvature for the gas is calculated using equation (17).

Next the exponent m_G needs to be solved iteratively using equations (18) and (19).

$$\left(\frac{R_G}{D_3}\right)_{\min} = \frac{\left[1 + \left(2\,\frac{D_1}{D_3}\,\sin\theta\right)^2\right]^{1.5}}{2\,\frac{D_1}{D_3}\,\sin^2\theta} \tag{17}$$

$$\frac{R_G}{D_3} = \frac{\left[1 + \left(m_G\,\frac{D_1\,b_G}{D_3\,D_1}\,\sin\theta\right)^2\right]^{1.5}}{m_G\,(m_G - 1)\,\frac{D_1\,b_G}{D_3\,D_1}\,\sin^2\theta} \tag{18}$$

$$\frac{R_G}{D_3} = \frac{\left(\frac{R_G}{D_3}\right)_{\min}}{\left(\frac{b_G}{D_1}\right)^{n_G}} \tag{19}$$

The angle between the gas and liquid streamlines at the point of impact (β) is computed using equation (20).

$$\beta = \tan^{-1}\left(m_G\,\frac{D_1\,b_G}{D_3\,D_1}\,\sin\theta\right) - \tan^{-1}\left(m_L\,\frac{D_1\,b_L}{D_3\,D_1}\,\sin\theta\right) \tag{20}$$

Compute γ from equation (21).

$$\tan\gamma = \frac{\sin\beta}{\cos\beta - \frac{\rho_G}{\rho_L}\left(\frac{\mu_G}{\mu_L}\right)^2 \frac{R_L}{R_G}} \tag{21}$$

Now the slip ratio can be calculated using equation (22).

$$S = \frac{\cos(\gamma - \beta)}{\cos\gamma} \qquad \textbf{(22)}$$

Compare the slip calculated with equation (22) with the slip corresponding to the velocities of each phase, as calculated in equation (6). If these values of slip are sufficiently different, a new guess for the depth of the zone of influence for the gas is required and the calculations are repeated, starting from equation (15). Once the values of slip agree, then from the knowledge of the mass distribution of the gas phase in the inlet and the depth of the zone of influence of the gas just computed, then the gas mass ratio in the arm (ω_{3G}/ω_{1G}) can be calculated since the "zone of influence" is specified by δ_G/D_1.

where:

A	=	area of pipe
b	=	distance
D	=	inside diameter of pipe
m	=	exponent
n	=	exponent
R	=	radius of curvature of streamline
S	=	slip ratio
u	=	velocity
β	=	angle between gas and liquid streamlines at point of impact
γ	=	angle of the velocity triangle
δ	=	depth of zone of influence measured from the pipe wall
Θ	=	angle between inlet and branch pipe
μ	=	viscosity
ρ	=	density
Φ	=	angle subtended at the center by the bounding chord of the zone of influence
ω	=	mass flow rate

Subscripts:

E	=	equal phase split
G	=	gas
L	=	liquid
1	=	inlet pipe
3	=	branch pipe

Chien and Rubel(4)

For mass extraction ratios (ω_3/ω_1) less than or equal to 0.5, use equation (23):

$$\frac{x_3}{x_1} = \left[a^{\frac{1}{(1-b)}}\right] * \left[\frac{\omega_3}{\omega_1}^{\frac{b}{(1-b)}}\right] \qquad \textbf{(23)}$$

For mass extraction ratios larger than 0.5 use equation (24):

$$\frac{x_3}{x_1} = \frac{1 - \left(a \left(1 - \frac{\omega_3}{\omega_1} \right) \right)^{\frac{1}{(1-b)}}}{\frac{\omega_3}{\omega_1}} \quad \textbf{(24)}$$

where:

$$a = (1.352)^{(1-x_1)} \quad \textbf{(25)}$$

$$b = (0.435)^{(1-x_1)} \quad \textbf{(26)}$$

x_1	=	mass fraction of gas in the inlet
x_3	=	mass fraction of gas in one arm
ω_1	=	total mass flow rate in inlet
ω_3	=	total mass flow rate in arm

Hong and Griston(5)

The liquid volume fraction in the branch is computed using equations (27) to (30).

$$F = V_{1L} * 0.3048\, u_{1G} \quad \textbf{(27)}$$

$$a = 0.2088 - 0.2541 \ln F + 0.3928\, F^{0.5} \quad \textbf{(28)}$$

$$b = -3.0662 - 0.4052\, F - 1.0381 \ln F + 2.7228\, F^{0.5} \quad \textbf{(29)}$$

$$V_{3L} = a \; V_{1L} \left(\frac{u_{3G}}{u_{1G}} \right)^{b} \tag{30}$$

where:

F	=	product of inlet liquid volume fraction and inlet gas velocity
u_{1G}	=	gas velocity in inlet (m/s)
u_{3G}	=	gas velocity in arm (m/s)
V_{1L}	=	liquid volume fraction in inlet
V_{3L}	=	liquid volume fraction in arm

To use this method, one must guess the gas velocity in the arm and then use the above equations to solve for the liquid volume fraction in the arm. Then the calculations are repeated for the second arm. Check the mass balance for consistency; otherwise, a new guess for the gas velocity in one arm is required.

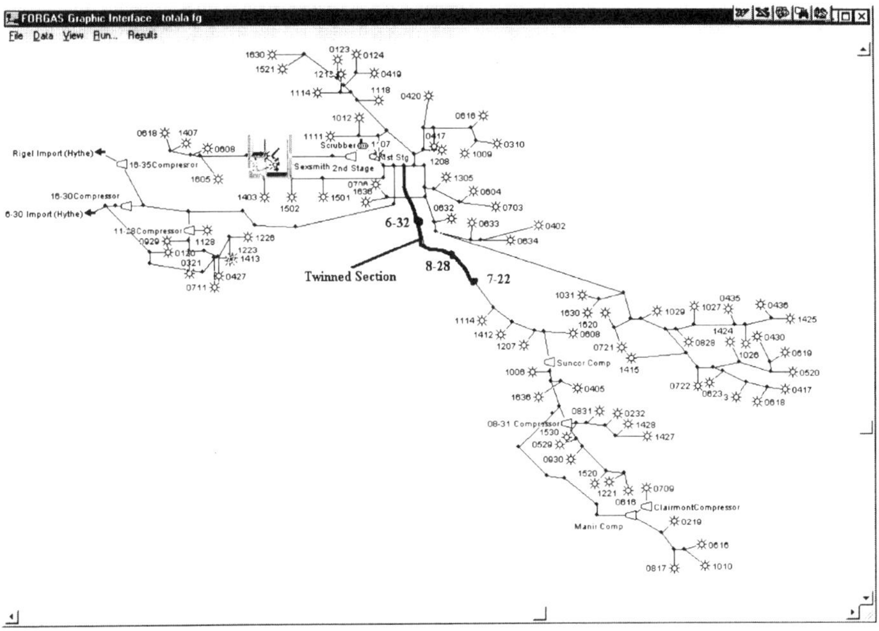

Figure 1
Location of Twinned Lines in Sexsmith Gathering System

	Hong(1)	*Azzopardi et al(2)*	*Hwang et al(3)*	*Chien & Rubel(4)*	*Hong & Griston(5)*
Fluid Type	*air/water*	*air/water*	*air/water*	*steam*	*air/water, steam*
Diameter (mm)	*9.5*	*31.8*	*38.0*	*49.3*	*19.0 air 50.8 & 101.6 steam*
Pipe Material	*lucite*	*acrylic resin*	*plexiglass*	*steel*	*lucite, steel*
Inlet Angle	*horizontal*	*vertical*	*horizontal*	*horizontal*	*horizontal*
Inlet Flow Pattern	*?*	*annular*	*bubbly, bubbly-stratified*	*annular, annular-mist*	*stratified, annular-mist*
Pressure (kPa)	*122*	*170*	*130 - 190*	*2758, 4137*	*?*
Mass Flux ($kg/m^2/s$)	*50.8*	*52.4 - 123*	*1350, 2050, 2700*	*230.5-1982.3*	*1.3 10^{-3} to 6.5 10^{-3} m^3/s air 6.6 10^{-6} to 4.2 10^{-4} m^3/s water ? steam*
Input Liquid Fraction	*0.0009*	*0.42-0.79*	*0.996-0.998*	*0.2-0.8*	*0.005-0.06 air/water 0.01-0.10 steam*
Arm Flow Split	*more than 0.15-0.85*	*entire*	*0.02-0.95*	*0.2-0.5*	*0.05-0.95*
Gas Velocity (m/s)	*27*	*?*	*?*	*12-40 superficial*	*4.6-22.9 air 1.5-21.3 steam*

Figure 2
Conditions Investigated at Impacting Tees

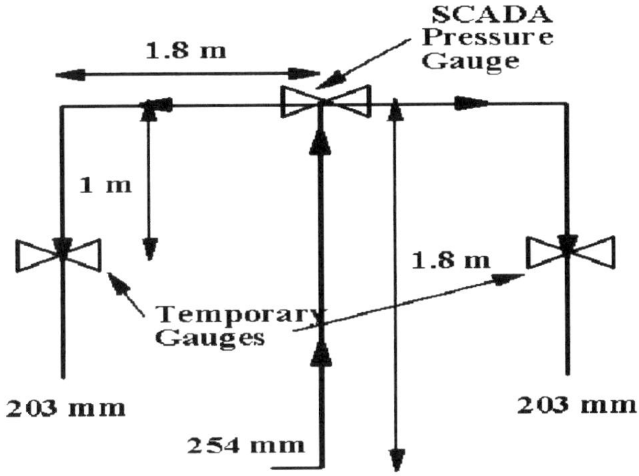

Figure 3
Schematic of Impacting Tee

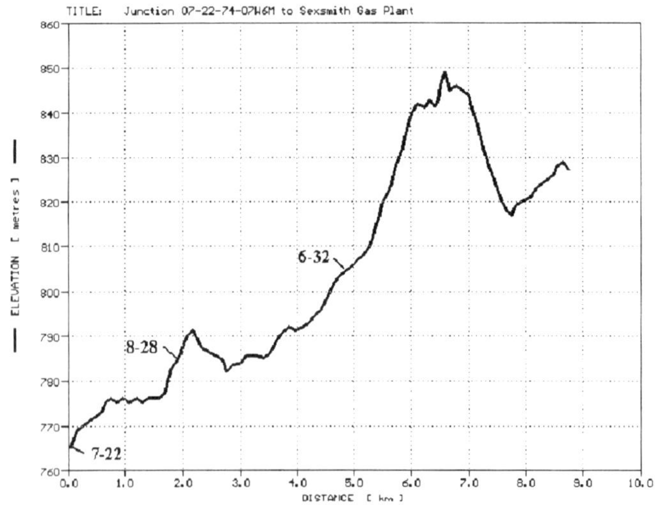

Figure 4
Elevation Profile of Twinned Pipelines

N_2	0.00468	**C_1**	0.79495	**nC_4**	0.00718
CO_2	0.00889	**C_2**	0.04625	**iC_5**	0.00205
H_2S	0.10297	**C_3**	0.02246	**nC_5**	0.00237
H_2	0.00040	**iC_4**	0.00323	**nC_6**	0.00169
Residual MW=101.0	.00090	**Pseudo 1 MW=98.66**	0.00123	**Pseudo 2 MW=128.3**	0.00069
Pseudo 3 MW=176.2	0.00007	**Pseudo 4 MW=283.4**	0.00000		

MW = Molecular Weight

Figure 5
Recombined Fluid Analysis

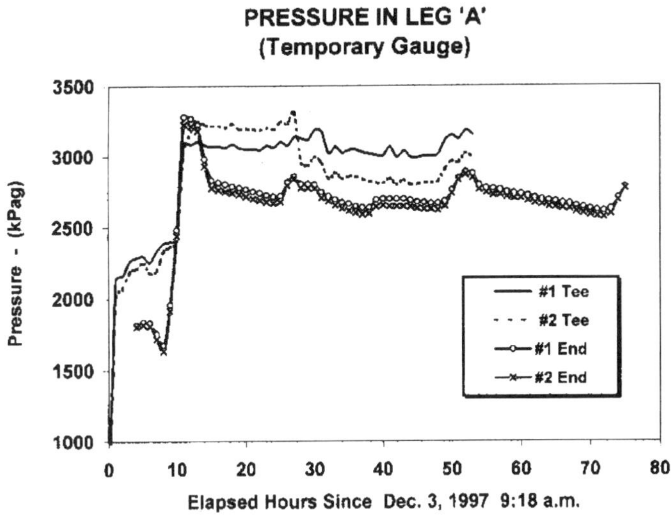

Figure 6
Measured Pressures In Leg A from Temporary Gauge

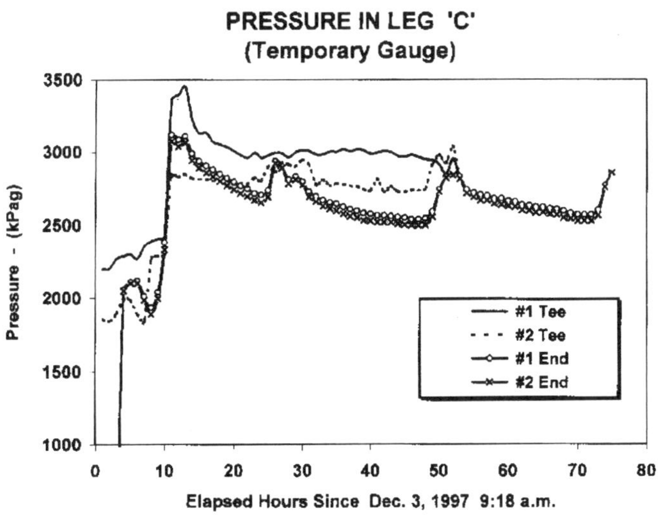

Figure 7
Measured Pressures in Leg C from Temporary Gauge

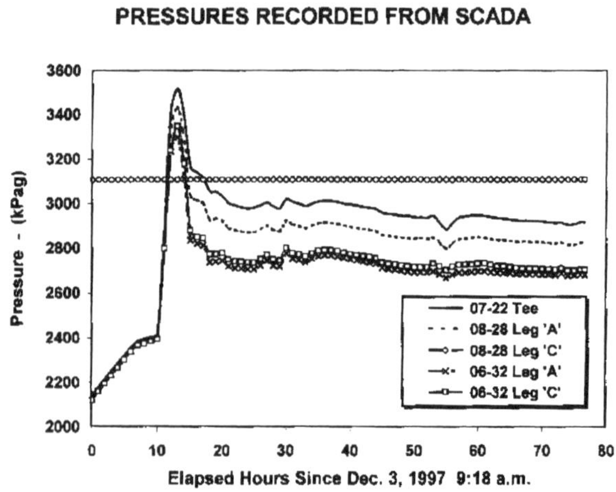

Figure 8
Pressures Measured by Permanent Gauges

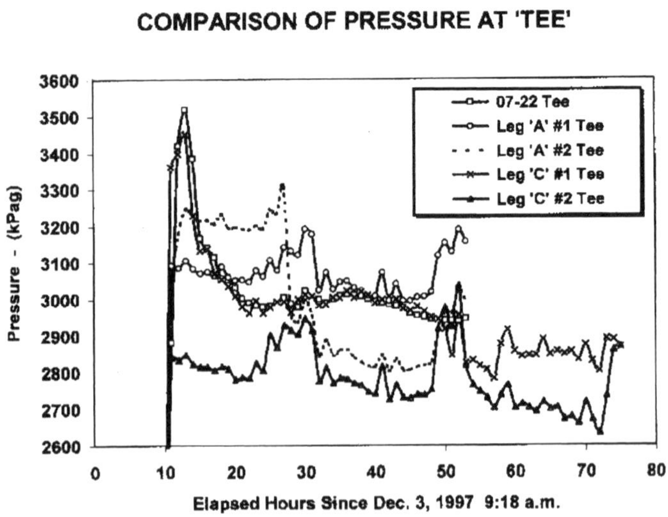

Figure 9
Comparison of Measured Pressures at Tee

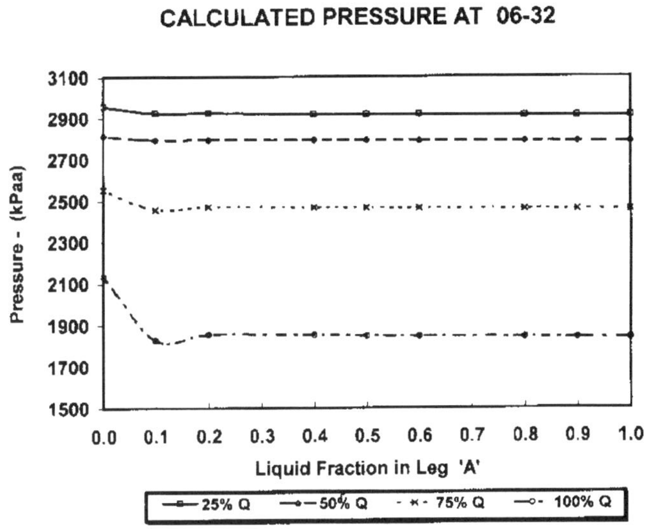

Figure 10
Calculated Pressure at 6-32 as a Function of Flow Rate and Liquid Split

	Pressure (kPaa) At 8-28	Pressure (kPaa) At 6-32	Pressure (kPaa) At End
Measured	2924 (Leg A)	2787 (Leg A) 2808 (Leg C)	
OLGAS	2928	2791	2617
Eaton-Oliemans	2933	2803	2638
Eaton-Oliemans-Flanigan	2912	2754	2537
Oliemans Mechanistic	2926	2785	2606
Beggs-Brill	2932	2817	2652

Figure 11
Comparison of Computed Pressures from Various Correlations, Assuming Equal Flow Rate and Equal Phase Split

Separation

Vertical zero net liquid flow: effects of high-pressure on holdup

R W DUNCAN and **S L SCOTT**
Louisiana State University, Baton Rouge, USA

1 ABSTRACT

The Gas-Liquid Cylindrical Cyclone (GLCC) Separator is a simple, compact, inexpensive vertical separator with a wide variety of potential applications. These compact separators are being examined for use in subsea separation, multiphase metering, well testing and to expand the capacity of existing separation facilities. Recently, a model has been proposed to predict the performance of the GLCC. However, this model has not been verified at the high pressures that are likely to be encountered in subsea separation and in many metering applications. One important component of this model is the treatment of the upper section of the GLCC above the tangential inlet. In this region, a peculiar flow, termed zero-net liquid flow (ZNLF) occurs where the gas phase flows out the top of the GLCC while the liquid undulates but remains in the separator. Determination of liquid holdup under ZNLF conditions is essential for prediction of liquid carry-over in the GLCC. This paper presents new experimental high-pressure measurements of liquid holdup for ZNLF. To investigate the influence of pressure on ZNLF, a 7.62 m (25ft) tall, 12.4 cm (4.897 in) ID experimental apparatus was constructed. Field-scale experiments were performed using methane and water at pressures ranging from 689.5 kPa to 6894.8 kPa (100 to 1,000 psig), and gas flow rates from 4,871 to 139,900 m^3/day (170 to 12,000 MSCFD). The ZNLF holdup is observed to decrease as pressures increase. Comparison with the TUSTP/Chevron Model (Arpandi, *et al* ,1996) shows a significant disagreement for pressures above 689.4 kPa (100 psig). A single dimensionless curve has been developed that incorporates the effects of pressure on liquid holdup. This new development is used to extend the Arpandi et al. ZNLF model to high pressure conditions.

2 INTRODUCTION

Multiphase fluid flow has become a normal part of offshore oil and gas operations. It is becoming even more common as these operations move further offshore into deeper water. Economic considerations are driving the search for better methods of transporting, separating and metering mixed streams of gas and liquid. Compact separation is an evolving new technology that provides an alternative to the large 'low technology' vessel-type separators that have been used in the past. Potential applications of compact separators range from subsea separation to multiphase metering and well testing. These separators are simple, compact, and lightweight, and utilize centrifugal forces rather than gravity segregation to obtain their increased efficiency. Two areas of application are of particular interest for the compact separator. These are for subsea separation and for subsea multiphase metering.

Subsea Separation

Surface separators are normally large, gravity-type knockout drums with limited operational envelopes. They have many characteristics unsuited to an underwater environment. Ideally, a subsea separator should be small, lightweight, and simple, since it must be placed and retrieved from inside a subsea manifold. It also must be extremely reliable and have a high pressure rating with a large range of operating conditions (high turndown ratio). Once placed on the seafloor, the manifold must operate undisturbed for many years due to the expense involved in moving a semi-submersible workover rig over the location and retrieving either individual parts of the manifold or, in extremity, the entire subsea production manifold. Thus all parts of the subsea manifold, including the separator, must be capable of operating without repairs or maintenance for long periods. This requires simplicity and extreme reliability. However, if maintenance or repairs are necessary, current design requires that components of the subsea manifold be lightweight and cylindrical with a very small footprint. Current vessel-type separators meet none of these criteria.

One type of compact separator that has generated a lot of interest is the Gas-Liquid Cylindrical Cyclone (GLCC) Separator (Kouba et al., 1995). This is somewhat similar to the hydrocyclone, which is well known to provide excellent separation of liquid and solids over a wide range of conditions. The hydrocyclone is also widely used in the food and chemical process industries, and its most common use in the oil and gas industry is as a desander and desilter on drilling rigs. The mixture enters the hydrocyclone tangentially, and the centrifugal force thus imparted forces the heavier solid particles to the outside of the cylinder and downwards. The lighter liquid components are forced to the inside and upwards. A compact GLCC separator works in much the same fashion, forcing the heavier liquid outwards and down, and the gas toward the center and upwards. Its basic design consists of a vertical piece of high strength pipe (the GLCC body), an inclined inlet and gas and liquid outlets. Due to the design, the GLCC can be made of relatively small diameter pipe which means a higher pressure rating, less wall thickness, less weight, and much less cost. Having no internals, it is simple and reliable and also has a large turndown ratio. The GLCC is currently being tested by Chevron at carefully selected land locations, and appears to be performing up to expectations. The University of Tulsa has been studying the GLCC for several years and has developed a complex computer model of the hydrodynamic processes occurring. Thus far, the model has been found to be quite accurate in predicting the performance of these separators and several dozen have been placed in service. However, the operating separators and the research models are for low pressure.

Multiphase Metering

Recently, Kouba (1998) observed that higher pressures act to reduce the uncertainty of the rate measurements provided by multiphase meters. These meters show great promise; however, their widespread acceptance is still some time away, due to concerns with accuracy and their acceptance for production allocation purposes. Unfortunately, in the Gulf of Mexico a number of regulatory and technical challenges have thus far prevented deployment of multiphase meters. Current Federal law requires that the determination of royalty payments be made using one of two methods. The first allows commingled production, but requires that each well be tested once a month to allocate the production. Therefore, wells must be shut-in in turn and the difference used to allocate production, or a separate test line must be run in addition to the flowline. The other method involves separating the phases and utilizing conventional single-phase meters. A compact separator may provide this type of convention metering and be more readily accepted than other types of multiphase meters.

This paper investigates the complicated two-phase flow behavior in the upper portion of the gas-liquid cylindrical separator at high pressures, and at rates comparable to those found in subsea and other higher pressure applications. First, a model of ZNLF is discussed and then experimental results are presented. A new method is developed that collapses pressure effects onto a single dimensionless curve. Using this approach, improvements to the Arpandi, et al. model are suggested.

3 MODELING VERTICAL ZERO NET LIQUID FLOW

The upper portion of the GLCC operates in a peculiar condition termed zero-net liquid flow (ZNLF). Under ZNLF conditions, the gas flows through the vertical pipe and out the top. The liquid phase moves up and down in the vertical pipe, but no liquid is produced out the top, i.e. zero-net liquid flow. Under these conditions the flow can assume the normal range of vertical flow patterns. This includes bubble, slug, churn and for high gas flow rates, annular flow. Low-pressure observations reveal that the slug flow pattern is rarely seen. This is due to the relatively short length of GLCC's, and, as noted by Taitel et al. (1980), the churn flow pattern is an entry region phenomena. Slug flow exists only after this entry region. A review of the literature reveals that few experimental or modeling studies have been performed to study vertical ZNLF, especially at high pressures. Studies performed by Kokal & Stanislav (1986) and Amaravadi et al. (1994) have considered ZNLF for inclined pipelines. Low-pressure vertical ZNLF data was presented by Kouba et al. (1995) and by Arpandi, et al. (1996) for 5.08 and 7.62 cm (2- and 3-inch) diameter pipes. The model presented by Arpandi et al. uses a modified Taylor bubble rise velocity method to estimate the liquid holdup for ZNLF conditions. The following is a presentation of the components of the ZNLF model which include calculations of: 1) bubble rise velocity; 2) liquid holdup; and, 3) liquid blowout velocity.

Taylor Bubble Rise Velocity

Davies and Taylor (1950), while studying the behavior of bubbles during an underwater explosion, derived an equation to calculate the gas bubble rise velocity in large tubes. This equation was later refined by other researchers. Harmathy (1960) studied large cylindrical bubbles which we now call "Taylor" bubbles. In the case of gas bubbles moving through a stagnant (non-flowing) liquid, he found that there was a relationship between bubble velocity and gravity, diameter, and density as given below:

$$v_D = 0.35\sqrt{\frac{g\,D(\rho_L - \rho_G)}{\rho_L}} \quad \text{.......... (1).}$$

Harmathy found that for $N_{E\ddot{o}} > 400$, the best fit curve appears to become asymptotic resulting in the 0.35 coefficient. Under stagnant conditions, Bendiksen (1984) gives a method to obtain this drift velocity for inclined pipes and Weber et al. (1986) show how to correct Eq. (1) for viscous fluids where the Eötvös number is low. For two phase flow, Taitel and Dukler (1980) represented the gas velocity by the equation:

$$v_G = 1.2v_m + v_D \quad \text{.......... (2).}$$

The first term of the right-hand side of the Eq. (1) uses a coefficient of 1.2, which is the approximate ratio of centerline velocity to average velocity for fully developed turbulent flow. Other values of this coefficient have been used by other researchers and in general form can be denoted as C_0. The mixture velocity in Eq. (2) is equivalent to the sum of the superficial gas velocity and the superficial liquid velocity ($v_{SG} + v_{SL}$). Under ZNLF conditions, the liquid velocity is zero, therefore only the v_{SG} term is used. Using Eq. (1) for the drift velocity yields:

$$v_{G0} = C_0 v_{SG} + 0.35\sqrt{\frac{g\,D(\rho_L - \rho_G)}{\rho_L}} \quad \text{.......... (3).}$$

At low pressure, Harmathy's dimensionless density fraction term becomes very close to unity. However, at higher pressures, gas density becomes appreciable and the term must be included.

Liquid Holdup

For the bubble flow regime, the definition of slip can be used to estimate the liquid holdup. Slip between the gas and liquid phases is defined as:

$$v_S = v_G - v_L = \frac{v_{SG}}{(1 - H_L)} - \frac{v_{SL}}{H_L} \quad \text{.......... (4).}$$

For ZNLF the liquid velocity is zero and v_G is determined from Eq. (3). Rearranged Eq. (4) yields:

$$H_{LO} = \left(1 - \frac{v_{SG}}{v_{GO}}\right) \quad \text{.......... (5).}$$

Arpandi's equation for liquid holdup is shown below.

$$H_{LO} = \left(1 - \frac{v_{SG}}{v_{GO}}\right)\left(1 - \frac{L_d}{L}\right) \quad \text{.......... (6).}$$

The second term is a correction to account for the fact that holdup will exist only in that portion of the GLCC that is in slug/churn flow. L is the total length of the pipe, or for a GLCC, the distance from the tangential inlet to the gas outlet. The variable L_d represents the length of the droplet region. The equation for the length of the droplet region was determined from a droplet ballistic analysis that included a drag coefficient, C_d, of 0.44 from Turner, Hubbard and Dukler (1969). This equation from Arpandi is shown below.

$$L_d = \frac{1}{\dfrac{2g}{v_{SG}^{2}} - \left(\dfrac{C_d}{2}\right)(\rho_G v_{SG})^2 \left(\dfrac{3}{32\rho_L \sigma g}\right)} \quad \text{.......... (7).}$$

The resulting model is a very weak function of pressure. Figure 1 shows a plot the holdup model performance over a wide range of superficial velocities for a 12.4 cm (4.897-inch) ID pipe at a pressure of 689.5 kPa (100 psi). At velocities over 6.1 m/sec (20 ft/sec), a discontinuity is observed. Arpandi et al. have used the concept of blowout velocity to limit the model to velocities before this discontinuity.

Liquid Blowout Velocity

The final relationship needed to model ZNLF is the gas velocity at which all the liquid above the inlet is blown out of the pipe. At blow-out conditions, the pipe will be entirely in annular flow, which means that the droplet region (L_d) will be equal to the entire height from inlet to outlet (L). Arpandi et al. rearrange Eq. (7) to solve for what was termed the blow-out velocity, v_{BO}. Thus, the blowout velocity can be determined from transition criteria for annular flow.

Turner, et al. (1969) determined the minimum gas velocity necessary to continuously remove liquids from gas wells. This was called the terminal velocity, v_t. They examined two physical models for the removal of gas well liquids; liquid film moving along the walls of the pipe, and liquid droplets being entrained in the high velocity gas core. The models were compared with data from actual gas wells producing at various rates to determine which model was the proper one to use. The droplet model showed a better match with field data when compared with the film model.

The liquid droplet model of Turner et al. is based on work by Hinze (1955). Hinze determined that "it is the antagonism of two pressures, the velocity pressure or inertia ($v^2\rho_G/g$), and the surface tension pressure (σ/D) " that determines the maximum droplet size. The ratio of these two pressures is the Weber number ($N_{WE} = v^2\rho_G D/\sigma g$). If the Weber number exceeds a certain critical value, a liquid droplet will shatter. Hinze stated that "Experiments have shown that the mechanism of breakup is quite simple when N_{WE} is equal to or slightly higher than $(N_{WE})_{crit}$." For a falling drop of water in air, he found that $(N_{WE})_{crit} \approx 22$. To indicate that the curve match was not exact, Turner, et al, used a range of values from 20 to 30. The larger the Weber number, the larger the droplet diameter; and the larger the droplet diameter, the higher the gas velocity required to remove it. Therefore, as a worst case scenario, they chose a Weber number of 30 as the value to use to determine the maximum droplet diameter at its terminal velocity.

$$D_{max} = \frac{30\sigma g}{\rho_G v_t^2} \qquad (8).$$

Substituting this into a general free settling equation rewritten in terms of drop diameter, using a constant drag coefficient of 0.44 resulted in their terminal velocity equation. Unfortunately, the calculated minimum gas flow rates were, in many cases, too low. Therefore, the coefficient was corrected upwards by about 20% to produce their final equation. Converting to SI units yields:

$$v_t = 0.566\frac{\sigma^{1/4}(\rho_L - \rho_G)^{1/4}}{\rho_G^{1/4}} \qquad (9).$$

This equation, although empirical, has been tested for many years and works extremely well. Figure 2 shows a comparison of the Turner et al. terminal velocity, v_t, with Arpandi's v_{BO} (Figure 2).

4 EXPERIMENTAL SETUP AND PROCEDURES

Field-Scale, high pressure experiments were performed at Louisiana State University's Petroleum Engineering Research and Technology Transfer Laboratory. This large facility has the capability to perform multiphase flow tests at pressures up to 34,474 kPa (5,000 psi) and liquid rates to 2,703 m^3/day (17,000 bbl/day). Natural gas is delivered to the facility at rates up to 991,200 m^3/day (35 MMSCFD) via an adjacent 20.3 cm, 4,445 kPa (8-inch, 630 psi) gas sales line. If necessary, the gas is compressed to 15,168 kPa (2200 psi) and stored in three gas storage wells located on the facility.

The ZNLF apparatus is shown in Figure 3 below. The main body is 15.2 cm (6-inch) XXS pipe with an ID of 12.4 cm (4.897-inch) and the liquid accumulation section is 30.4 cm (12-inch) Schedule 160 pipe with an ID of 25.7 cm (10.126-inch). Most of the connecting piping is 7.6 cm (3-inch) XXS pipe. The distance between the gas inlet and outlet on the main body is 7.37 m (24.17 ft). A Rosemount Model 3051T 'Smart' absolute pressure transmitter is mounted on the top of the apparatus and a Foxboro Model IDP10 differential pressure transmitter is connected to the base. The 4-20 mA output of these transmitters was directed to a data acquisition system using Labview from National Instruments. For pressures of 3,447 kPa (500 psig) and lower, the gas flow was measured with a 10.2 cm (4-inch) Daniel junior orifice meter and a Daniel Model 2231 Flow Computer. The 6,894.8 kPa (1000 psig) tests utilized a 5.08 cm (2-inch) Daniel senior orifice meter.

Water and pipeline quality methane were used throughout the testing. The water used was city water with a density of 1.013 g/cm^3 (8.45 ppg). The composition of the gas is shown in Table 1. The experimental procedure entailed filling the main body of the device with water. The automatic Warren choke was set to hold backpressure equal to the test pressure, and the outlet line was 'pre-loaded' with gas at the test pressure. This ensured that the test was begun at the test pressure, thus reducing the chance of a large Taylor bubble removing too much water during the test startup. The large needle valve on the gas inlet was then slowly opened until the flowrate equaled that selected for the test. The flowrate, pressure, and holdup were allowed to stabilize for approximately 30 seconds, and the needle valve was closed ending the test. The choke was then opened and the pressure was slowly bled off the system. The valve upstream of the differential pressure sensor was opened and the sensor was used to measure the height of the remaining water. This height was converted to volume and used to calculate holdup H_L. The gas flow rate in standard cubic feet was converted to *in situ* conditions and divided by the flow area to obtain the superficial gas velocity (v_{SG}). Tests were conducted at pressures ranging from 689.4 kPa (100 psig) to 6,894.7 kPa (1000 psi).

5 DISCUSSION OF RESULTS

Figure 4 shows a summary of the results obtained for the high pressure testing. For a given superficial gas velocity, higher pressures act to decrease the liquid holdup for ZNLF. Figure 5 shows the same data plotting liquid holdup as a function of pressure for fixed superficial gas velocities. The data indicate that pressure does have a significant effect on the liquid holdup for vertical ZNLF.

Comparing these experimental results with the model of Arpandi et al. (1994) finds a good match for the lower range of gas velocities at a pressure of 689.4 kPa (100 psig) (Figure 6). The model does, however, show a discontinuity at higher superficial gas velocities as was noted earlier. At higher pressures the match is unsatisfactory. Figure 7 shows a plot of the differences in the experimental data and the model as a function of v_{SG}. At higher velocities, the model performance is poor. This would indicate that the modified Taylor bubble holdup method performs reasonable well at low velocities where some type of bubble/slug/churn flow exists. However at high gas velocities, holdup for the churn and pseudo-annular flow regimes are poorly predicted.

Extrapolation of the data to zero holdup shows a good agreement with the model of Turner et al. (1969) for liquid loading in gas wells. The data is best described with a N_{WE} = 37 and when the 20% correction factor utilized by Turner et al. is <u>not</u> applied.

This changes the coefficient of Eq. (9) from 0.566 to 0.512. It can be observed on Figure 8 that normalizing v_{SG} with v_t has the effect of collapsing all the data onto a single curve. This figure shows a straight line behavior from liquid holdups less than 0.2 to blowout conditions.

Based on these observations, the following modification to the Arpandi, et al. model is suggested. First, the Arpandi et al. model is utilized to predict the low pressure ZNLF performance. This model is observed to perform well at low pressures and low to moderate velocities. Since it is a physical model it has the capability to consider differing pipe diameters, lengths, and fluids and, therefore, is a versatile design tool. At higher gas velocities, when the value of $H_{L0} < 0.2$, the performance of the modified Taylor bubble model is poor and discontinuities are observed. Experimental data indicate a straight line function of H_{L0} versus (v_{SG}/v_t) at these low values of liquid holdup. It is suggested that the Arpandi model should be used to calculate H_{L0} until a value of v_{SG} is reached where the following condition is met:

$$-\frac{dH_{L0}}{dv_{SG}} = \frac{H_{L0}\,|_{v_{SG}}}{v_t - v_{SG}} \quad \text{.............(10).}$$

That is, the Arpandi model should be used at values of v_{SG} from zero until a point where a line drawn tangent to the curve can be extended to equal the terminal velocity, v_t, as defined by Eq. (9) (with the 0.512 coefficient). This limiting superficial gas velocity, $(v_{SG})_{LIMIT}$, corresponds to a limiting valve of liquid holdup, $(H_{L0})_{LIMIT}$. For higher gas velocities, the following straight line equation can be used to predict H_{L0}.

$$H_{L0} = (H_{L0})_{LIMIT}\left(\frac{v_t - v_{SG}}{v_t - (v_{SG})_{LIMIT}}\right) \quad \text{.............(11).}$$

Secondly, the low pressure prediction of this modified model must be corrected for pressure effects. As demonstrated by Figure 8, corrections for high pressures can be performed by normalizing superficial gas velocities by the value of v_t that corresponds to the pressure desired. The curve generated in the first step represents this curve for low pressures (P = 100 psig). Therefore, using a value of v_t for 100 psig will define a single dimensionless curve similar to that shown in Figure 8. This curve is valid for both low and high pressures. To calculate the high-pressure values of H_{L0} at a given v_{SG}, the v_{SG} is divided by a value of v_t that is calculated based on the pressure level desired.

6 CONCLUSIONS

Field-scale experimental data indicate that pressure has a significant impact on liquid holdup for vertical ZNLF. The existing model of liquid holdup poorly predicts behavior for pressures greater than 689.4 kPa (100 psig). A methodology has been proposed to improve both the stability of the existing model and its ability to predict high pressure liquid holdup behavior. By normalizing the superficial gas velocity by the minimum lift/annular flow transition velocity, the effects of pressure collapse onto a single dimensionless curve. The data is best fit using the Turner et al. (1969) model to predict this transition velocity with a slight modification to the Weber number.

7 ACKNOWLEDGEMENTS

This work was sponsored by Chevron, the State of Louisiana Board of Regents and the U.S. Minerals Management Service (MMS). The authors wish to thank Dr. Gene Kouba and Dr. Ovadia Shoham for their encouragement and the many useful discussions regarding these experiments. We would also like to thank Dr. Jinghai Yi and Mr. James Tidwell for their assistance in constructing the test apparatus and in performing these experiments.

8 NOMENCLATURE

D = diameter
g = acceleration of gravity
H = holdup
N = dimensionless number
v = velocity
ρ = density
σ = surface tension
μ = viscosity

Subscripts and Superscripts
BO = blow out
crit = critical
d = drag or droplet
D = drift
Eö = Eötvös
G = gas
G0 = gas, zero net liquid flow
L = liquid
L0 = liquid zero net liquid flow
m = mixture
s = slip
SG = superficial gas
SL = superficial liquid
t = critical gas velocity
WE = Weber

9 REFERENCES

1. Amaravadi, S. K. Minami and O. Shoham: "The Effect of Pressure on Two-Phase Zero-Net Liquid Flow in Upward Inclined Pipes," SPE paper number 28544 presented at the SPE 69th Annual Technical Conference, New Orleans (1994).

2. Arpandi, I, A.R. Joshi, O. Shoham, S. Shirazi, and G.E. Kouba: "Hydrodynamics of Two-Phase Flow in Gas Liquid Cylindrical Cyclone Separators," *SPEJ*, 427-436 (1996).

3. Bendiksen, K.H.: "An Experimental Investigation of the Motion of Long Bubbles in Inclined Tubes," Int. J. Multiphase Flow, **10**, 467-483 (1984).

4. Davies, R. M., and G. I. Taylor: "The Mechanics of Large Bubbles Rising Through Extended Liquids and Through Liquids in Tubes," *Proc. Roy. Soc.*, **A200**, 375 (1950).

5. Harmathy, T. Z.: "Velocity of Large Drops and Bubbles in Media of Infinite or Restricted Extent," *AIChE J.*, **6**, 281 (1960).

6. Hinze, J. O.: "Fundamentals of the Hydrodynamic Mechanism of Splitting in Dispersion Processes," *AIChE J.*, **1**, 289 (1955).

7. Kokal, S. and J.F. Stanislav: "An Analysis of Two-Phase Flow in Inclined Pipes with Zero Net Liquid Production," J. Can. Pet. Technology, 60-64 (1986).

8. Kouba, G.E.: "A New Look at Measurement Uncertainty of Multiphase Flow Meters," *ASME J. Energy Resources Tech.*, **120** (March 1998).

9. Kouba, G. E., O. Shoham, and S. Shirazi: "Design and Performance of Gas-Liquid Cylindrical Cyclone Separators," Proceedings of the BHR Group 7th International Meeting on Multiphase Flow, Cannes, France, 307-325 (1995).

10. Taitel, Y., D. Barnea, and A. E. Dukler: "Modeling Flow Pattern Transitions for Steady Upward Gas-Liquid Flow in Vertical Tubes," *AIChE J.*, **26**, 345-354 (1980).

11. Turner, R. G., M. G. Hubbard, and A. E. Dukler: "Analysis and Prediction of Minimum Flow Rate for the Continuous Removal of Liquid from Gas Wells," *J. Petroleum Tech.*, **21**, 1475, (1969).

12. Weber, M.E. , A. Alarie, and M.E. Ryan: "Velocity of Extended Bubbles in Inclined Tubes," Chem. Eng. Science, **41**, 2235-2240 (1986).

Gas Composition

COMPONENT	Mole %
Hydrogen	2.195583
Nitrogen	0.959395
Methane	94.53269
Carbon Dioxide	0.965665
Ethane	1.277686
Propane	0.068979

Gas gravity - 0.564516

Table 1

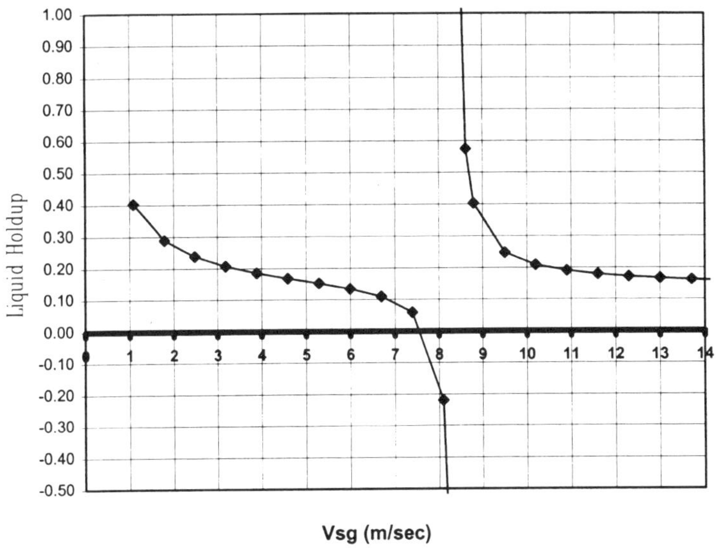

Figure 1
Discontinuous Holdup Function - 689 kPa (100 psi)

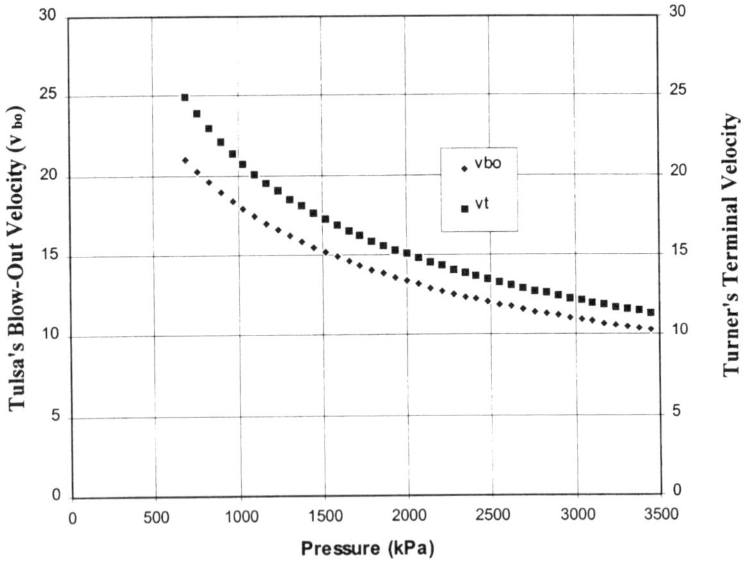

Figure 2
Blow-Out vs Terminal Velocity

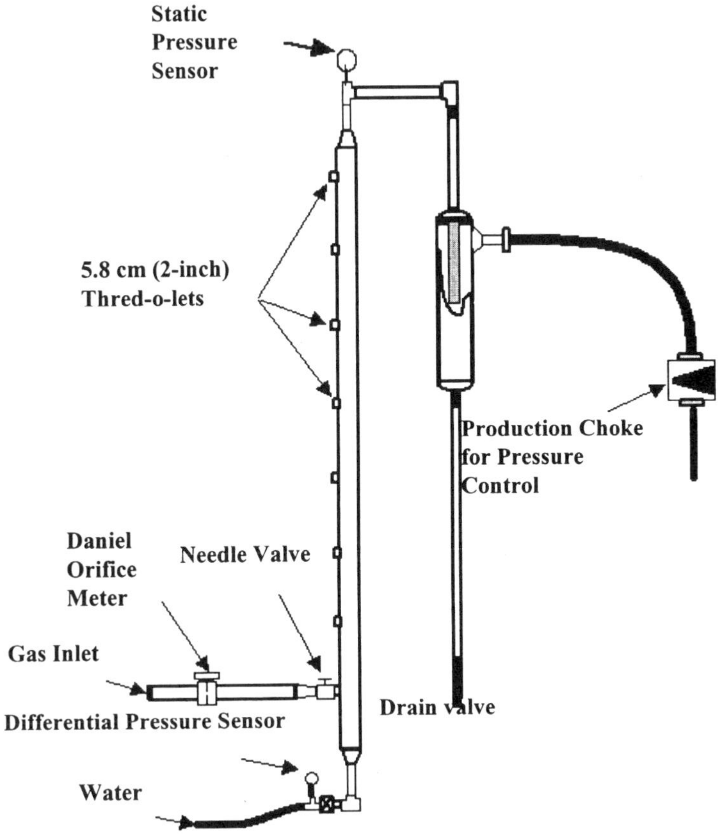

Figure 3
High Pressure Zero Net Liquid Flow Test Apparatus

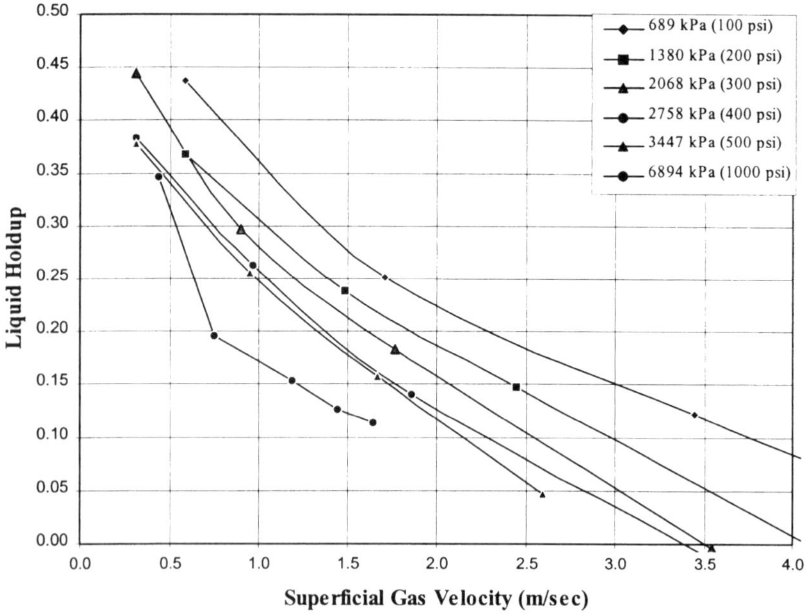

Figure 4
High Pressure Experimental Results

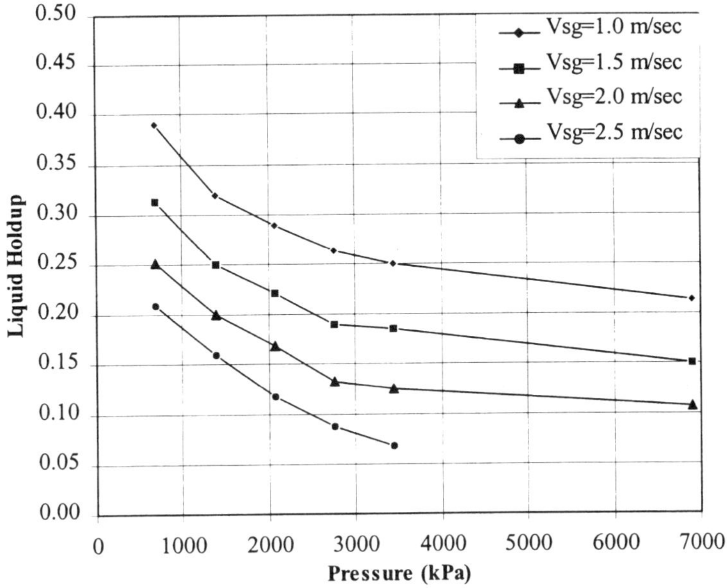

Figure 5
Effects of Pressure on Holdup

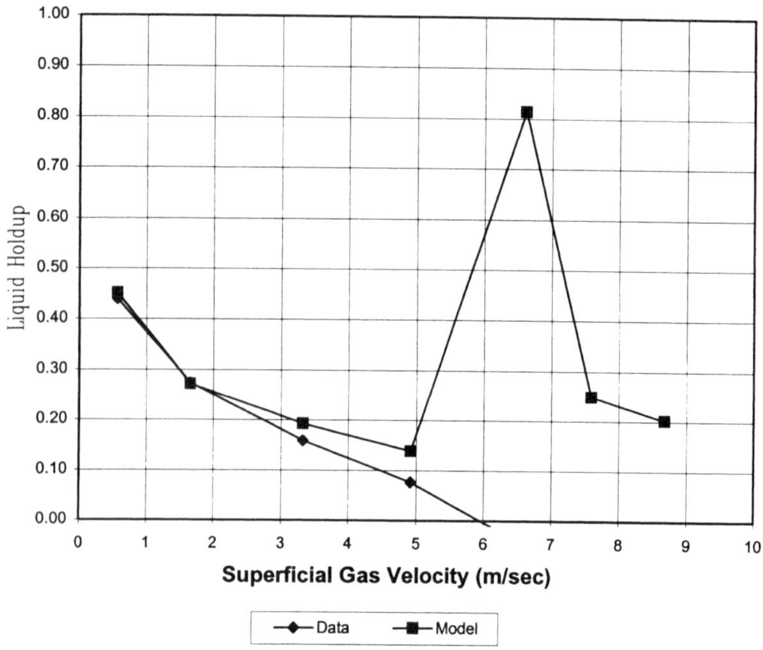

Figure 6
Comparison of Arpandi Model with Experimental Data
689 kPa (100 psi)

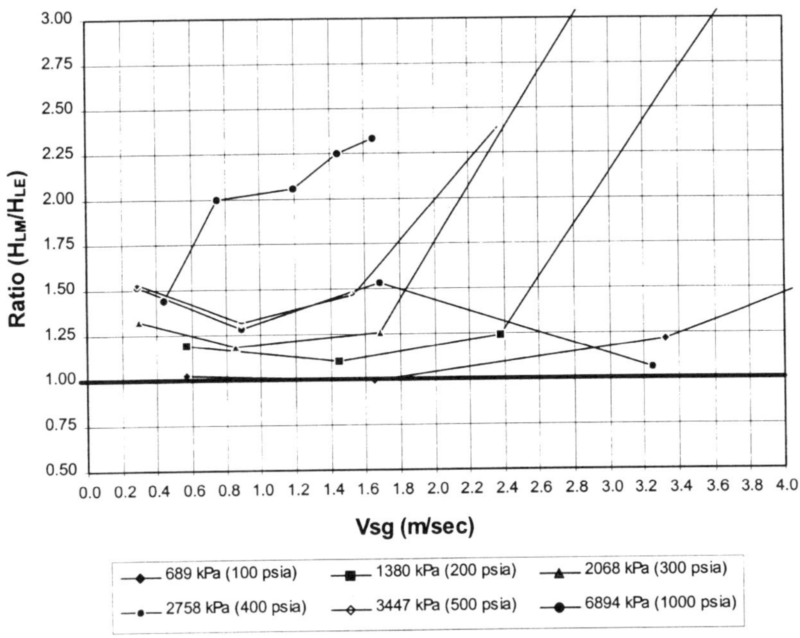

Figure 7
Error in Model Holdup Predictions

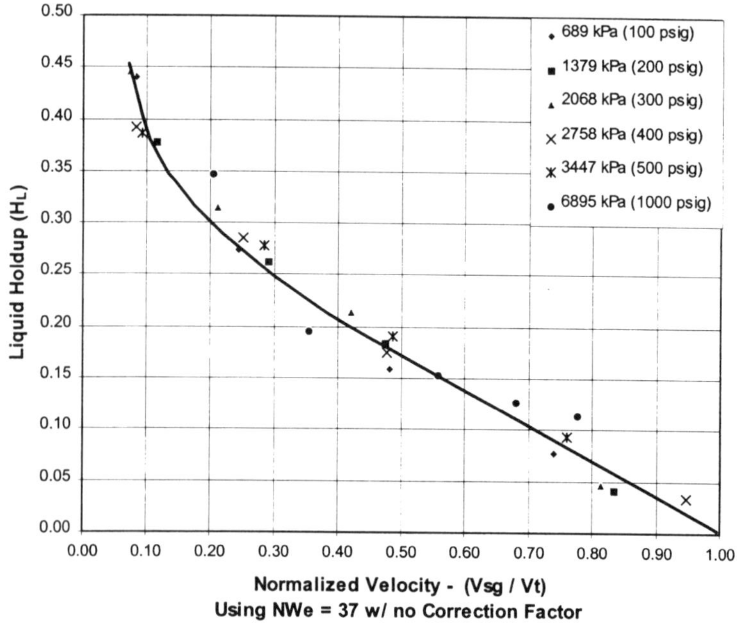

Figure 8
Normalized Velocity vs Liquid Holdup

Primary separation of oil–water mixtures: an analysis of plate separators

A M DAHL and **A HALLANGER**
Christian Michelsen Research A/S, Norway
H LINGA
FRAMO Engineering, Norway
G A DAVIES
University of Manchester Institute of Science and Technology (UMIST), UK

In order to increase the coalescence rate and hence the separator efficiency, parallel plates can be installed in gravity drum separators. The paper addresses the performance of a plate separator operating in cross-flow mode and with two-phase liquid-liquid flow. Selected results are shown for the plate separator performance as a function of plate angle, plate distance, plate material and plate location in the separator volume. Special attention has been put on the separator performance as function of hydraulic conditions upstream to the plates as affected by the distance to the inlet distributor baffle and the length of the separator volume.

1 INTRODUCTION

Primary separation of oil-water dispersions is associated with the two processes, droplet settling and droplet coalescence. In ordinary gravity separators, both the settling rate and the coalescence are low. Thus in order to gain acceptable performance of gravity separators the separator residence time and hence the size need to be relatively large. In order to increase the settling rate the dispersion can be provoked to electrostatic fields or centrifugal forces. Introducing solid surfaces at which droplets may coalesce with other droplets or with a liquid film may increase the coalescence rate. This paper addresses the latter method, performance of a gravity separator whenever a collection of plates parallel to the separator axis is installed in a part of the separator volume.

In principle, plate separators may be classified in three systems in terms of flow direction and draining, see Fig. 1:

1) Cross-flow.

The main flow within the plates is parallel to the horizontal. Perpendicular to the main flow direction the plates are tilted at an angle relative to the horizontal so that the coalesced dispersed droplets are allowed to drain off the plate surface.

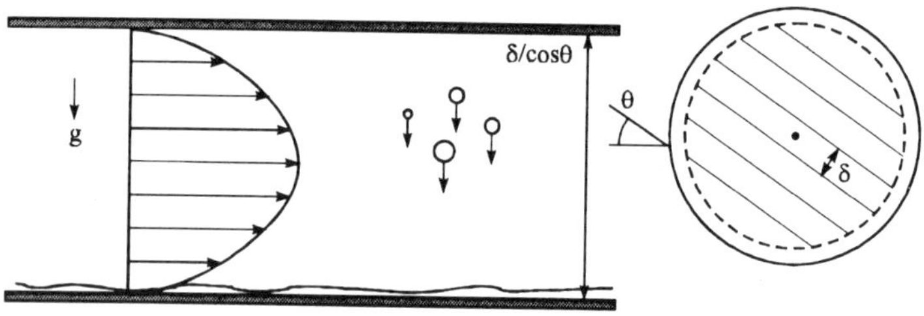

Figure 1a: Plate separator systems. Cross flow. Channel in vertical cut plane parallel to separator axis.

2) Counter-current flow.

In the direction of the main flow, the plates are inclined relatively to the horizontal. For example, with oil dispersed in water continuous flow the plates are inclined downward. Thus the oil droplets rise to the upper plate surfaces on which they coalesce and an oil film moves upward counter-currently to the main flow direction.

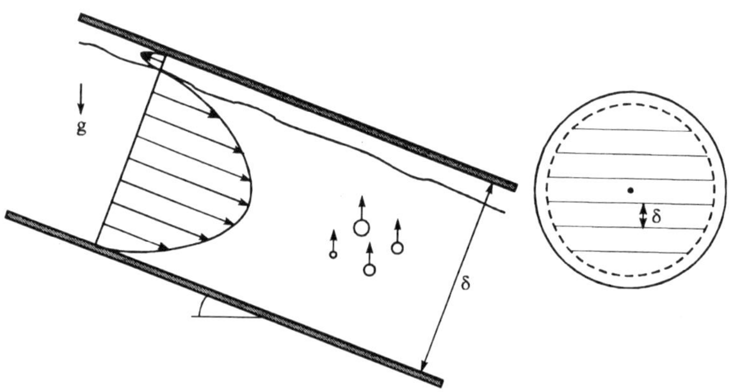

Figure 1b: Plate separator systems. Counter-current flow. Channel in vertical cut plane parallel to separator axis.

3) Co-current flow.

Similarly to the counter-current flow system, the plates are inclined in the direction of the main flow with an angle to the horizontal. For example, with water dispersed in oil continuous flow the plates are inclined downward. Thus the water droplets settle to the lower plate surfaces on which they coalesce, and a water film moves downward co-currently to the main flow direction.

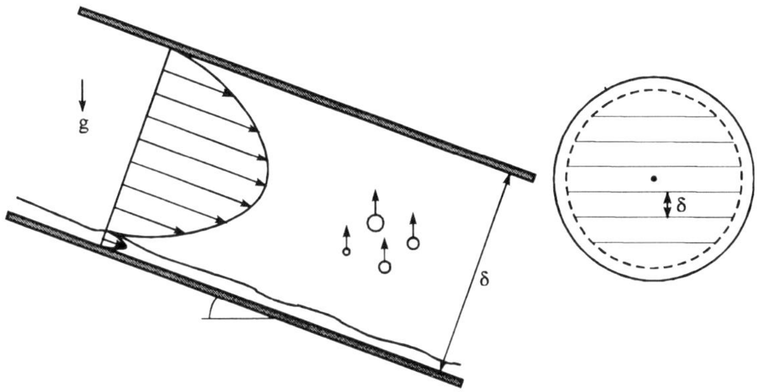

Figure 1c: Plate separator systems. Co-current flow. Channel in vertical cut plane parallel to separator axis.

Additionally, the counter-current and co-current flow separators can be configured to yield cross-flow of the coalesced phase.

In this paper, we address the performance of a cross flow separator for liquid-liquid flow of water dispersed in oil continuous flow. Experimental findings concerning plate separator efficiency as a function of plate angle, plate distance and plate material will be shown. Moreover, the importance of the flow field upstream to the separator will be demonstrated with experimental results as well with CFD-simulations obtained with the 3-D three-phase flow simulator GALIFA (1).

2 THEORY

2.1 Terminal settling-velocity

From the balance between the forces from friction, buoyancy and gravity, the terminal settling velocity u_s of a single droplet of diameter d_d and density ρ_d in a continuous phase of density ρ_c and viscosity μ_c can be expressed;

$$u_s = \left[\frac{4}{3} \cdot \frac{g \cdot d_d}{f_d}\right]^{\frac{1}{2}} \cdot \left[\frac{\rho_c - \rho_d}{\rho_d}\right]^{\frac{1}{2}} \tag{1}$$

where f_d is the friction factor between the droplet and the continuum. In the Stokes regime, $2 \cdot 10^{-5} < \mathrm{Re}_d < 2$, where Re_d is the droplet Reynolds number;

$$\mathrm{Re}_d = \frac{\rho_d \cdot u_s \cdot d_d}{\mu_d} \tag{2}$$

The settling velocity u_s is expressed as,

$$u_s = \frac{g \cdot d_d^2 \cdot (\rho_c - \rho_d)}{18 \cdot \mu_d} \tag{3}$$

2.2 The cut-off diameter

Standard methods for predicting the plate separator potential as a function of droplet size are based on the concept of the cut-off diameter d_{co}. For a given dispersion droplets greater than d_{co} will be separated, whereas droplets smaller than d_{co} will be entrained.

The mean residence time T of the liquid within the plate separator is given by,

$$T = \frac{L}{U} \tag{4}$$

where L is the separator length and U is the mean velocity in the main flow direction within the channel. The cut-off diameter is associated with droplets entering the channel just below the upper plate. Thus this droplet will be captured at the lower plate if the droplet propagates the vertical spacing between the plates $\delta/\cos\theta$ within the residence time T. Here δ is the plate distance and θ is the angle between plates and the horizontal. From the considerations above the following expression for the cut-off diameter can be expressed;

$$d_{co} = \sqrt{\frac{18 \cdot \delta \cdot \mu \cdot U}{L \cdot (\rho_c - \rho_d) \cdot g \cdot \cos\theta}} \tag{5}$$

For rough estimates, the separator efficiency can be calculated from Eq. (5) if the droplet distribution is known. Equation (5) relies on simplifications as no droplet-droplet interaction, no-slip in the main flow direction and uniform flow field, droplet distribution and concentration at the plate separator inlet. However, Eq. (5) will give some idea of what droplet sizes can ideally be expected to separate.

All plate separators are designed to operate with the continuous phase in laminar flow. Consider a single channel in a plate pack. Particles entrained in the feed will settle and a vertical separation will take place in the channel. The forces acting on a particle, viz. gravity, buoyancy, drag and inertia, can be expressed by the equation:

$$\frac{\pi \cdot d_p^3}{6} \cdot (\rho_p - \rho_c) \cdot \mathbf{g} + c_D(\text{Re}) \cdot \frac{\pi \cdot d_p^2}{4} \cdot \frac{\rho_c}{2} \cdot \mathbf{u}' \cdot \mathbf{u}' - \frac{\pi \cdot d_p^2}{6} \cdot \rho_p \cdot \frac{d\mathbf{u}_p}{dt} = 0 \tag{6}$$

The particle is assumed to be spherical and the diameter is d_p. c_D is the drag coefficient, which is a function of the particle Reynolds number, *Re,* where

$$\text{Re} = \frac{u' \cdot d_p \cdot \rho_c}{\mu}$$

And **u**' is the relative velocity of the particle to that of the continuous plane fluid. The trajectory of an oil droplet at any instant is shown in Figure 2.

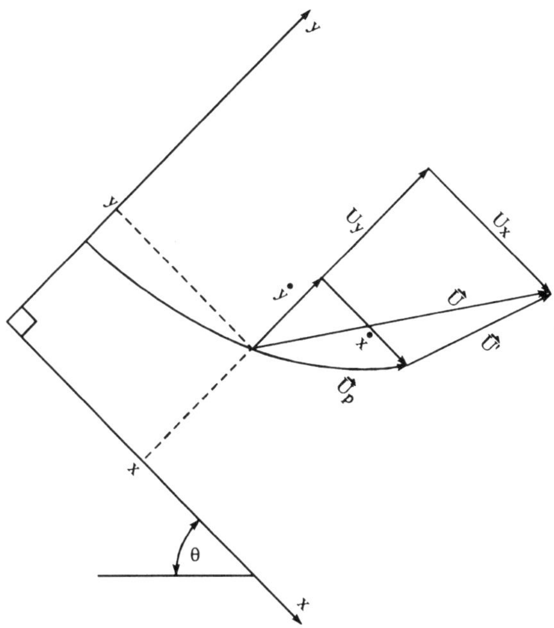

Figure 2: Forces acting on a particle.

This trajectory clearly is dependent on the local fluid velocity **u**, components, u_x *and uy.* If the effect of the dispersed phase on this flow is ignored then **u** can be readily evaluated, If steady laminar flow is assumed then generally **u** = **u**(*x, y*). In the entrance region of the channel this has been evaluated by Wang and Longwell (2) and Schlichting (3). When the velocity profile is fully developed **u** = **u**(*y*) and

$$u_x = \frac{-P' \cdot Y^2}{2 \cdot \mu} \cdot \left[1 - \left(\frac{y}{Y} \right)^2 \right] \tag{7}$$

where

$$P' = \left[\frac{\Delta P}{L} + \rho_c \cdot \mathbf{g} \right]$$

For cross-flow, Figure 1a $\mathbf{g}$=0 and $P' = \Delta P/L$
For co- or counter-current flow, Figure 1b and Figure 1c

$$P' = \left[\frac{\Delta P}{L} + \rho_c \cdot g \cdot \sin(\theta) \right]$$

For both cases, $u_y = 0$.

If we return to equation (6), $\mathbf{u}_p = u_p = u_p(x, y)$ and equation (6) can be simplified

$$\frac{-du_p}{dt} + a \cdot \mathbf{g} + b \cdot c_D(\mathrm{Re}) \cdot \mathbf{u} \cdot \mathbf{u}' = 0 \tag{8}$$

where

$$a = \frac{\rho_p - \rho_c}{\rho_c} \tag{8i}$$

and

$$\mathrm{b} = \frac{3}{4} \cdot \frac{\rho_c}{\rho_p} \cdot \frac{1}{d_p} \tag{8ii}$$

When equation (8) is expressed as components resolved in the direction of the axes, x, y, z, the terms depend on the type of system. If cross flow is considered then

$$\mathbf{g} \begin{bmatrix} 0 \\ -g\cos \\ -g\sin \end{bmatrix}, \mathbf{u} \begin{bmatrix} u_x \\ u_y \\ u_z \end{bmatrix}, u_p \begin{bmatrix} \dot{x} \\ \dot{y} \\ \dot{z} \end{bmatrix}, u' \begin{bmatrix} u'_x \\ u'_y \\ u'_z \end{bmatrix}$$

and

$$u'_x = u_x - \dot{x}\, etc.$$

$$u' = \left[(u_x - \dot{x})^2 + (u_y - \dot{y})^2 + (u_z - \dot{z}) \right]^{\frac{1}{2}} \tag{9}$$

Equation (8) must be solved to compute the trajectory of the particle in the channel. The conditions for capture or entrainment being for co-counter current flow

$y = +Y$, $x \leq L$ $\rho_p < \rho_c$
$y = -Y$, $x \leq L$ $\rho_p > \rho_c$

and for cross flow

$y = +Y,\ x \leq L$ and $z \leq Z$

The solution of equation (8) requires an expression for c_D. Under normal conditions the particles separated are small, $d_p \sim 100$ μm, and settling takes place in Stokes' or Newton's ranges. In Stokes' region $Re_p<1$

$$c_D = \frac{24}{\mathrm{Re} \cdot \lambda} \tag{10}$$

λ is a correction factor[4] to allow for the fluidity of the particle

$$\lambda = \frac{2 \cdot \mu_c + 3 \cdot \mu_p}{3 \cdot (\mu_c + \mu_p)} \tag{10i}$$

λ = 1 for a solid particle and for a liquid drop 0.66< λ <1.0.

If equation (10) is substituted into the equation describing the forces acting on the particle equation (8) becomes:

$$\frac{-du_p}{dt} + a \cdot g + \frac{18 \cdot \mu}{\rho_p \cdot d_p^2 \lambda} \cdot \mathbf{u}' = 0 \tag{11}$$

The solution of equation (11) depends on two factors, the particle diameter d_p and the initial boundary condition, the entry point to the channel, (0, y, z). Since a counter-current flow system is considered this becomes (0, y).

The solution of equation (11) will provide the locus of the trajectory of a particular particle in the channel leading to capture or entrainment. Indeed the conditions outlined in the simple cut-off analysis will be one particular trajectory. The mixture entering a separator is polydispersed so that no single trajectory can define the overall process and a way must be found to link the motion of particles to the inlet and outlet concentrations. This will then give an immediate expression for the separation efficiency.

The mass concentration of particles at any point in the separator may be defined by C. In general

$$C = C(x, y, z, t) \tag{12}$$

The concentration of particles of a given size d_p in the suspension is

$$C_D = C(x, y, z, t) f_{x,y,z,t}(d_p) \tag{13}$$

where $f_{x,y,z,t}(d_p)$ is the mass density or distribution function with respect to particles of a diameter d_p. If the minimum particle diameter in the suspension at *x, y, z, t* is d_{min} and the maximum diameter d_{max}, then

$$\int_{d_{min}}^{d_{max}} f_{x,y,z,t}(d_p)\,d\,d_p = 1 \tag{14}$$

The mass conservation equation can now be written for a particular particle diameter

$$\frac{\partial C_d}{\partial t} + \nabla \cdot (C_d \cdot \mathbf{u}') = 0 \tag{15}$$

If this equation is combined with equations (7), (9) and (11) the concentration at any point in the channel can be computed. Assuming steady flow

$$C(x,y,z) = \int_{d_{min}}^{d_{max}} C_d(x,y,z)\,dd_p \tag{16}$$

At the outlet x = *L* the exit concentration, C_E, is:

$$\frac{\int_{-Y}^{+Y}\int_{-Z}^{+Z} C(L,x,y)\,dz\,dy}{\int_{-Y}^{+Y}\int_{-Z}^{+Z} dz\,dy} = C_E \tag{17}$$

Then the separation efficiency η is

$$\eta = \frac{C_0 - C_E}{C_0} \tag{18}$$

This formal procedure must be followed if the concentration of particles in the channel is to be related to the trajectory of particles. It requires that the mass density distribution function is defined for the inlet conditions.

The procedure can be applied to convert the calculations of cut-off size, and the assumptions inherent in this, to an exit concentration. Thus from the simple analysis

$$C_E \cong C_0 \int_{d_{min}}^{d_{co}} f_{0,y,z}(d_p)\,dd_p \tag{19}$$

Before considering a method of solution of the equations it should be emphasised that in deriving an expression for **u** no account was made for the effect of the coalesced phase and the flow of this in the channel. For example in counter-current separators the oil must flow out of the channel and if the concentration in the feed, C_0, is large this will influence **u**. This will be considered later.

In order to solve equation (17) to calculate C_E the mass distribution function $f_{0,y,z}(d_p)$ must be known. It will be assumed that the feed is well mixed so that at the inlet

$$f_{x,y,z}(d_p) = f_0(d_p) \tag{20}$$

but beyond this point

$$f(d_p) = f_{x,y,z}(d_p)$$

A stochastic solution is applied to this problem using a Monte Carlo procedure to link the particle trajectories to the free stream concentration, *C*.

The inlet conditions to the separator must be defined in terms of flow, concentration of dispersed phase and mass density function $f_0(d_p)$. The mean velocity of the liquid entering each channel is easily determined from the feed flowrate *V*. Thus

$$\bar{u}_c = \frac{\upsilon}{(m-1)\cdot 4\cdot Y\cdot Z} \tag{21}$$

where *m* is the number of plates.

In applying a stochastic solution to the equations, individual particles in the inlet distribution will be selected, in doing this the diameter density distribution function $f'(d_p)$ rather than the mass density distribution $f(d_p)$ is used. $f'(d_p)$ can easily be transposed into $f(d_p)$. In this work normal and log normal distribution function were used to define $f'_0(d_p)$. These only require two parameters to define them, d_p and σ. Plate separators, relying on gravity settling should only be used to separate primary dispersions hence a limit on d_{pmin}>30 μm is appropriate. Thus $\bar{d}$ is chosen such that $\bar{d}_p > d_{min} + 2\sigma$.

To compute the exit concentration from this point a particle is selected from $f'_0(d_p)$. The trajectory is computed and the outcome determined, captured in the channel or entrained .By repeating the calculations successively for a large number of particles, C_E can be determined. The details are as follows. A particle diameter d_i is selected from the distribution $f'_0(d_p)$ using the central limit theorem. Thus

$$d_{pi} = \bar{d}_p + \sigma \cdot \left[\frac{\sum_{1}^{M} r_j - (M/2)}{\sqrt{(M/12)}} \right] \tag{22}$$

where r_j is a uniformly distributed random variable between 0 and 1 and *M* is an integer $M \geq 12$. The point of entry into the channel is then determined, assuming equation (20) holds. For cross flow systems *y* and *z* are selected at random such that $-Y \leq y \leq Y$ and $-Z \leq z \leq Z$, for co- and counter-current systems the flow can be considered as 2-dimensional and only *y* needs to be considered. To simplify the discussion the latter cases will be described. The channel is broken down into a series of equal increments in the *y* direction, Δy. Starting at the inlet at (0,*y*) first the terminal settling velocity of the particle is determined using equation (8). The particle Re is then determined and the appropriate equation for calculation of the drag coefficient used. The particle velocity is then computed, $\mathbf{u}_p$, using equation (8). The first increment of the trajectory is then calculated. If the particle is less dense than the continuous phase fluid the settling will take place in the +ve y direction and the position x_1, when the particle has travelled from y_0 to $y_0 + \Delta y$, is computed. The average velocity u_x between y_0 and $y_0+\Delta y$ is calculated to provide x_1. The second increment is then computed $x_2 = x_1+\Delta x$ for $y_1+\Delta y$ where $y_1= y_0 + \Delta y$. The procedure is repeated until either:

(a) y_n=Y and $x_n \leq L$

or

(b) y_n<Y and $x_n= L$

In the first condition the particle is determined to be captured in the separator at x_n. In the second condition, the particle is entrained through the separator. In each case d_i the condition (a) or (b) is recorded and if (a) pertained then x_n is recorded. The process is repeated for between 1 x 10^4 and 1 x 10^5 drops. The total mass of particles entering and leaving the system is computed.

In this example $\rho_p<\rho_c$, this corresponds to the separation of oil from water in a T.P.I. For the system to work oil coalesced on the underside of the upper plate, $y=Y$, must then flow counter-currently up the channel. In the calculations this has been ignored and may be justified at low oil concentrations C_0. If however C_0 is high, say >10% wt/v then the thickness of the oil film and its effect on the velocity profile in the channel must be taken into account. At any point in the channel the velocity profile will be modified, the change is shown on Figure 3.

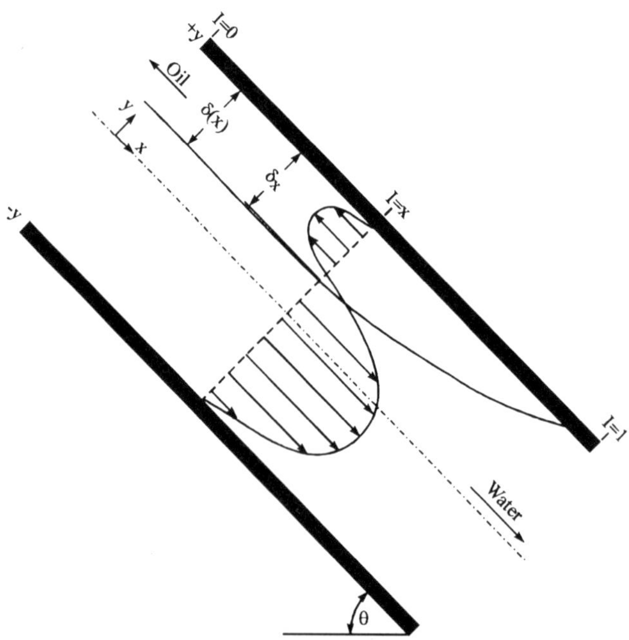

Figure 3: Velocity profile in a counter-current, T.P.I., plate separator.

The analysis of the flow to give $u_x = f(y)$ is straightforward if the film thickness, $\delta=\delta(x)$, is known. δ depends on the droplet trajectories and the mass density of captured droplets along the channel. This will not be known a priori. In the simulation, however, $\delta=\delta(x)$ can be computed assuming coalescence takes place at $y = +Y$ and (in the 1st iteration) neglecting the effect of C_0 on the velocity profile. The point where each captured droplet coalesces is computed. The volume is known and assuming cumulative flow in the channel the profile can be computed. If the sample is large enough then the number, mass and coalescence flux along the channel can be calculated for a given inlet mass concentration C_0. The profile $\delta=\delta(x)$ can then be calculated. The calculations are then all repeated for the same number of droplets fed into the channel but now the velocity profile is modified to allow for $\delta=\delta(x)$ from the 1st iteration. The outlet concentration from this second iteration C_{E2} is compared with C_{E1}. If the difference is less than an acceptable error the calculations are terminated and the separation efficiency computed. Otherwise the velocity profile is recalculated for $\delta=\delta(x)_2$, from the second iteration, and the whole procedure repeated.

From the analysis data on C_E, $\delta(x)$ and the mass density distribution function of droplets entrained from the channel $F_{Ly}(d_p)_E$, can be computed. The latter data is particularly important if, as is usually the case in offshore processing, a secondary separator is used in series with a plate system.

A wide range of design conditions, plate length L, spacing $2Y$, flowrate defined as loading per unit cross section (free) area of plate pack, m^3/m^2hr, have been studied. Fluid properties ρ_c, ρ_p and μ have also been varied over ranges typical of offshore operations. The performance has been computed in terms of the mass of dispersed phase leaving the separator per unit mass entering the separator C/C_0, These concentrations refer to the sample size used in the simulation. A whole range of data will be provided. In addition to the parameters referred to above, the distribution function $f'_0(d_p)$ is introduced. To simplify the review of results the data for $f'_0(d_p)$ is defined in terms of two-dimensionless variables:

$$\alpha = \frac{\bar{d}_p}{d_{co}}$$

$$\beta = \frac{\sigma}{\bar{d}_p}$$

It is convenient to express the 1st moment of the distribution to the cut-off size since it allows ready comparison to the region of operation v.v. present design ranges and d_{co} takes in plate design, flow and physical property data.

Cross flow and co-counter-current systems have been analysed. The appropriate equations for d_{co} and $\mathbf{u}_p$ have been used.

Results to show the effect of increasing the flowrate, keeping all the other conditions constant are first shown in Figure 4. This simply illustrates the fact that as expected the entrainment increases with increasing flow. Superimposed on this figure are results for two plate spacings $Y_1 > Y_2$ and again the entrainment increases with increasing plate spacing. This is as expected, the model predicts the trend. If results were displayed in this way examining each of the parameters in equations (8) and (9) influencing the trajectory of particles passing through the separator, the conclusions confirmed are that C_E/C_0 increases as the plate length and density ratio, $(\rho_c - \rho_p)/\rho_c$, decreases and as the continuous phase viscosity increases. The important extra feature of the stochastic model is the ability to examine the effect of the particle size distribution in the feed on the separation efficiency. The parameters α and β are used to present the results.

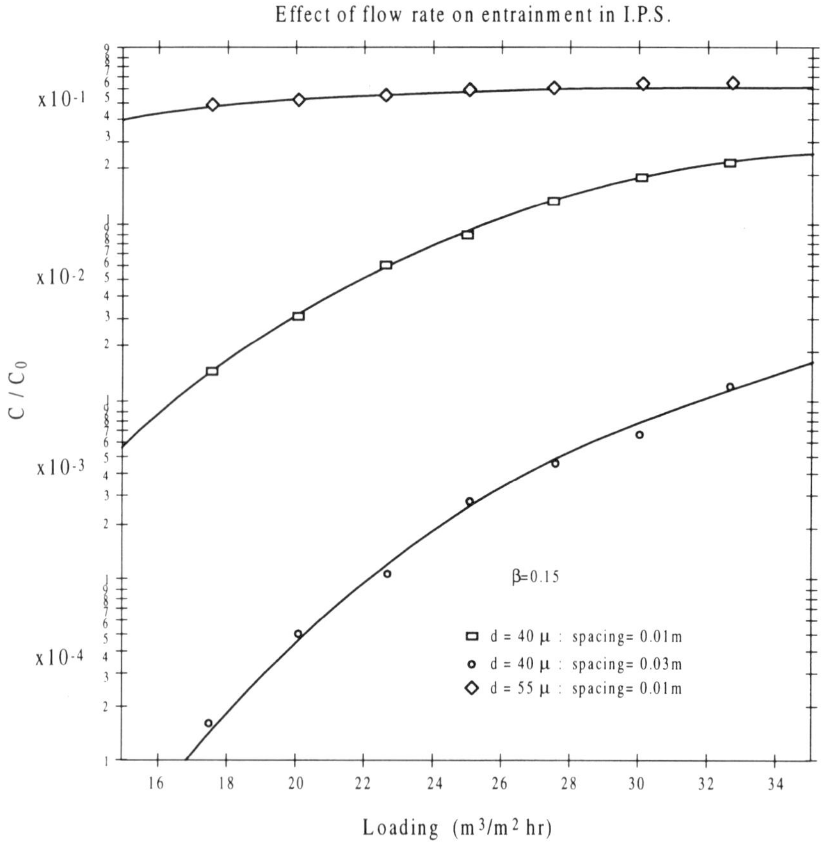

Figure 4: Variation of the outlet concentration with flowrate.

A set of results is shown in Figure 5. These were obtained for an I.P.S. in counter-current flow. The relevant process data are examined, at constant β, then entrainment decreases with increasing drop diameter. The scale for C_E/C_0 is logarithmic, the concentration of oil in the outlet stream increases rapidly as α approaches 1. The effect of the second moment σ on the performance of the separator can best be recognised at larger values of α, $\alpha>1.3$. In this region as σ increases and the dispersion becomes more diffuse and the entrainment increases.

It was emphasised that a process specification normally calls for a minimum entrainment in the effluent from a separator, C_{Emax} for a range of flow conditions. Present design techniques produce a cut-off size and field experience is quoted to back up the entrainment data. The solution described here produces a method of predicting C_{Emax}, but this by necessity requires a definition of $f_0(d)$ or $f'_0(d)$. The application of this solution may then be restricted if data on distribution functions in the feed are not known or cannot be characterised. A method of using the solution can however be found since the data can be represented as a set of "universal" curves, Figure 5. A design requirement usually specifies a maximum C_E or (C_E/C_0).

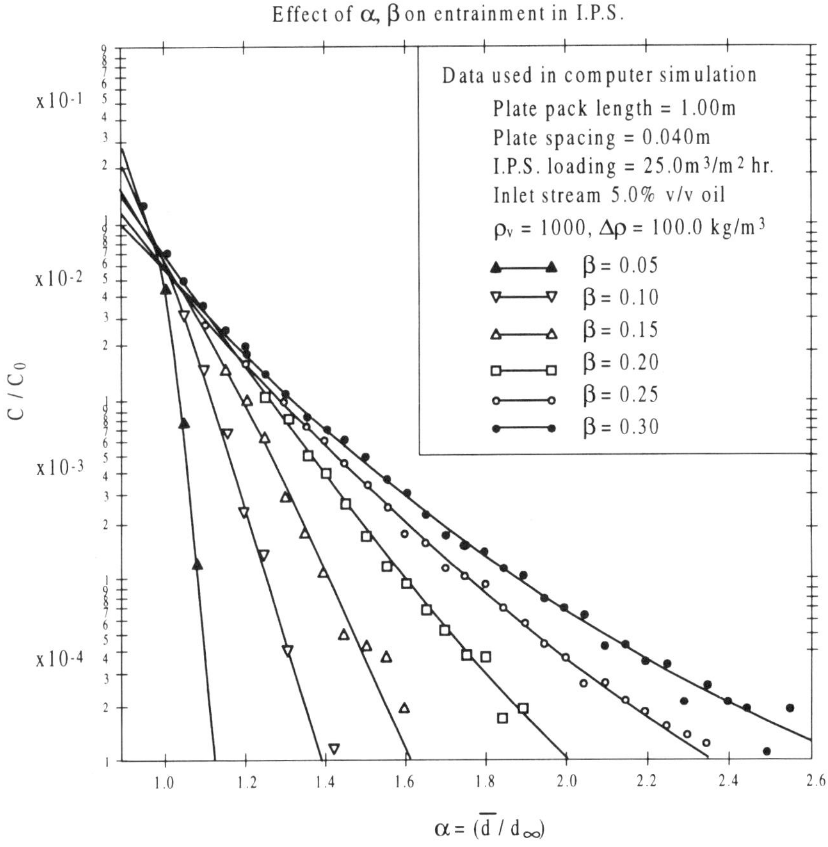

Figure 5: Predicted outlet concentration for a T.P.I. separator.

If we consider Figure 6 a value for $(C_E/C_0)_{max}$ specified for a design is shown. Values of α in the feed < α_1 could not apply since the entrainment will always exceed the permitted value, except at very low values of β. As α increases then the entrainment decreases but even at α_2

the entrainment exceeds that allowed over part of the dispersion range. If $f'_0(d)$ is not known we cannot assume to have any control of the variance ratio!

However at α_3 the predicted entrainments are less than the set value over the whole range of drop size. Therefore this represents a possible operating region. The critical condition of course is where the concentration limit becomes tangent to the β curve. This will give the minimum value of α possible to achieve the separation. A selection of design parameters must then be made, L, $2Y$, V, to determine a value for d_{co}. Practical considerations may restrict the minimum value of $2Y$ which can be used such as silting, wax formation in the separator, cleaning etc. The value of d_{co} must also rely to known conditions of upstream equipment and the primary-secondary dispersion transition. For the value of α used to ensure that C_{Emax} is not exceeded the value of d can be calculated and this must relate to upstream equipment. A check can also be made to the penetration of the dispersion in the separator in line with an agreed value. When all these factors are satisfied then a design with the predicted entrainment can be achieved. The outlet drop size distribution can then be computed for the worst case, lowest α value and β value according to the maximum condition on Figure 6.

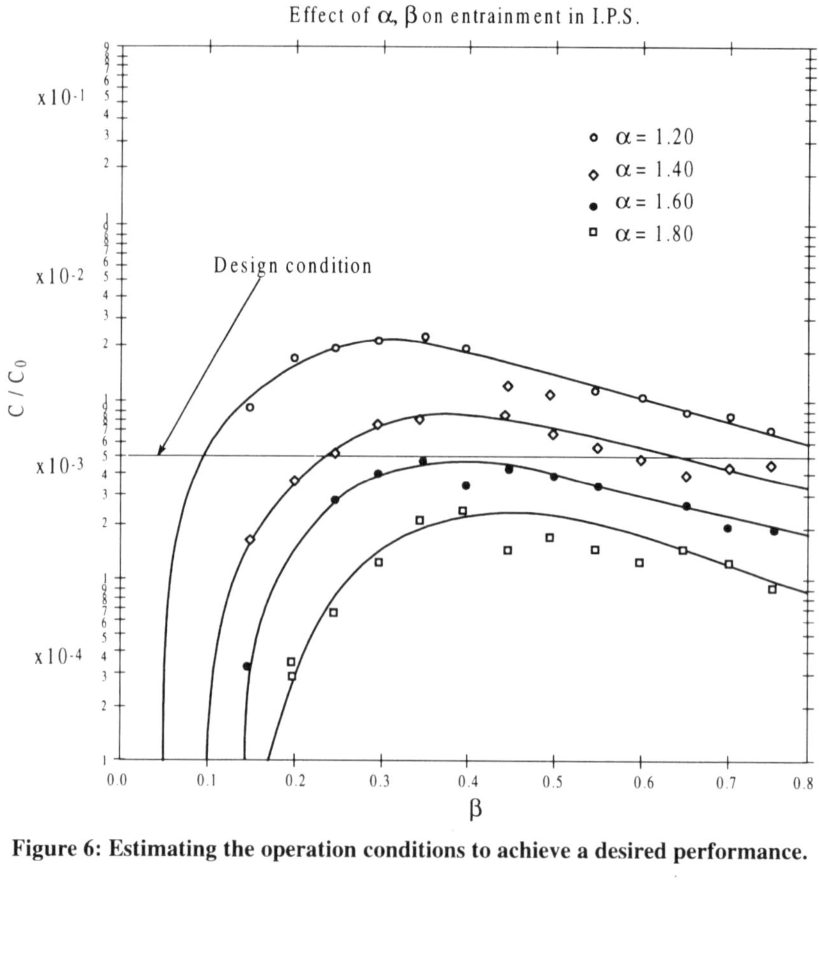

Figure 6: Estimating the operation conditions to achieve a desired performance.

3 EXPERIMENTAL SET-UP AND PARAMETER RANGE

3.1 Separator design

The separator including inlets and outlets has been designed to isolate the plate pack performance as far as possible from the hydraulic flow conditions and coalescence effects within the volumes surrounding the plate pack. The plate pack has been enclosed in a cylindrical separator, and only liquid-liquid separation has been studied within the liquid filled cylinder. The multiphase flow entering the separator has been made up by oil-water flow only, and during start-up of the multiphase flow the initial gas volume in the separator is vented.

Experiments have been run with two different lengths of the separator. The layout of the longest separator is shown in Figure 7. With the short separator, 1.3 m shorter, the separator performance was significantly affected by non-homogeneous flow through the separator, see Chapter 4. Hence, it was not possible to study the effects from the plate pack characteristics separately. However, with the long separator this was achieved.

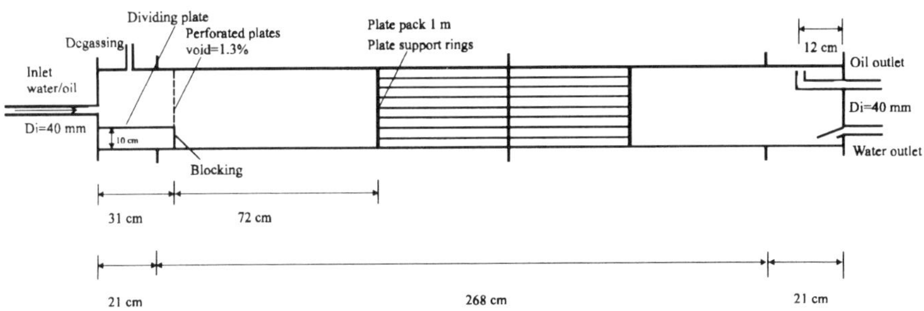

Figure 7: Simplified sketch of the long separator configuration.

3.2 Plate-pack design

A total of four plate materials have been tested in order to investigate the qualitative and quantitative differences in the plate separator performance due to plate surface properties. The materials used are glass, steel, Teflon and plywood. Here Teflon and plywood represent the extremes in terms of hydrophilic (plywood) and hydrophobic (Teflon) properties. The plates were supported by grooved PVC rings, which were reinforced with two steel rods. The plate distances used were 20 and 40 mm, corresponding to 6 and 12 plates respectively. The plate pack length was kept constant at 1 m. Figure 3 shows a cross-sectional view of the plate separator with 6 plates installed. Water drainage from the tilted sides of the plates was allowed in about 1 cm space between the plate sides and the separator wall. The plate pack was allowed to be installed at any point between the perforated plate and the outlet, see Figure 7.

3.3 Flow conditions

The oil-water flow entering the separator is mixed in a screw pump and then in a choke valve located 2.0 m upstream to the separator. In general, the choke was set in a position that gave the least separable liquid-liquid flow dispersion. The water level was kept at 25% of the separator diameter (water-oil interface located 8 cm over the bottom). A perforated plate (1.3 % void) is installed near the inlet to ensure an even flow distribution over the separator cross-section. This plate is blocked in the lower section corresponding to the water zone. A horizontal dividing plate is also installed at the inlet to avoid flushing from the inlet flow into the bulk water phase. By this, the amount of water droplets being entrained from the water bulk phase into the oil continuous zone is reduced.

The flow conditions as a function of volumetric flow rate are summarised in Table 1. Here U_{mix} is the superficial mixture velocity calculated for the separator diameter (cylinder with internal diameter 32 cm). The oil residence time corresponds to the average residence time for the oil in the separator. The limits of the intervals indicated represent water cuts 2 (left) and 20 % (right), respectively.

Table 1: Flow conditions as a function of volumetric flow rate.

Q [l/min]	U_{mix} [cm/s]	**Reynolds number [-]**				**Oil residence time [s]**
		Pipe	**Cylinder**	**Plate**		
				d=2cm	**d=4cm**	
150	3.1	21800-25900	2000-1700	195-160	390-320	79-97
200	4.1	29000-34000	3000-2300	260-210	520-420	59-73
250	5.2	36000-43000	3500-2800	320-260	650-530	48-58

3.4 Fluid properties

For the oil phase, EXXOL D100S was chosen, and the water phase was filtered tap water. To simulate formation water 30 ppm salt (Na_2SO_4) was added to the water phase. Dye was added to the dispersed phase (water) in order to obtain better contrast when observing droplet behaviour. Table 2 shows the fluid properties of the fluids at the relevant conditions.

Table 2: Fluid properties.

Fluid	Density [kg/m³]	Viscosity [cP]
Exxsol D100S	820	3.2
Water	1028	1.0

4 RESULTS

The major goal for the experiments is to reveal the mechanisms and performance of the plate pack for different configurations. Hence, the separator performance will in general be compared to the performance with the reference case; no plate pack in the separator.

Since the experiments represent oil continuous flow, the volumetric fraction of water entrained in the oil flow at the oil outlet, *WiO* [%] is defined as the measurement of the separator performance. The value of *WiO* is determined by volumetric sampling of the oil outlet flow by directing the flow into a settling tank.

The plate pack parameters studied are the plate distance, the plate angle, the plate material and the plate pack position within the separator. Typically, the flow rates are run at 150-200-250 l/min, and the water cut values 2-5-10-20 % are covered. See Chapter 3.3 for resulting flow conditions.

4.1 Plate angle

Figures 8 and 9 illustrate the entrained water *WiO* as a function of plate angle for the two plate distances 20 and 40 mm. The values as obtained for the reference case, no plate pack installed, are for convenience indicated at plate angle 90°. All the results are obtained with plywood plates located near the separator outlet. Apart from plate angles close to the vertical, a significant improved separator performance is achieved with plate packs installed.

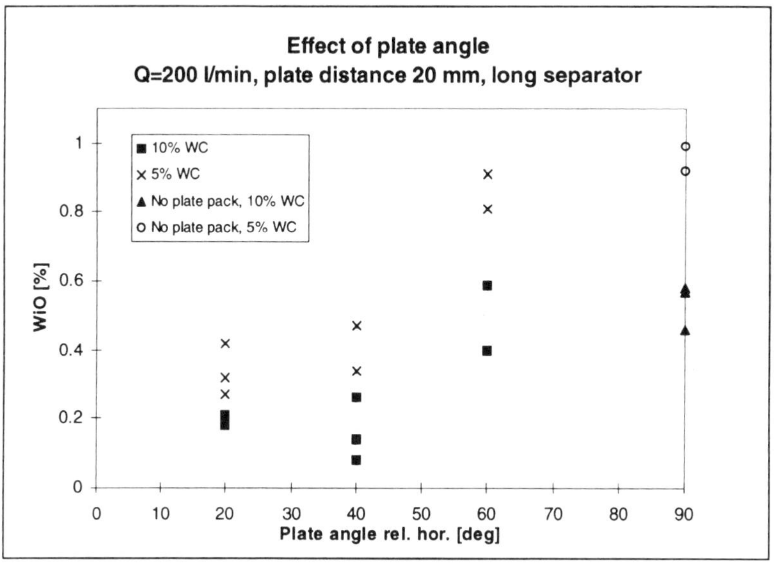

Figure 8: Water entrainment vs. plate angle.

This is also shown in Figures 10 and 11 where we have introduced the plate pack efficiency η;

$$\eta = 1 - \frac{WiO_P}{WiO_R} \tag{23}$$

where WiO_P is the results with plates installed and WiO_R is results without plates. Thus the η-value 0 % corresponds to no effect from the plate pack and 100 % corresponds to no entrained water in the oil outlet flow. The results indicate that the optimal plate angle is close to 40°.

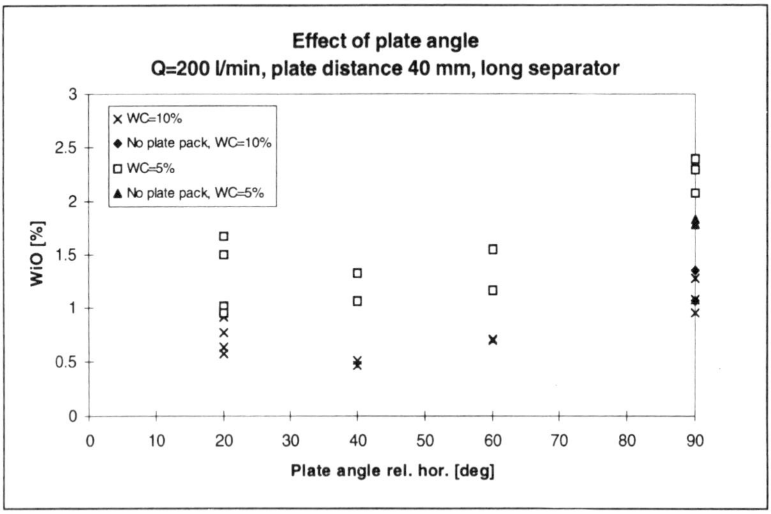

Figure 9: Water entrainment vs. plate angle.

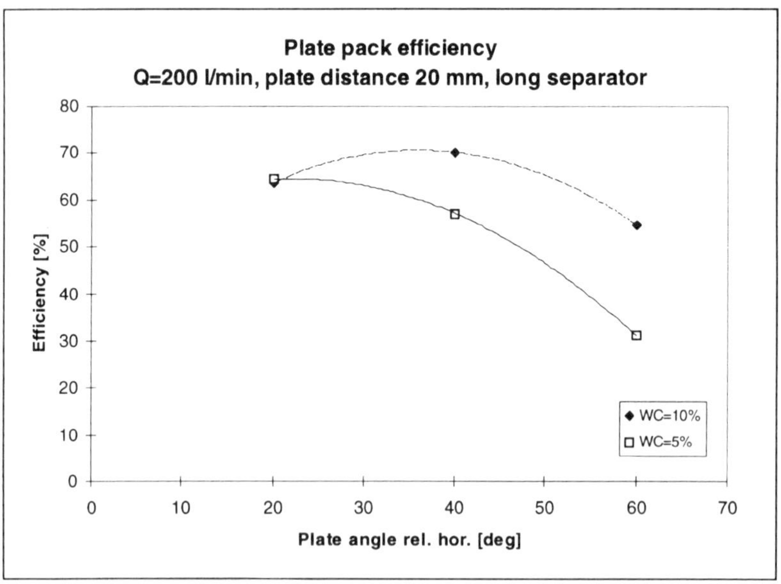

Figure 10: Plywood plate pack efficiency relative to no plate pack installed in separator.

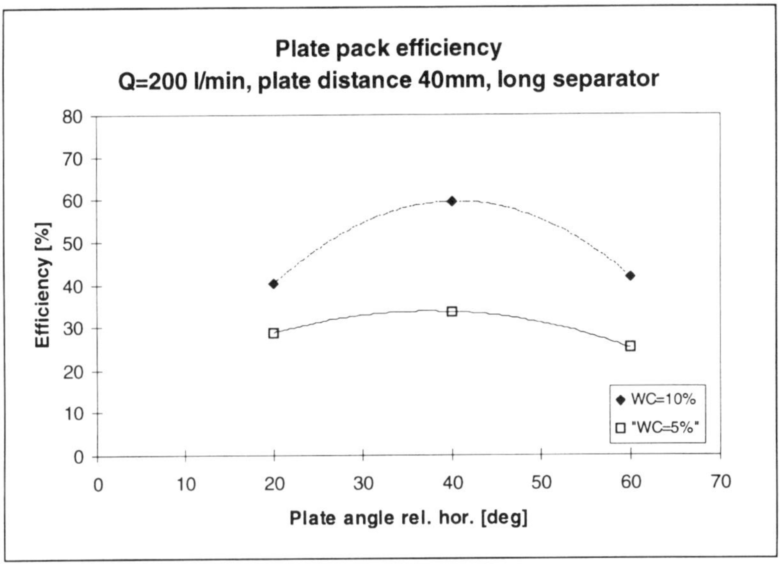

Figure 11: Plywood plate pack efficiency relative to no plate pack installed in separator.

4.2 Plate distance

The effect of plate distance is illustrated by comparing Figures 8 to 9, or alternatively Figures 10 to 11. With water cut 5 %, the plate pack efficiency is increased from typically 33 % for plate distance 40 mm to 58 % for plate distance 20 mm. The numbers are taken for plate angle 40°. Similarly for the water cut 10 % the efficiency increases from 60 % to 71 % when reducing the plate distance from 40 to 20 mm.

4.3 Plate location

The effect of plate pack location is illustrated in Figure 12. Data for three plate pack locations are indicated; close to the inlet section, in the middle of the separator and close to the outlet of the separator. The poorer performance is obtained with the plate pack close to the inlet section. On the average, the better performance is obtained with the plate pack close to the outlet section.

4.4 Plate material

Entrained water *WiO* is plotted versus water cut in Figure 13 with the plate material as the parameter. At water cut 5 % the entrained water is reduced with some 25 % when changing from the hydrophobic material (Teflon) to the hydrophilic material (plywood). At water cut 10 % the differences are within the repeatability of measurements.

Figures 14 and 15 illustrate the different droplet coalescence and draining mechanisms observed for the plywood and Teflon plates respectively. The plywood plates were subjected to pure water for more than 30 minutes prior to the oil-water experimental in order to obtain an initial water film on plate surface. This initial film serves as a 100% hydrophilic surface.

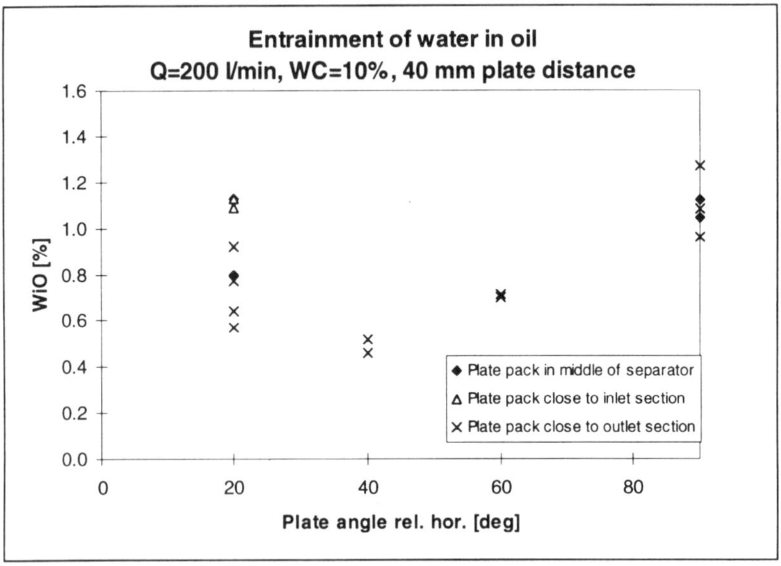

Figure 12: Water entrainment vs. plate angle for different plate pack positions.

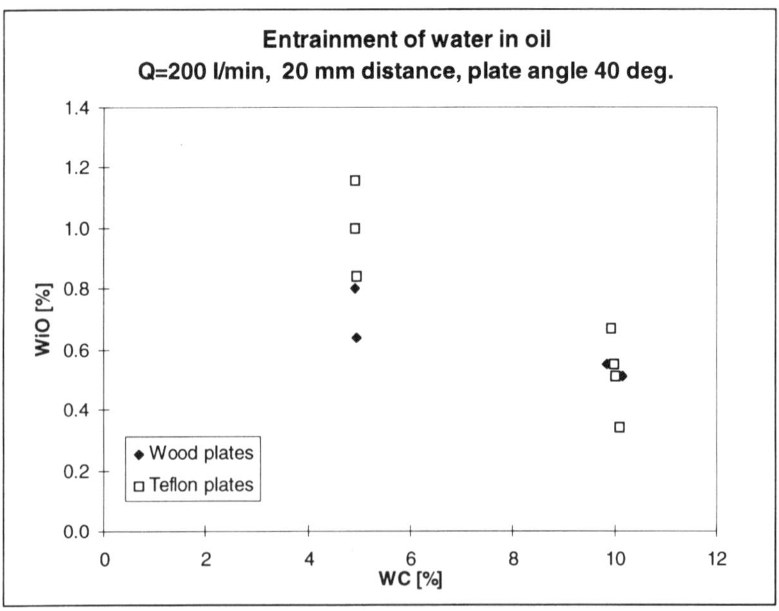

Figure 13: Effect of plate surface material. Long separator.

As the momentum exchange between the sedimented droplets and plate surface strongly depends on droplet diameter, large droplets are observed to cause significant disturbances of the water film as they hit the plate.

For the smaller droplets, the coalescence with the film is observed to be almost instantaneous and only small disturbances of the film are observed. The water film drains down to the plate edge, breaks up into large water fragments and falls down to the underlying plate. The fragments coalesce with the film on this plate similarly as do the large settling droplets. Finally the water fragments arrive and coalesce with the water bulk phase. For the plate pack with the smallest spacing, 20 mm, water fragments occasionally form continuous water "bridges" between the plate edges.

The plate pack "activity"; water draining as observed visually, is observed to be far more evenly distributed along the plate pack length for the lower water cuts than for the higher water cuts. For higher water cuts, the main activity is in the first part of the plate pack.

The mechanisms for droplet coalescence and draining off the Teflon top coated plates are quite different from the mechanisms observed for the plywood plates. No water film was formed, and hence the sedimented droplets (deform if they are large) roll down on the plate surface in a direction determined by the main flow velocity and the plate angle. The rolling droplets fall over the plate edge and down to the underlying plate similarly as settling droplets. The overall plate pack activity consists of a "swarm" of quite equally medium sized droplets falling from one plate edge to the other until the bulk water phase is reached.

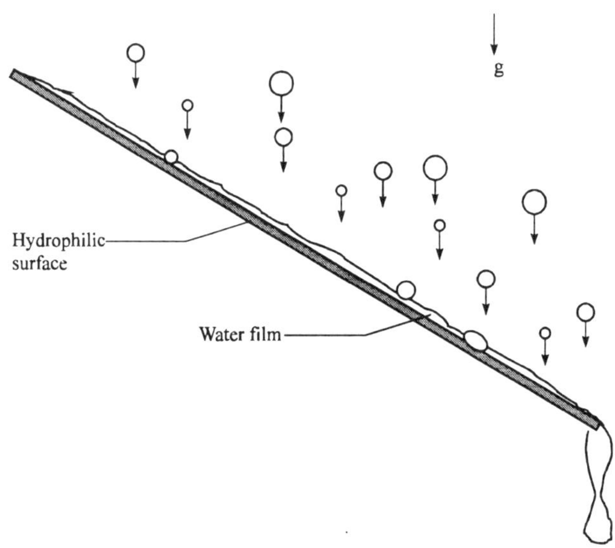

Figure 14: Droplet behaviour at hydrophilic surfaces. Plate surface wetting mechanisms.

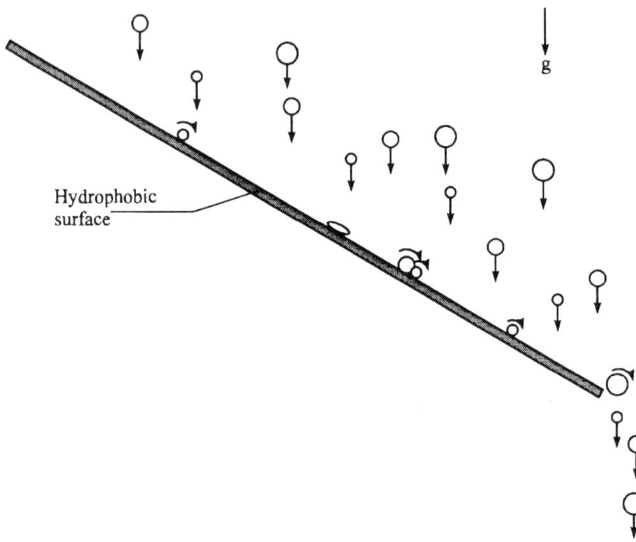

Figure 15: Droplet behaviour at hydrophobic surfaces.

4.5 Hydraulic performance of the plate pack installation

The plate pack may cause short-circuiting of the flow if the separator inlet installation distributes the volumetric liquid flux non-homogeneously over the plate pack. This may be the case with improper design of the porous baffle at the inlet or with a relatively short separator length compared to the plate pack length. An effect of the latter is shown in Figure 16 for the short separator, refer Chapter 3.1. Here the separator performance is reduced significantly when the plate pack is installed. The effect is most pronounced for the higher water cut where the entrained water is shown to increase by a factor of about 4 when the plate pack is installed.

The flow simulator GALIFA has also confirmed the effect of short-circuiting of the flow in the short separator, see Ch. 5.

4.6 Separation properties of the fluid system

The separation properties of the oil-water mixture for a given flow rate are illustrated in Figure 17. At the water cut 2 % very little of the water flow is separated for the actual residence time in the separator. At the nominal water cuts 10 % and 20 % the majority of the water flow is separated. For this oil-water system the *critical water cut* (4) is lying somewhere in-between 2 and 5 %. The results are obtained with no plate pack installed.

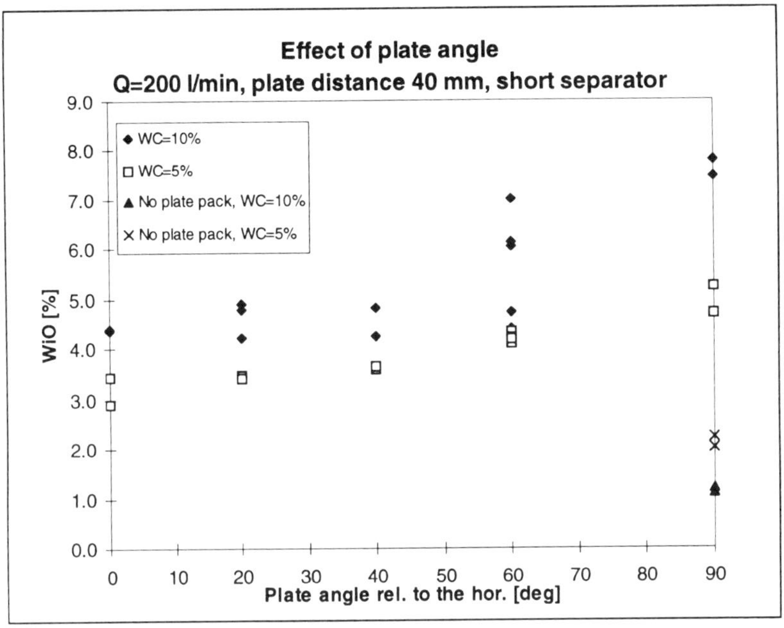

Figure 16: Water entrainment vs. plate angle. Short separator.

5 HYDRAULIC PERFORMANCE – CFD SIMULATIONS

The effect of short-circuiting of the flow in the short separator has also been confirmed from CFD simulations with the GALIFA multiphase code. In the computations, the flow is assumed single phase, and the water-oil interface is modelled as a solid surface. Although this is a simplification of the actual flow, it is a decent approximation of the experiments with low water cut. Both the short and long separator configurations are simulated.

In the short separator, the plate pack is positioned close to the distributor plate. For the long separator, two cases are investigated. First, the plate pack is centred between the distributor and the outlet. Secondly the plate pack is removed, and this will show how the plates are influencing the flow. For all cases with plate packs installed a vertical orientation of the plates is chosen. Plots in a vertical cut plane through the middle of the plate packs are shown in Figure 18. The top section of Figure 18 shows simulations for the short separator, cut through the middle of the plate pack (90^0 angle with horizon). The middle section of Figure 18 shows simulations for the long separator, cut through the middle of the plate pack (90^0 angle with horizon). The lower section of Figure 18 shows simulations for the long separator, without plate pack installed.

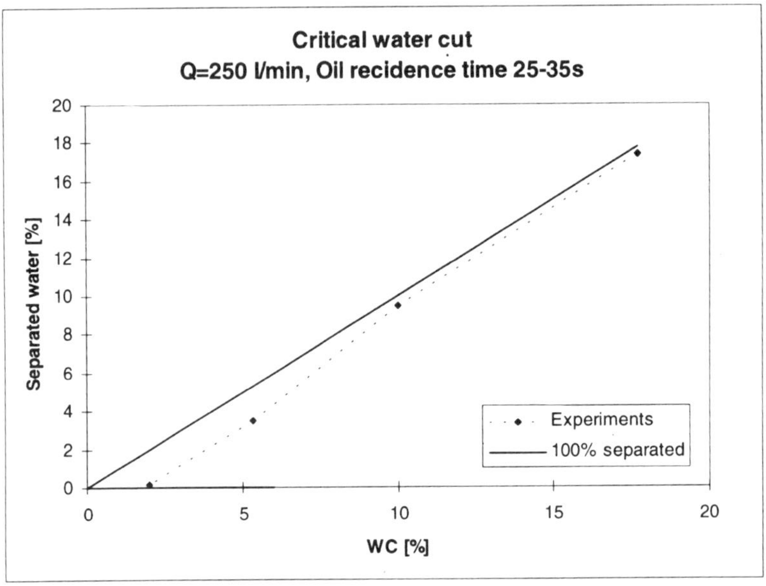

Figure 17: Separation properties: critical water cut. Long separator without plate pack installed.

When the plate pack is positioned near the distributor plate, the axial velocity profile is strongly inhomogeneous. The highest axial velocity is between the two central plates, and the velocity falls off rapidly between the other plates when approaching the wall. There is a strong cross-flow parallel with the plates, and again the highest velocities are found between the central plates. For the long separator with the plate pack centred in the cylinder, the results are different. The axial velocity is now uniformly distributed between the plates. The magnitude of the cross flow is in most of the cross-section low compared with the axial velocity. Only in the space between the plate edges and the cylinder wall there are flow accelerations. This is consistent with visual observations in the experiments. When the plate pack is removed the flow is even smoother, and with two nearly symmetrical vortices in the cross-section.

The CFD-calculations are supported by the experimental results shown in Chapter 4.5. When the plate pack is too close to the distributor plate to obtain a sufficient pressure recovery the performance of the plate separator is poor and can even have an adverse effect. The main reason for this is the mal-distribution of the axial velocity profiles in combination with strong secondary cross-flows. In order to obtain good results with plate separators considerable attention should be given to the hydraulic conditions at the inlet to the plate separator and to the performance of the flow distributor.

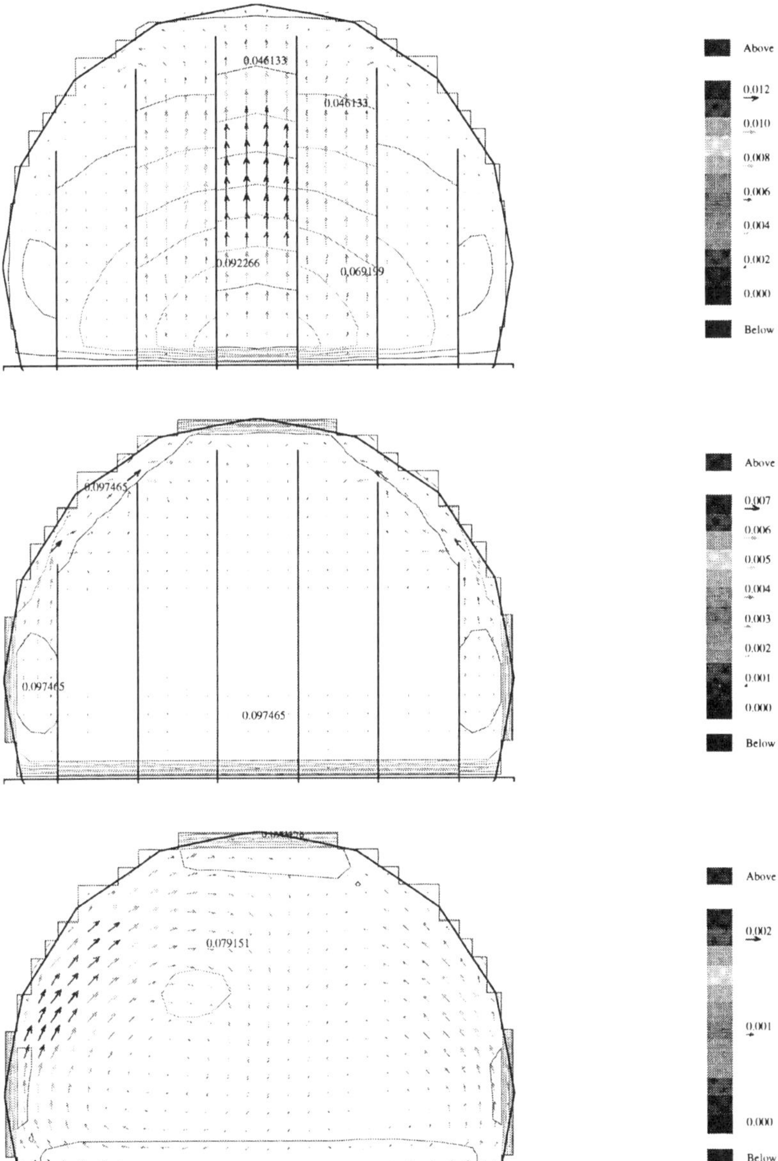

Figure 18: Single phase flow field simulations from GALIFA.

6 CONCLUSIONS

The water-oil experiments run with a horizontally mounted cylindrical separator show that the separator performance obtained by installing plate packs is very sensitive to the homogeneity of the flow field upstream to the plates. Using an improperly designed inlet section or a relatively short separator compared to the plate length, short-circuiting of the flow through the plate pack may occur. Consequently, the installation of a plate pack may even reduce the separator performance. The flow simulator GALIFA has shown to be an efficient tool for designing separator and plate pack lengths which combined yield an acceptable hydraulic performance.

For the experiments run with an acceptable hydraulic performance of the separator a positive effect on the separator performance is clearly observed when installing the plate pack. Within the separator volume, the better plate pack performance is obtained with the plate pack located close to the outlet section. The poorer results are obtained with the plate pack located close to the inlet distributor baffle.

The optimal plate angle concerning separator efficiency is close to 40° relative to the horizontal. A significantly improved separator performance is observed when reducing the plate distance from 40 to 20 mm. Compared to the reference case with no plate pack installed, the water entrained in the oil outlet can be reduced by a factor 2 with the plate distance 40 mm, whereas the reduction factor is about 3 for the plate distance 20 mm.

The results indicate that the plate material is of some importance for the lower water cuts. Thus for water cut 5 % the water entrainment in the oil outlet is on the average reduced by some 25 % when changing from hydrophobic (Teflon) to hydrophilic material (plywood). At the higher water cut 10 % no significant difference was observed.

Concerning the separation properties of the liquid-liquid system tested; water-Exxsol D100S, the critical water cut for the relevant flow conditions and residence time is observed to be between 2 and 5 %.

7 ACKNOWLEDGEMENTS

The experimental results have been achieved during the project "Development of Plate Separators for Oil and Water Continuous Flow" sponsored by Norsk Hydro ASA and Kværner Oilfield Products a.s. The theoretical work was sponsored by the members of the DIPPS consortium, Amoco Oil UK Ltd., BP E&P Ltd., HSE, Merpro Ltd., Norsk Hydro a.s., Petrobras, Shell I.O.P, Texaco UK, EPSRC. The sponsoring companies are greatly acknowledged.

8 REFERENCES

(1) Hallanger A., Soenstaboe F., Knutsen T. (1996) `A Simulation Model for Three-Phase Gravity Separators` *The 1966 Annual Technical Conf. Denver, Colorado,* 1996, special print SPE 36644.

(2) Wang, Y.L., Longwell, P.A. (1964), *AIChE J,* **10**: 323.

(3) Schlichting, H., (1934), *ZAMM,* **14**: 368, also (1960), *Boundary Layer Theory*, 4th ed., p. 169 (McGraw-Hill Book Co., N.Y.).

(4) Davies G.A., Nilsen F. P. and Gramme P. (1996) `The Formation of Stable Dispersions of Crude Oil and Produced Water: The Influence of Oil Type, Wax & Asphaltene Content` Presented at SPE Annual Technical Conference and Exhibition, Denver, Colorado, Oct. 1996. Special prints SPE 36587.

The effect of tubing and choke valve on oil droplet break-up

M J VAN DER ZANDE and **W M G T VAN DEN BROEK**
Delft University of Technology, The Netherlands

1. SYNOPSIS

When oil and water are produced simultaneously -and the oil/water ratio is low- break-up of oil droplets will occur, notably in the choke valve. This paper describes laboratory work on break-up of oil droplets dispersed in water, in a number of model chokes in the form of circular restrictions and venturis. It is shown that the size of the droplets downstream of the choke valve is a function of, among other parameters, the interfacial tension between oil and water, the dimension of the choke and the energy dissipation rate of the flow in the valve. Tentative statements are given on the implications for actual oil production.

2. SYMBOLS

A	: Cross-sectional area of the pipe	[m²]
C	: Constant	[-]
D	: Diameter	[m]
K	: Constant	[-]
Re	: Reynolds number	[-]
U	: Mean velocity	[m/s]
We	: Weber number	[-]
d	: Droplet diameter	[m]
p	: Pressure	[Pa]
u	: Velocity fluctuation	[m/s]
x	: Distance	[m]
β	: Diameter ratio	[-]
Δ	: Difference	[-]
ε	: Energy dissipation rate	[W/kg]

μ	: Viscosity	[Pa·s]
ρ	: Density	[kg/m^3]
σ	: Interfacial tension	[N/m]
τ	: Disturbing stress.	[Pa].

Subscribts

c	: Continuous phase
crit	: Critical
d	: Dispersed phase
dr	: Droplet
o	: Orifice
p	: Pipe
perm	: Permanent
vc	: Vena Contracta
0	: 0% Break-up
95	: 95 Volume %
100	: 100% Break-up

3. INTRODUCTION

During the production of oil, often water is produced as well. In practice the amount of produced water increases in the course of time. This associated production has a negative influence on the total production costs, because oil and water have to be separated and, subsequently, the treated water has to be discharged, e.g. in the sea. Economical production, however, can remain possible even for oil/water ratios as low as, say, 5-10 vol. %. Under these circumstances water is the continuous medium, and usually the oil is present in the form of droplets. From laboratory work (1) it is known that the size of these droplets at the bottom of the well is roughly in the range between tens of micrometers and a few millimetres. At the surface, however, upstream of the separators, smaller to much smaller sizes are encountered. Consequently, between the bottom of the well and the surface facilities, oil-droplet break-up occurs, which has an additional negative effect on the production costs. As in separation equipment small droplets are more difficult to remove than larger ones, more sophisticated (and thus more expensive) separation equipment may be needed. Break-up takes place in both the tubing, connecting the well with the surface equipment, and the choke valve at the end of the tubing. In the tubing the break-up process is not intense (2), thus it is expected that most of the oil droplets are broken up in the choke. This paper describes experiments to investigate the break-up processes taking place in a small-scale choke valve. The results are compared with general theoretical models on droplet break-up.

4. THEORY

The theory of droplet break-up can, generally speaking, be described using only two forces; an external disturbing force tries to deform the droplet and an internal restoring force counteracts this motion. In 1955, Hinze (3) used this approach to study droplet break-up and many others have followed his example since.

Hinze assumed that the disturbing force, τ, can either be viscous or inertial, depending on the nature of the flow around the droplet. The interfacial tension, σ, gives rise to a restoring force equal to σ/d, where d is the droplet diameter. In this approach the contribution to the restoring force of the dispersed phase viscosity, μ_d, is neglected. The ratio between the disturbing and the restoring force is given by the Weber number, We, which is equal to $\tau d/\sigma$. If for a certain droplet We is larger than a critical value, We_{crit}, the droplet will eventually break up. The droplet diameter will decrease and so will We. Break-up takes place until for all droplets $We < We_{crit}$.

To be able to calculate the disturbing force acting on a droplet, the flow around the droplet has to be known. The work described in this paper concerns break-up of oil droplets in a choke valve. It is therefore necessary to look in detail into the flow occurring in a choke valve.

4.1 **Flow through a restriction**

A great variety of choke valves is being used in actual oil-production systems. Many different sizes and designs such as for example Multiple Orifice Valves, Plug and Cage Valves and External Sleeve Valves are installed in such systems to create a pressure drop and to control the production rate. All choke valves have one thing in common: the fluid flow is forced through a restriction where energy is dissipated and a pressure drop occurs. In order to understand the basic principles of droplet break-up in a choke valve, we have in first instance simply modelled a choke as a circular orifice in a circular pipe.

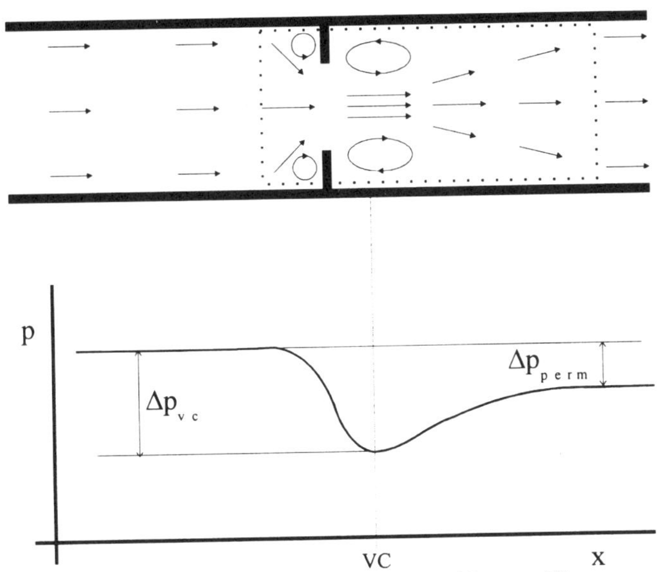

Figure 1: Flow and pressure field inside an orifice

In Fig. 1 schematic representations of the flow and pressure fields inside an orifice are given. Upstream of the orifice a fully developed turbulent flow is present. The flow is accelerated through the restriction and a jet is formed. Upstream of the orifice plate a small recirculation zone is generated and downstream a larger recirculation zone is generated. The minimum diameter of the jet is smaller than the orifice diameter, D_o. The reason for this extra contraction is the inwardly directed momentum of the fluid at the position of the orifice. Due to the acceleration the pressure decreases and reaches a minimum at the position of minimum jet diameter. This position is called the Vena Contracta, *VC*. In this region the velocity is high and energy is dissipated at an increased rate. Downstream of the Vena Contracta, the fluid decelerates, and part of the pressure is recovered. At a sufficient distance from the orifice the flow returns again to a fully developed turbulent pipe flow.

The maximum pressure drop, Δp_{vc}, can be calculated using the orifice equation:

$$\Delta p_{vc} = \frac{1}{C_o^2}\frac{1}{2}\rho_c U_p^2\left(\beta^{-4} - 1\right) \tag{1}$$

where C_o is the orifice coefficient, ρ_c the density of the continuous phase, U_p the mean velocity in the pipe and β the ratio between the orifice diameter, D_o, and the pipe diameter, D_p. In general, C_o depends on U_p and β, but for sufficiently high Reynolds number ($Re > 10{,}000$) C_o becomes constant and is approximately 0.62 (4).

The recovery factor, which is the ratio between the permanent pressure drop, Δp_{perm}, and Δp_{vc} is given by (5):

$$\frac{\Delta p_{perm}}{\Delta p_{vc}} = 1 - \beta^2 \tag{2}$$

Based on experimental data (6, 7) the following characteristic distances can be used. The fluid flow starts to accelerate at a distance approximately 0.6 D_p upstream of the orifice plate. The position of the Vena Contracta is 3.3 dam heights downstream of the orifice plate. The dam height is equal to $\frac{1}{2}(D_p-D_o)$. The distance at which pressure recovery is completed, lies between 4.5 D_p (6) and 8.5 D_p (7) downstream of the orifice. In this paper the position at which full pressure recovery is achieved is assumed to be 6.5 D_p.

4.2 **Break-up due to acceleration**

In 1983 Percy and Sleicher (8) wrote an article on the subject of droplet break-up in the flow through an orifice in a pipe. In their experiments they released 20 droplets of a certain size upstream of the orifice. Downstream the droplet sizes were measured. In this case that all 20 droplets broke up (100% break-up), the initial droplet diameter was called d_{100}. For those runs that no droplets broke up, it was called d_0. To provide a physical basis for correlating the data, Percy and Sleicher made the following assumptions.

The disturbing stress acting on the droplets is the stress induced by the acceleration of the fluid in the entrance region of the orifice. This stress is equal to the pressure drop across the droplet diameter:

$$\tau = \frac{\partial p}{\partial x} d \approx \frac{\Delta p_{vc}}{\Delta x} d \tag{3}$$

where Δx is the distance over which the maximum pressure drop takes place. Percy and Sleicher assumed that Δx is proportional to D_o. The expression for the Weber number becomes therefore:

$$We = \frac{\tau d}{\sigma} = \frac{\Delta p_{vc} d^2}{\sigma D_o} \tag{4}$$

Considering that a droplet can only break up if $We > We_{crit}$, the following correlation for the droplet diameter can be derived from Eq. 4:

$$d = K\left(\frac{D_o\sigma}{\Delta p_{vc}}\right)^{0.5} \tag{5}$$

where K is a constant. To fit their data for d_0, Percy and Sleicher found K to be 0.32; for d_{100}, K is equal to 3.1.

4.3 **Break-up due to turbulent motion**
In the previous section break-up is solely contributed to the acceleration of the fluid in the entrance region of the orifice. Another approach is to consider the turbulent forces which are present in the jet zone of the orifice. The velocity in this zone is much higher than the mean velocity in the pipe, and therefore the turbulence is more intense. Droplets which are stable in the turbulent pipe flow can break up due to the increased turbulent intensity at the orifice. In general, droplet break-up due to turbulence can be described using the Weber number. To be able to calculate the disturbing force acting on a droplet some characteristics of the turbulence have to be known.

In a turbulent flow, the action of many eddies of varying sizes results in a random and chaotic movement of the fluid. If a droplet of diameter d is released in a turbulent flow, a velocity difference, Δu, across the droplet will be present. The inertial stress induced by this velocity difference is equal to $\rho_c(\Delta u)^2$. The viscous stress on the droplet is equal to $\mu_c\Delta u/d$, where μ_c is the viscosity of the continuous phase. The ratio between the inertial and the viscous stress acting on the droplet is called the droplet Reynolds number, Re_{dr}, which is equal to:

$$\mathrm{Re}_{dr} = \frac{\rho_c \Delta u\, d}{\mu_c} \tag{6}$$

For the droplet sizes and the level of turbulence we are interested in, it appears that $Re_{dr} >> 1$, which means that the inertial force is the dominant force causing droplet break-up. In this case the Weber number is given by:

$$We = \frac{\rho_c(\Delta u)^2 d}{\sigma} \tag{9}$$

For the case of homogeneous isotropic turbulence a relation between Δu and the energy dissipation rate of the flow, ε, is given by:

$$\Delta u \propto \left(\varepsilon d\right)^{1/3} \tag{10}$$

It is clear that the assumption of homogeneous isotropic turbulence is not valid for the entire flow field in the orifice. On the scale of the droplets, however, this assumption can be used. Eddies of a size larger than the droplets tend to translate the droplets through the flow field; eddies of the same size and smaller than the droplets make the droplet oscillate and eventually break up. It is commonly known that the energy distribution of the smaller eddies is independent of the geometry of the boundaries of the flow (9). The distribution mainly depends on the energy dissipation rate of the flow, which basically makes the turbulence isotropic at this small scale. Furthermore, if one considers a turbulent zone where ε is approximately constant, the requirements for homogeneous isotropic turbulence are met.

By substituting Eq. 10 in Eq. 9, and considering that droplets will break up for $We >> We_{crit}$, the following relation for the maximum stable droplet diameter, d_{max}, in a turbulent zone of energy dissipation rate, ε, can be derived:

$$d_{\max} \propto \left(\frac{\sigma}{\rho_c}\right)^{0.6} \varepsilon^{-0.4} \tag{11}$$

It is in our set-up, however, not possible to determine the energy dissipation rate locally. Therefore the mean energy dissipation rate of the entire orifice zone, $\bar{\varepsilon}$, is used for correlating d_{max}. To determine $\bar{\varepsilon}$, the following approach has been used. Consider a fluid element enclosed by the dotted line sketched in Fig. 1. The energy needed to pass this element through the orifice zone is $\Delta p_{perm} \cdot A \cdot \Delta x_{perm}$, where A is the cross-sectional area of the pipe and Δx_{perm} is the length of the orifice zone, which is equal to the distance between the point where the pressure first starts to decrease and the position where the pressure is recovered. The mass of the fluid element is equal to $\rho_c \cdot A \cdot \Delta x_{perm}$. The average residence time of the fluid element in the orifice zone is proportional to $\Delta x_{perm}/U_p$. Combining the expressions given above, the mean energy dissipation rate of the orifice zone, which is the energy dissipated per unit mass and time, can be calculated:

$$\bar{\varepsilon} \propto \frac{\Delta p_{perm} U_p}{\rho_c \Delta x_{perm}} \tag{12}$$

All quantities in Eq. 12 can easily be measured.

5. EXPERIMENTAL SET-UP

A schematic representation of the set-up for the investigation of droplet break-up in a choke valve is given in Fig. 2. The set-up mainly consists of three sections. In the first section pumps and valves are situated, which are needed to inject the oil. The second contains a choke valve where oil droplets break up, and in the last section the oil-droplet-size distribution is measured.

A Verder gear pump [1], V108-05, with a maximum flow rate of 1.8 l/min, pumps water out of a container. Part of the main water flow is side-tracked and directed through a needle valve [2] where oil is injected [3]. The size of the injected oil droplets depends upon the water flow rate through the needle valve. Therefore the flow rate is measured [4] and kept constant, independent of the main flow rate. This is achieved by adjusting the valves in the pipe lines [5]. The oil-water mixture flows through a restriction [6] where the oil droplets break up. The pressure drop over this restriction is measured [7]. The mixture flows through the Malvern Particle Sizer, type 3600E, where the droplet-size distribution is measured [8]. The sizer is able to measure droplets from 1.2 to 564 μm. Before the dispersion is disposed in a separator vessel, the main flow rate is measured [9]. To be able to measure the size distribution of the

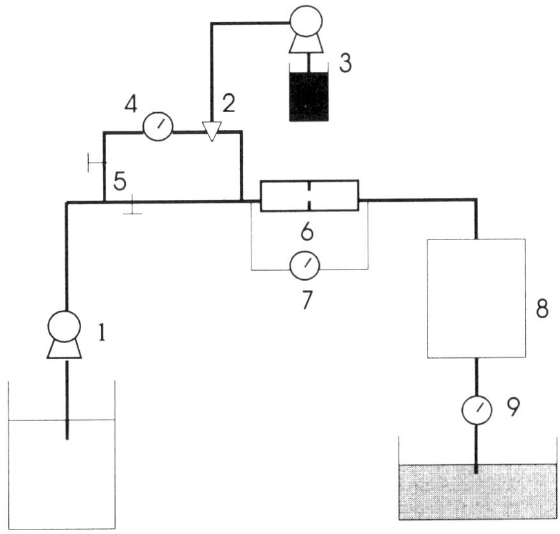

Figure 2: Outline of the set-up for the investigation of droplet break-up in a choke valve. The set-up is described in the text.

injected droplets, the restriction can be replaced by an undistorted pipe. The internal diameter of the pipelines is 4.5 mm. Two types of restrictions have been used for the experiments. Firstly experiments have been conducted with square-edged orifices. The orifice diameter varies from 4.0 mm down to 0.5 mm in steps of 0.5 mm The orifice plate is 0.5 mm thick.

The second type of experiments have been carried out with venturi shaped restrictions. A schematic representation of these venturis is given in Fig. 3. Venturi 1 has a gradual inlet and a gradual outlet. Venturi 2 has the same gradual inlet as Venturi 1, while at the outlet the channel diameter increases stepwise to the tube diameter. Venturi 3 has a stepwise restriction at the inlet and the same gradual outlet as Venturi 1. The flow is directed from left to right. If water is flowing from right to left the venturi is called reversed, e.g. "Venturi 1 reversed".

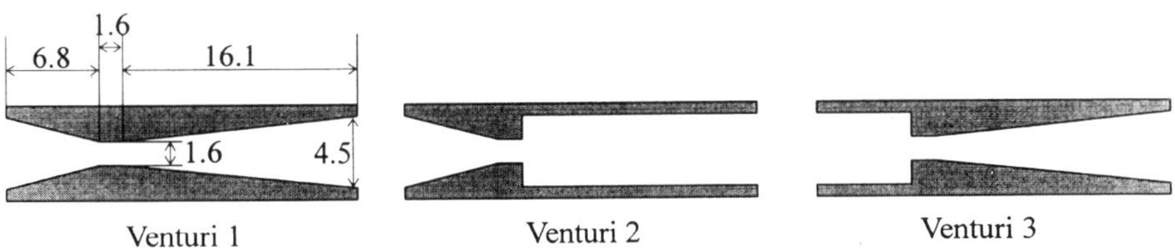

Figure 3: Outline of the venturi shaped restrictions. All sizes are in mm.

The experiments are carried out by varying the main water flow rate. The flow through the injection valve is kept constant. The droplet-size distribution is measured downstream of the restriction for varying orifice sizes and venturi shapes.

RESULTS

For the orifice experiments, two types of mineral oils have been used: Shell Vitrea 9 and 46. The numbers indicate the kinematic viscosity of the oil in mm^2/s at 40 °C. The interfacial tension of the oils is $30.9 \cdot 10^{-3}$ and $36.5 \cdot 10^{-3}$ N/m, respectively. The temperature of the continuous phase, de-mineralised water, during the experiments was 22±1 °C. Since the oil droplets are small, they immediately adapt the temperature of the continuous phase. The actual dynamic viscosity of Vitrea 9 and 46 is therefore $1.8 \cdot 10^{-2}$ and $1.4 \cdot 10^{-1}$ Pa·s, respectively. The volume concentration of the dispersed phase is typically 0.5 vol. % for all experiments. To obtain droplet-size distributions which lie within the Malvern range, the orifice diameter has been varied between 3.0 and 1.0 mm. In Fig. 4 the results for these experiments are shown.

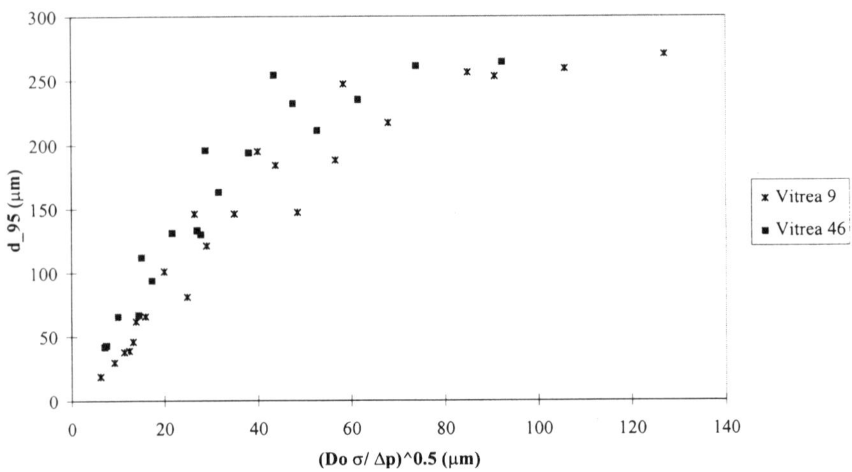

Figure 4: For orifice sizes varying from 3.0 down to 1.0 mm, d_{95} is plotted versus the parameter derived by Percy and Sleicher (dispersed phase: Vitrea 9 and 46).

The droplet diameter after break-up, below which 95 volume percent is present, d_{95}, is plotted versus the parameter derived by Percy and Sleicher. From Fig. 4 it can be seen that our break-up data seem to follow Eq. 5 rather well, especially for $(D_o\sigma/\Delta p)^{0.5} < 60\ \mu m$. At higher values, the data seem to deviate from a straight line through the origin. This result indicates that the approach of the break-up process, which Percy and Sleicher used, seems to be correct. This would mean that droplets break up due to acceleration of the fluid in the entrance zone of the orifice. As a consequence of this, restrictions with similar flow and pressure fields in the entrance zone (i.e. same geometry of the entrance) result in the same droplet sizes. In Fig. 5, however, it is shown that this assumption is not correct.

It is for example shown in Fig. 5 that for the same flow rate d_{95} for Venturi 1 is larger than for Venturi 2. Based on the assumption made by Percy and Sleicher one would expect these values to be equal, since the geometry of the entrance zone of these restrictions is identical. In Fig. 5b it is shown that for identical flow rates Venturi 3 results in the largest droplets, and Venturi 2 reversed gives slightly larger droplets than the 1.6 mm orifice. Again, based on Percy and Sleicher one would expect these restrictions to produce the same droplet sizes.

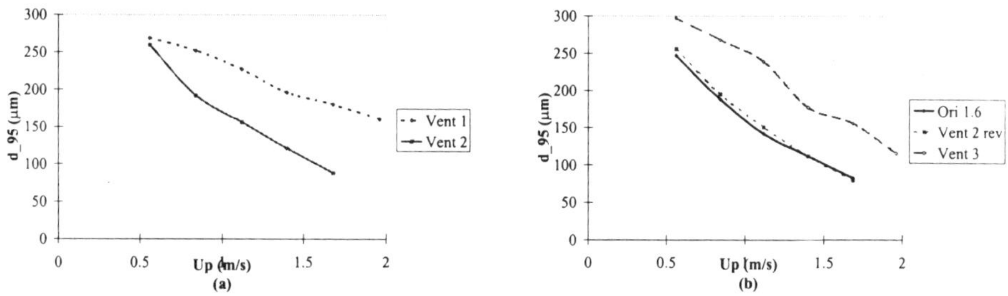

Figure 5: A comparison between droplet break-up in different choke geometries. In (a), Venturi 1 and 2 are compared, in (b) the orifice with diameter 1.6 mm, Venturi 2 reversed and Venturi 3 are plotted. Vitrea 9 has been used as the dispersed phase.

The results shown in Fig. 5 indicate that droplets do not break up due to the acceleration of the fluid in the entrance zone. Therefore we now consider droplet break-up due to the increased turbulent intensity downstream of the restriction. For the analysis of the data we use Eqs. 11 and 12. The results for the venturis are shown in Fig. 6.

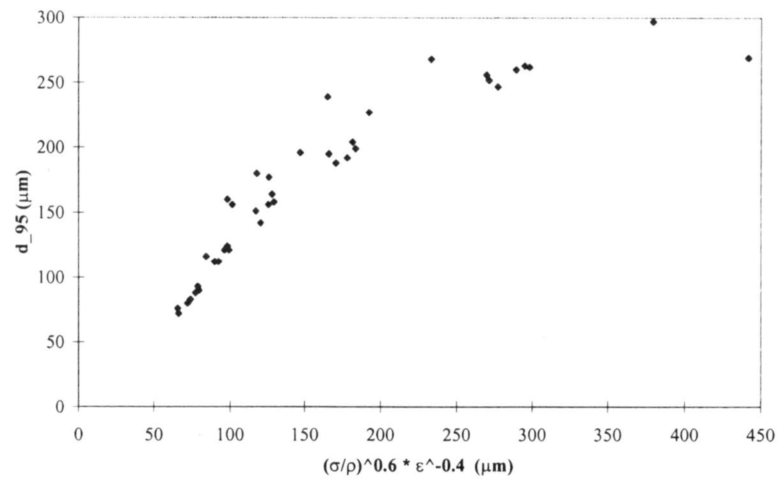

Figure 6: Break-up data obtained with the 1.6 mm orifice, Venturi 1, 2 and 3. Equation 11 has been used to correlate the data.

The break-up data of the venturis seem to correlate rather well with the parameter given in Eq. 11. For $(\sigma/\rho)^{0.6}\varepsilon^{-0.4} < 250\ \mu m$ the data can be fit with a straight line through the origin, where for larger values the data seems to deviate from this line.

In Fig. 7, the data from the experiments with the orifices are also plotted according to Eq. 11. Note that the data shown in this figure are the same as that shown in Fig. 4.

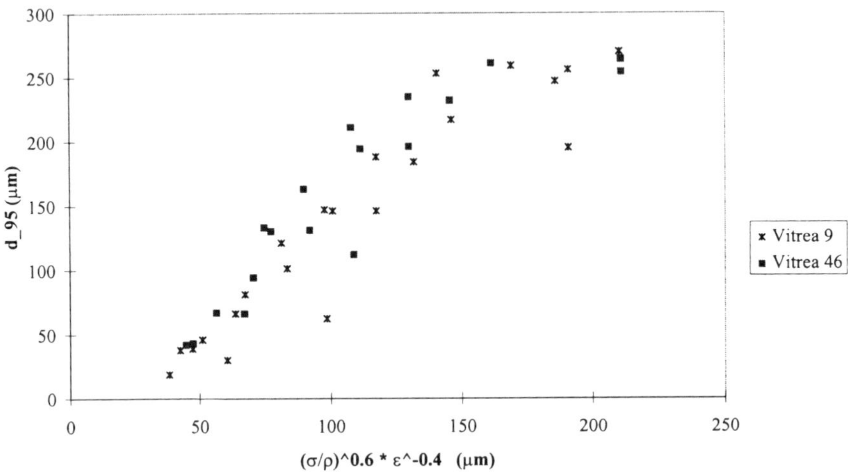

Figure 7: Data obtained with the orifices, plotted according to the turbulent break-up model (Eq. 11).

7. DISCUSSION AND CONCLUSIONS

7.1 Experiments

The break-up data obtained with the orifices can be predicted rather well with the expression given in Eq. 5. It is shown, however, in Fig. 5 that the physical basis for break-up which has been used to derive this relation is not correct. Based on the theory provided by Percy and Sleicher it would be expected that restrictions with an identical geometry of the upstream side result in the same droplet size distributions. The data in Fig. 5 show, however, that although the entrance is identical, the restriction with a stepwise expansion downstream results in the smallest droplets. This indicates that the energy dissipation -which leads to the permanent pressure drop- and not the pressure gradient is responsible for the droplets to break up. The maximum stable droplet size can, therefore, be predicted using Eq. 11. As can be seen in Fig. 6, the data obtained with the venturis show that this relation can indeed be used to describe the break-up data. In Fig. 7 it is shown that this relation is also suitable to describe the data obtained with the orifices.

The effect of the dispersed phase viscosity on the maximum stable droplet size is shown in Fig. 7. As expected, an increase in μ_d results in an increase of d_{95}. It is surprising, however, that, although there is a large difference in viscosity for Vitrea 9 and 46 ($1.8 \cdot 10^{-2}$ and $1.4 \cdot 10^{-1}$ Pa·s, respectively), the difference in d_{95} is small (less than 10%). Break-up data for Vitrea 9 and 46 obtained in a turbulent pipe flow (2) show an increase of d_{95} of approximately 60% for a similar dispersed phase viscosity ratio. This indicates that the droplet break-up mechanism in a restriction is different from the mechanism in a turbulent pipe flow. The effect of the dispersed phase viscosity on the droplet diameter is one of the topics which we will investigate further in the near future.

Another phenomenon which can be observed in Figs. 6 and 7 is that the data does not follow a straight line through the origin. Probably this is caused by the rather rough approximation of the break-up process. Firstly the expression for $\bar{\varepsilon}$ can be improved. If one considers the recirculation zones as stagnant water which do not contribute to the energy dissipation, the effective volume of the orifice decreases. As a consequence, the residence time in the orifice zone of the moving fluid also decreases. The result of this adjustment is that $\bar{\varepsilon}$ increases: the smaller the orifice the larger the increase. Current research on this topic shows that the effect of this adjustment on Fig. 7 is a reduction in the scatter of the data. The deviation of the straight line, however, still remains.

A second approach to improve the description of the break-up process is to take into account that the energy dissipation rate is not distributed homogeneously over the orifice region. It is not unlikely that a small region of very high turbulent intensity is situated immediately downstream of the orifice. Presumably not all droplets pass through this region, or stay long enough in this region to break up. By refining the description of the break-up process in this way, hopefully an explanation can be found for the difference in d_{100} and d_0 as found by Percy and Sleicher. A similar phenomenon has been observed in our experiments: the droplet-size distribution after break-up in the orifice depends on the distribution of the injected droplets upstream of the orifice. We are currently investigating this problem.

Furthermore, the effect of the geometry of the restriction on the droplet-size distribution needs to be investigated; the data for the orifices and venturis cannot be fitted with the same line.

7.2 **Production system**

Finally, when we extend the conclusions drawn above to an actual oil-production system, the following observations can be made. Although the residence time of the production fluids in the tubing is much larger than in the choke valve, droplets will predominantly break up in the choke. In the choke the pressure decreases over a much smaller distance and in a shorter period than in the tubing. Consequently, the energy dissipation rate is higher and according to Eq. 11 this leads to smaller droplets. The effect of the tubing, however, cannot be neglected. The droplet-size distribution of the droplets downstream of the choke valve is influenced by the distribution as present upstream. In order to predict this distribution, droplet break-up in the tubing has to be taken into account.

A topic which needs further research is the influence of choke valve design on droplet break-up. Until now, research on choke design has mainly been focused on the reduction of wear and the increase of lifetime.

8. REFERENCES

1: Janssen, P.H., Broek, W.M.G.T. van den, Dijk, K.M. van, Cras, R.S.A., and Klauer, H.S., 1996, "Characterisation of Near-Wellbore Oil/Water Morphology," Proceedings of the 2nd International Seminar in Practices of Oil and Gas Exploration, Lima, Peru, Vol. 2, pp. 1289-1303.

2: Zande, M.J. van der, Broek, W.M.G.T. van den, 1998, "Break-up of Oil Droplets in the Production System," to be published in the proceedings of the Energy Sources Technology Conference and Exhibition in Houston, February 2-4.

3: Hinze, J.O., 1955, "Fundamentals of the Hydrodynamic Mechanism of Splitting in Dispersion Processes," AIChE-Journal, Vol. 1, No. 3, pp. 289-295.

4: McCabe, W.L., and Smith, J.C., 1976, "Unit Operations of Chemical Engineering", 3rd edition, McGraw-Hill Chemical Engineering Series.

5: "Fluid Meters, Their Theory and Application," 1959, 5th edition, The American Society of Mechanical Engineers.
6: Teyssandier, R.G., and Husain, Z.D., 1987, "Experimental Investigation of an Orifice Meter Pressure Gradient," Transactions of the ASME, Journal of Fluids Engineering, Vol. 109, pp. 144-148.
7: Morisson, G.L., DeOtte, R.E., Nail, G.H. and Panak, D.L., 1993, "Mean Velocity and Turbulence Fields Inside a β=0.50 Orifice Flowmeter," American Institute of Chemical Engineers Journal, Vol. 39, No. 5, pp. 744-756.
8: Percy, J.S., and Sleicher, C.A., 1983, "Drop Breakup in the Flow of Immiscible Liquids Through an Orifice in a Pipe," American Institute of Chemical Engineers Journal, Vol. 29, No. 1, pp. 161-164.
9: Hinze, J.O., 1959, "Turbulence, an Introduction to Its Mechanism and Theory," McGraw-Hill Book Company.

Operational experiences with a gas–liquid cyclone

D KVAMSDAL, B CHRISTIANSEN, and **I HJELKREM**
Kværner Process Systems A/S, Norway

SUMMARY

Kværner Process Systems have, together with Statoil, developed a novel gas blockage arrangement with applications for gas liquid cyclones. The gas blockage arrangement allows high centrifugal forces to be applied to a gas liquid mixture without severe gas carry under taking place.

Gas liquid cyclones utilizing the gas blockage arrangement has been installed in a number of gravity separators in the North Sea during the last 18 months mainly to reduce foam problems but also in some cases to increase the separator throughput. The experiences gained so far from the offshore testing have been positive, the gas liquid cyclone has shown excellent performance in foam supression.

The gas liquid cyclone has also been tested as the key component in a multi-phase meter. The test data for an API 19 and an API 28 real crude are presented. The test data are very promising showing an overall accuracy normally within ±5% for both gas and liquid through the whole operational area.

1. BACKGROUND

Free gas will often limit the separator throughput in oil production. In order to prevent liquid carry over and foam build-up in the gravity separators the gas velocity and the inlet momentum often has to be within a limited range. A gas liquid cyclone with a novel gas blockage arrangement used as an inlet for gravity separators has shown promising results in allowing increased inlet momentum into the gravity separator. This usually allows for an existing separator to be up-rated or a new separator to be made smaller.

A cyclone is a static piece of equipment that makes use of the centrifugal force as the driving force for separation. The fluid mixture is set into rotation inside the cyclone body using a tangential inlet. The heavy phase, being the liquid for a gas liquid cyclone, is then thrown towards the wall because of the centrifugal force, while the lighter gas phase is driven towards the center of the cyclone. The clean gas is collected through the top of the cyclone (or the overflow) while the liquid is collected at the bottom of the cyclone (the underflow=. The main challenge in designing a gas liquid cyclone is to prevent the gas from following the liquid through the underflow of the cyclone. To prevent gas following the liquid, the cyclone is equipped with a gas blockage arrangement that directs the gas toward the overflow of the cyclone. The general working principle of a gas liquid cyclone is illustrated in Figure 1.1.

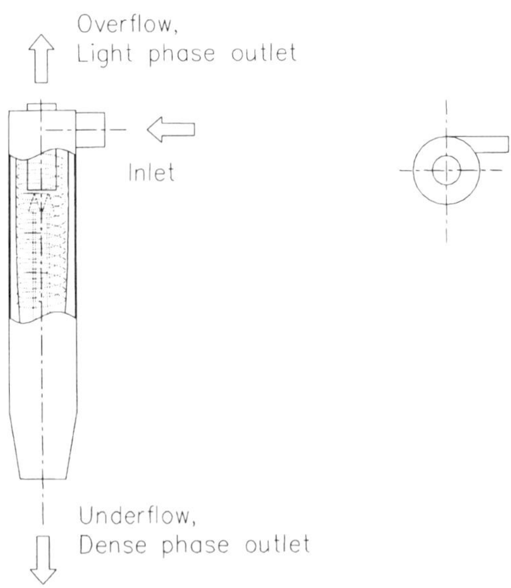

Figure 1.1 Operational principle of a gas liquid cyclone.

New compact equipment for liquid-liquid separation such as cyclones and electrostatic coalescers presents new opportunities to reduce size of separators. In the present howeve there are few compact gas liquid separators available that are suitable over a wide range of gas liquid compositions. Kvaerner Process Systems (KPS) in cooperation with the Norwegian oil company Statoil have togheter developed a gas liquid cyclone that utilizes centrifugal force to enhance the gas liquid separation. The gas liquid cyclone can handle actual gas liquid ratios ($GLR_{actual}= Q_{gas}/Q_{liquid}$) ranging from 0 - ~40. The cyclone is pressure balanced, so the pressure at overflow equal the pressure at underflow, which makes it suited as an inlet for gravity separators.

The cyclone might also be used as a self-standing unit for two-phase separation. As a two-phase separator, the cyclone is used in the KPS multiphase meter, and as a selfstanding compact two-phase separator. The cyclone allows high centrifugal forces to separate the gas from the liquid, which has been a limitation for most other gas liquid cyclones. A strong centrifugal field allows a compact cyclone design.

The product streams will be a liquid phase almost free of gas, and a gas phase with a small amount of liquid. The liquid phase will include some micro-bubbles that are not separated in the cyclone, in addition to the gas that eventually flashes off through the end part of the cyclone. The pressure drop over the cyclone should usually be less than 0.5 bar, depending on the application. Typically the liquid will include 0-4 volume percent gas, dependent upon water-cut, pressure level and oil density/-viscosity. The gas phase will under no circumstances contain more than 10 volume % liquid and normally below 1 volume %. The amount of liquid carry over (from the cyclone) is not critical to the gravity separator performance, while severe gas carry under (from the cyclone) might degrade the separator performance. The main design criterion for the cyclone inlet is for this reason to eliminate any possibility for severe gas carry under instead accept that some of the liquid follows the gas.

The cyclone has been installed in more than 20 gravity separators in the North Sea the last 18 months as an inlet device primarily to eliminate foam, but it has also been used to increase the liquid handling capacity of gravity separators. The commercialization of the cyclone has included a verification and acceptance work done in cooperation with Statoil. The experiences from these offshore tests will be presented here. References are made to earlier publications on the subject (1-3).

In addition to being an inlet in gravity separators the cyclone is used as a self-standing two-phase separator, and is the main component in a multiphase meter. The multiphase meter has been tested in the real crude test-rig operated by Norsk Hydro in Porsgrunn Norway. Some of these test results will be presented to further explain the operational characteristics of the gas liquid separator and the multiphase meter.

2 THE GAS LIQUD CYCLONE USED AS AN INLET FOR GRAVITY SEPARATORS

2.1 Background

An application where the gas liquid cyclone has been used with success is as an inlet in gravity separators. A gas liquid cyclone should be ideal for foam break down and thereby allow a high inlet momentum in to the gravity separators where the production are limited by foam production.

It was early identified that the high liquid velocity out from the cyclone could cause a degradation of liquid-liquid separation when used in a three-phase separator. To smooth out the high velocities, perforated plates are mounted at the underflow of the cyclone in addition to a perforated plate downstream the inlet area. Figure 2.1 shows a twin inlet cyclone mounted inside a gravity separator.

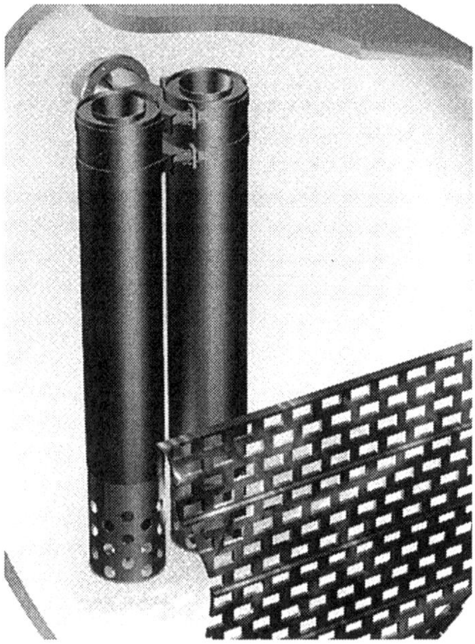

Figure 2.1 Twin inlet cyclone mounted inside a gravity separator. To reduce secondary flow in the gravity separator, the cyclones are equipped with perforated plates at the underflow. In addition there is a perforated plate in the separator downstream the inlet area.

There are currently three solutions for handling foam problems:

1. Since all foams are thermodynamically unstable they will break down, if left alone long enough. This will unfortunately cost a lot in terms of space, time and money, which are generally of little interest.
2. The stability of the film may be attacked by adding chemicals. These chemicals usually dissolve or displace the stabilizing substances, thereby collapsing the foam structure. Unfortunately, these chemicals may give problems in other parts of the separation and refining process, they are a menace to the environment and they cost money. However, the savings given by the foam reduction usually make them attractive anyway, and this is the usual approach to break foam today.
3. The gas can be separated from the liquid prior to exposing them to the mixing energies in the inlet. This has two significant advantages. Out of the cyclone there will be a liquid phase with only some dissolved gas and a gas phase with insignificant liquid amounts in terms of foam formation. Thus the mixing is inflicted on two clean fluids that do not have a foaming capability. In addition, the mixing energy at the gas side where the foam is created are reduced relatively to the inlet of the cyclone. The reduced mixing energy is due to both reduced mixture density and a reduced gas velocity into the separator.

A gas liquid cyclone inlet utilizes the mechanism as described in item 3. Any foam (or gas liquid dispersions) created upstream of the cyclone will collapse in the cyclone because of both the increased “weight” in the centrifugal field and the increased thinning velocities. A gas phase with minute concentration of liquid is rejected from the gas side and a liquid phase with very little gas is introduced at the liquid side. For gravity separators, a large reduction of foam destabilizer spending should be achieved, and often the need of foam breaking chemicals is eliminated.

2.2 Foam handling as tested on Statoil-Gullfaks C.

Statoil has tested the gas liquid cyclone as an inlet device in their test separator at Gullfax C. The test-conditions and procedures are described in (2)

The conditions for the tests are summarized in Table 2.1.

Table 2.1. Test conditions at Gullfaks C, foam tests.

Diameter	2000 mm
Tan to Tan	9500 mm
Weir height	400 mm
Inlet devices	Deflection plate (conventional inlet) Gas liquid cyclone with gas blockage arrangement
Internals	One full area vane pack One 2/3 area vane pack One 2/3 area demister

The cyclone’s ability to break foam was tested with reference to the previous inlet configuration, a circular deflection plate. To detect the foam level inside the gravity separator, a gamma ray measuring technique were used. The results from these tests are shown in Figure 2.2.

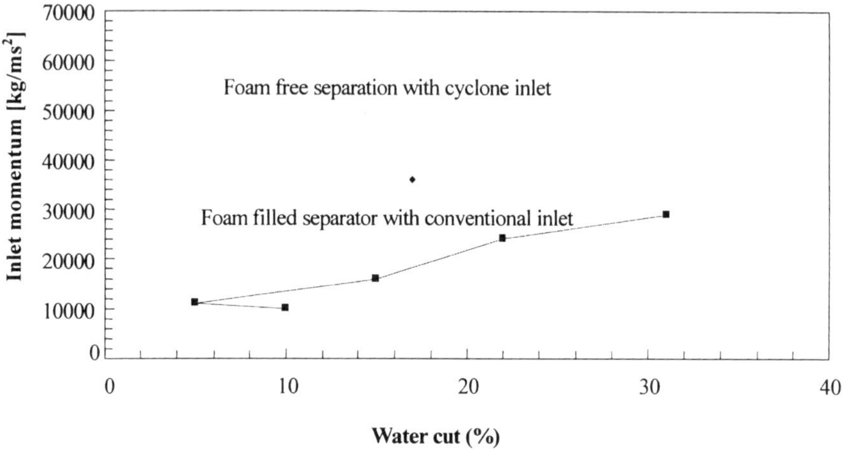

Figure 2.2. Foam production as measured at Gullfaks C. The lower line shows when the separator was completely filled with foam for the reference inlet. For the cyclone no foam was detected in the gravity separator, the two single points illustrate two of the testing rates. The inlet momentum is defined as: $\rho_{mix} u^2_{mix}$. (3)

From the tests done with the conventional inlet, we see that the foam production decreases with increasing water cut. This is expected because the oil that contains the foam stabilising chemicals. There is a connection between inlet momentum and foam in the separator. High inlet momentum will create more foam both due to increased mixing energy and increased inlet flow.

The gas liquid cyclone was used at inlet momentum up to 62000 kg/ms^2 without detecting any foam in the separator. The conclusion from the tests done at Gullfaks C is that the cyclone used as inlet in gravity separator shows excellent foam suppression abilities. Based on these test results, a production separator at Gullfaks C where refitted with the gas liquid cyclone (3).

2.3 Liquid-liquid separation utilizing the gas liquid separator as inlet device.

The high outlet liquid momentum from the cyclone may degrade liquid-liquid separation in three-phase gravity separators if precautions not are take. The high-velocity swirl flow out from the cyclone will set up secondary flows in the gravity that will reduce the liquid-liquid performance of the separator. In order minimize the negative effect of the high liquid velocities, the inlet cyclone is equipped with perforated liquid diffusers at the liquid outlet, as well as a perforated plate downstream the inlet separator. These are seen in Figure 2.1. The secondary flow and the effect of perforated plates mounted in the gravity separator are discussed by Kvamsdal (1). Generally, it can be said that the cyclone should have no negative effect on the liquid-liquid separation in three phase separators if precautions are taken.

At the Statfjord field operated by Statoil in the North Sea, the objective was to increase the liquid handling capacity of their current separation train. A test program in the gravity separator of Statfjord C was conducted. The tests looked at the influence of the inlet cyclone on liquid-liquid separation. The test conditions are summarized in table 2.2.

Table 2.2. Test conditions at Statfjord C, for the rate tests.

Diameter	2410 mm
Tan to Tan	9000 mm
Weir height	805 mm
Inlet devices	Deflection plate (conventional inlet) Gas liquid cyclone with gas blockage arrangement
Internals, gas liquid cyclone	One gas cassette partly submerged Perforated plate covering liquid side
Internals, deflection inlet	Two gas cassettes partly submerged

The test results are shown in the chart in Figure 2.4.

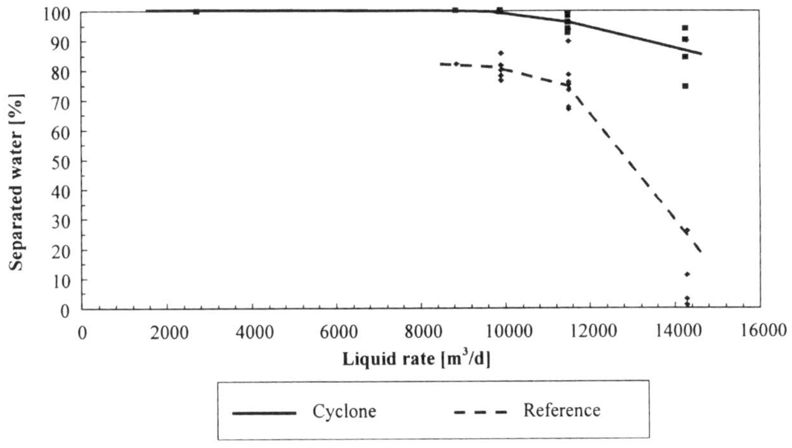

Figure 2.4 The water removal from oil as function of the liquid rate through the separator. (3)

Figure 2.4 shows that the water in oil quality where improved especially at high liquid rates. Some of the improvement with regards to separation can be explained by the installation of a perforated plate downstream the cyclone, giving better flow conditions for the settling part of the gravity separator. The results indicate that under the right conditions the gas liquid cyclone may increase the liquid-liquid separation performance.

3 THE GAS LIQUID CYCLONE USED IN A MULTIPHASE METER SYSTEM

3.1 Introduction

The gas liquid cyclone is also used as the key component in a self-standing two-phase separator. Three different configurations are currently being investigated.

1. A compact self-standing two-phase separator.
2. A multiphase meter where the liquid and gas stream are separated, and then measured before they are commingled downstream the measuring point.
3. The cyclone used as a gas knock out device before two or three stage liquid-liquid cyclone separation in a compact water knock out system.

The basic configuration of the separator will be the same for all three configurations. The difference will be in the control system and the scrubber and polishing stage for the gas stream. The basic configuration of the two-phase compact separator is shown in Figure 3.1.

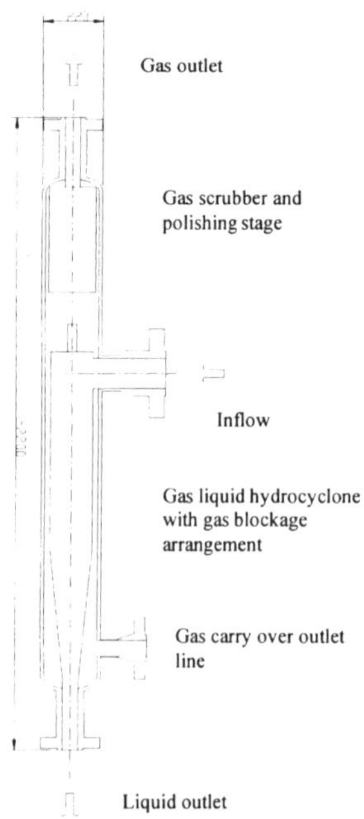

Figure 3.1 The gas liquid separator as a selfstanding two phase separator. The dimensions are shown for the multiphase meter prototype handling 40 m^3/h liquid and 200 m^3/h gas at 90 bar.

3.2 Experiences with the multiphase meter used at offshore conditions.

A prototype multiphase meter has been developed and tested at real crude conditions at Norsk Hydro's test facility in Porsgrunn (Southern Norway). The test facility is a closed loop high-pressure multi purpose test rig where the streams are separated and measured individually before entering the multiphase meter.

The multiphase meter can be supplied in various configurations, depending upon the required measuring precision and operational envelope. The tested meter is a high precision meter with the most sophisticated control system available. The multiphase meter is based upon two-phase separation and conventional single-phase measurements. The bulk of the separation is achieved in the cyclone, a second stage polishing stage are added to clean up the gas stream as well as for control purposes.

The separated gas is measured using a conventional gas flow meter (usually a v-cone differential pressure meter). The separated liquid is routed through a Phase Dynamics full-range microwave water cut meter, before being measured with any suited liquid flow meter (usually a coriolis meter). The physical layout of the system is shown in Figure 3.2.

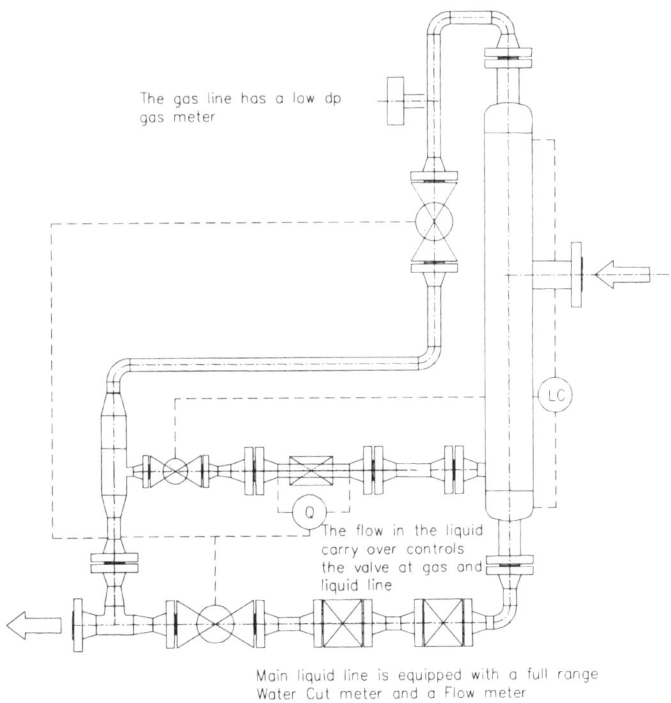

Figure 3.2. Multiphase meter lay out. The control system is indicated. The flow in the liquid carry over line controls the valves located on the main lines, while the liquid level inside the vessel controls the valve at the liquid carry over line.

The main advantages of the multiphase meter can be listed as:

- The meter is simple and compact, being based on a compact two-phase separator, plus proven single-phase measurements instrumentation, the meters should be easy to understand and operate/troubleshoot.
- On site calibration is normally not required.
- The meter is flexible; dependent on the physical layout and the flow meters, the operational envelope and the accuracy can be tailored to suit the different needs and overall system costs.
- No radioactive meters are needed.

The multiphase meter prototype tested includes a control system. There are three regulating valves: one on each of the main lines that assures that the cyclone is always inside its operational envelope. In addition, a valve on the liquid line from the scrubber that regulates the liquid level inside the pressure casing of the cyclone. The amount of liquid that follows the gas from the cyclone is measured as a basis for regulating the cyclone.

The meter has been tested under variable conditions. The results here are for the oil from Norsk Hydro's Grane Field that has quite heavy oil, with an API of 19, and Norsk Hydro's Troll C field with an 28 API crude oil with unusually strong emulsifying properties. The test conditions are summarized in Table 3.1.

Table 3.1 Test data for the multiphase meter tests.

Oil Type		Grane	Troll C
API		19	28
Density oil	[kg/m^3]	940	
Pressure	[BarG]	10, 20	23, 35
Flow Liquid	[m^3/h]	5-20	5-50
Flow Gas	[m^3/h]	40-120	20-200
Water Cut	[%]	30-55	

3.3 General observations

Gas carry under

The gas carry under in the main liquid line is estimated based on known densities. The mixture density measured in the Coreolis meter. The equlibrium densities are calculated based on the PVT data. The amount of gas in the underflow is shown in Figure 3.3 for the Grane crude.

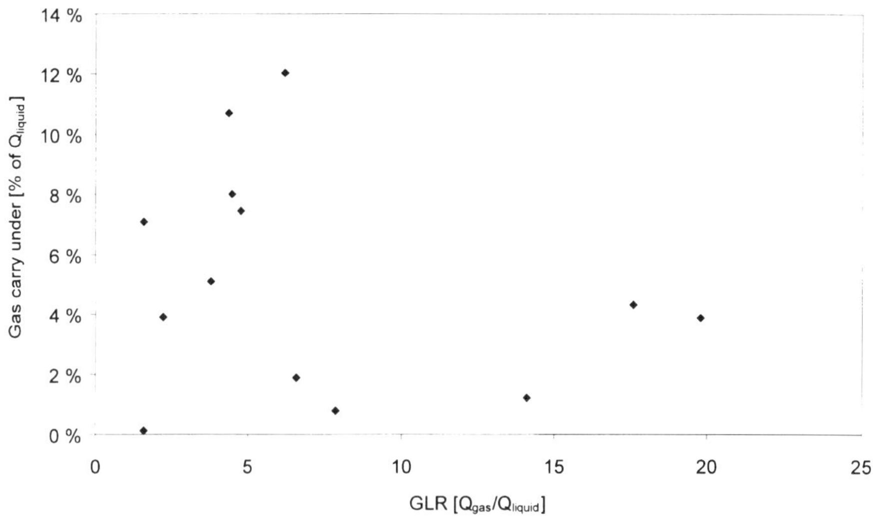

Figure 3.3 Gas carry under in the main liquid line as measured for the Grane Oil.

There are some more gas than expected, this is due to the viscous oil and the fact that the CCM meter are configured for higher flow. An increase in GLR gives better separation due to higher centrifugal forces.

Pressure drop.

The pressure drop from inlet to outlet is measured. The pressure drop will mainly be a function of the total inlet volumetric flow that decides the inlet velocity. The pressure drop over the cyclone for the case with Grane crude is shown in Figure 3.4.

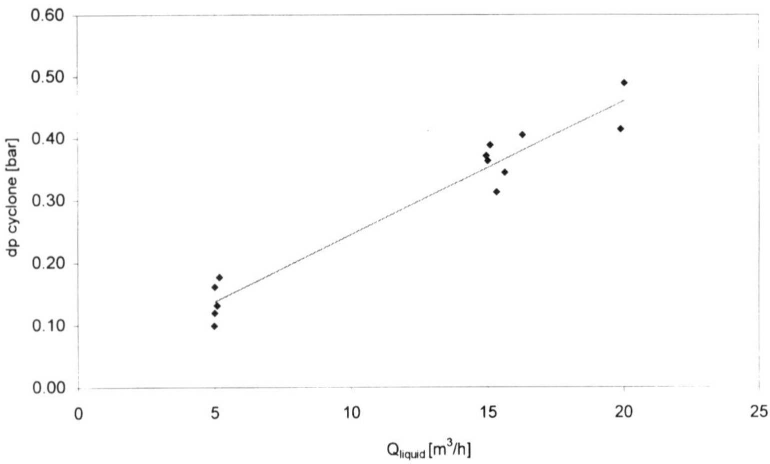

Figure 3.4 Pressure drop over the cyclone as measured for a 19 API oil.

3.4. Measurements done on the API 28 crude

The test data for the 28 API crude are presented. The ± 3 standard deviation curves are shown as stapled lines.

Test matrix

The test matrix for the tests run with Troll C crude oil are shown in Figure 3.5.

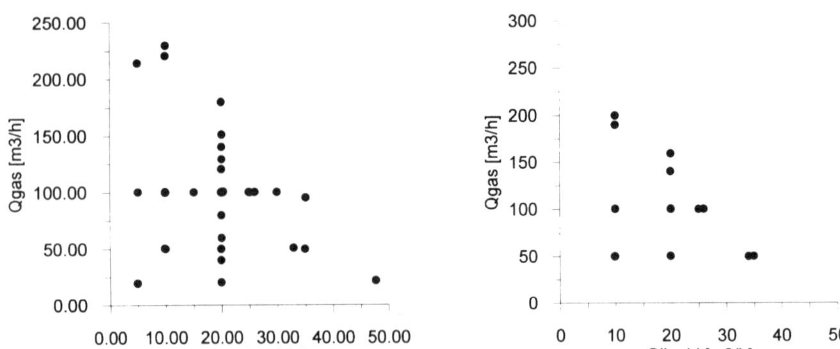

Figure 3.5 Test matrix at 23 and 35 bar for the 28 API crude.

Gas measurements.

The gas is measured with a V-cone differential pressure meter. The meter is based on pressure drop over a body. The main advantage is the low pressure-drop and the ability to handle wet gas.

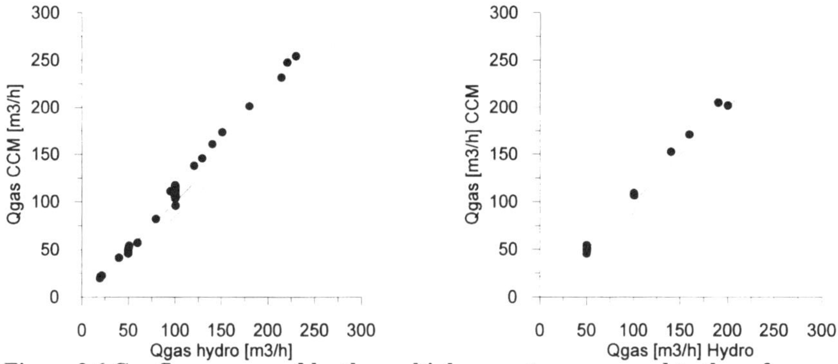

Figure 3.6 Gas flow measured by the multiphase meter, compared to the reference measured in the closed test loop for 23 and 35 bars, respectively.

The gas measurements show a strange discontinuity at 100 m^3/h where the CCM overpredicts the gas. The reason for this discontinuity is currently not understood.

Liquid measurements

The liquid measurements are done with a Khrone 1.5" straight through coreolis meter. The main advantages with the coreolis meter are high turndown and that it measures the density of the flowing media.

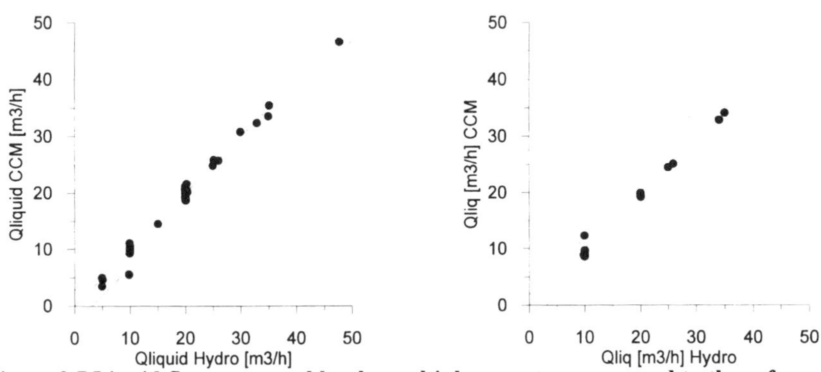

Figure 3.7 Liquid flow measured by the multiphase meter, compared to the reference measured in the closed test loop for 23 and 35 bars, respectively.

The liquid measurements show differences at low liquid flow. This is probably due to some liquid follows the gas, which has larger influence at low flow than at high flow. Overall the liquid measurements are well recreated.

Water cut measurements

A full range Phase Dynamics microwave water cut meter is used to decide the water cut.

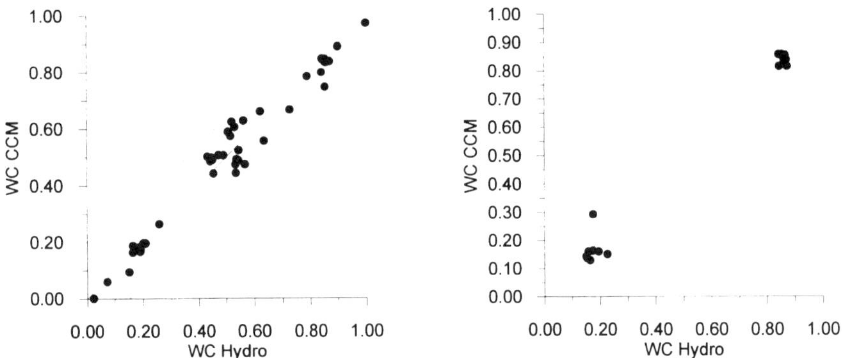

Figure 3.8. Water cut measured by the multiphase meter, compared to the reference measured in the closed test loop for 23 and 35 bars, respectively.

The water cut is relatively well predicted at the end of the specter where the flow goes towards single phase. At water cut around 0.5 the measurements gets poorer probably because of the property problems that occur around inversion.

Oil Flow measurements

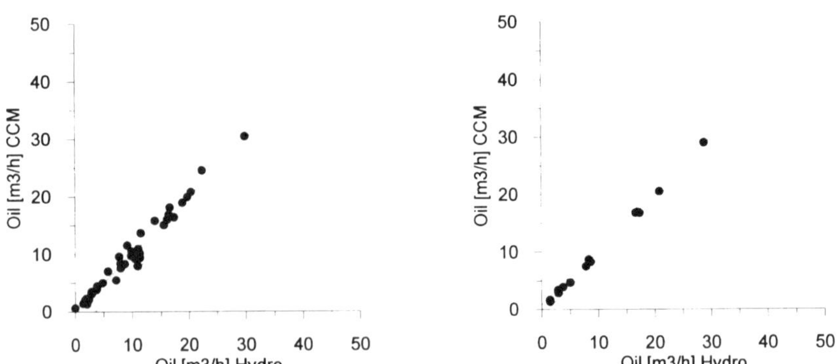

Figure 3.9 The oil flow measured by the multiphase meter, compared to the reference measured in the closed test loop for 23 and 35 bars, respectively.

Water Flow

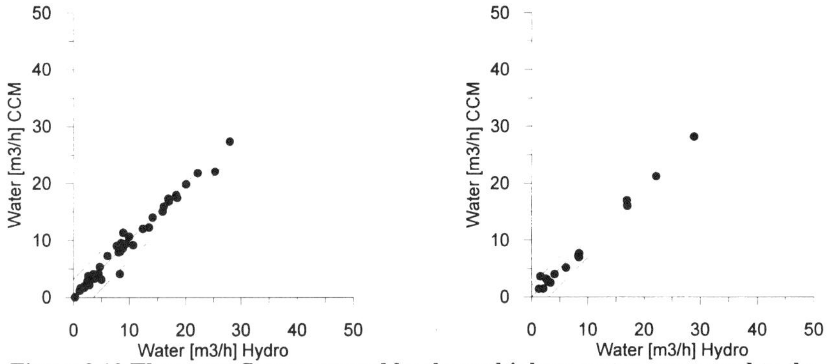

Figure 3.10 The water flow measured by the multiphase meter, compared to the reference measured in the closed test loop for 23 and 35 bars, respectively.

3.5 Liquid and gas flow for a 19 API crude

Just to give an impression of the ability to measure heavy oils, the liquid and gas measurements for the 19 API crude are shown in Figure 3.11.

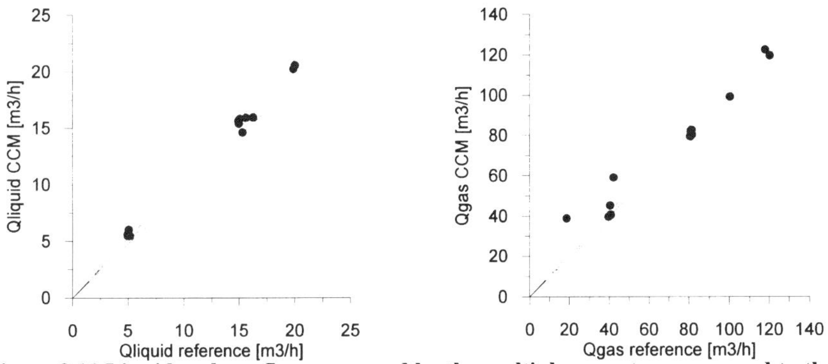

Figure 3.11 Liquid and gas flow measured by the multiphase meter, compared to the reference measured in the closed test loop. The stapled line shows ± 5 % error bands.

3.6 Multiphase meter configured for small land based wells.

A simple alternative for small on-shore facilities is developed and a prototype is made together with Phase Dynamics. The meter should remove the gas to make exact liquid measurements before combining gas and liquid downstream the meter.

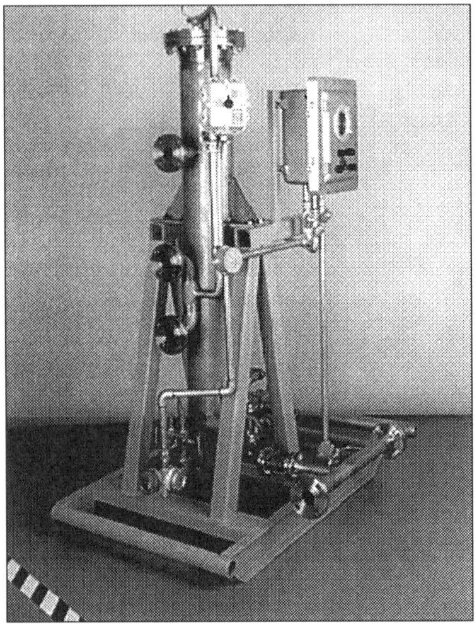

Figure 3.7 The compact gas liquid cyclone meter prototype for onshore liquid metering.

The regulation system is simplified and there is no gas metering. On the liquid side a Coreolis meter is used for the flow and a Phase Dynamics full range water cut meter is used to decide the water cut. The full skid is approximately 2 m high and a picture of the skid without the Coreolis meter are shown in Figure 3.7. The meter is a prototype and includes a regulating valve on the main gas line. The control function is integrated in the Phase Dynamics control system.

4 CONCLUSIONS AND FURTHER WORK

The gas liquid cyclone has showed good results as an inlet used in gravity separators. The cyclone's ability to suppress foam build up in the separator is well documented. For the liquid side the cyclone presents new opportunities to introduce the liquid into the separators. Current "conventional" inlets put the liquid on top of the gas liquid interface. With the cyclone the liquid may be introduced deep into the liquid layer. Experiences so far shows that the liquid separation in the gravity separator may be altered by where the liquid is introduced, and this will be looked further into.

The multiphase meter is well proven in laboratory. It will now be introduced on the American onshore market licensed by Phase Dynamics Inc.

The selfstanding two-phase separator has potential as a compact replacement for two-phase separators in separator processes. Compact equipment in the process may alter the control system and the available volumes at shutdowns. The regulation system and the influence of the reduced volume on shutdown times will be objectives in the work on the compact selfstanding separator

REFERENCES

1 Kvamsdal D, Christiansen B, Arntzen R, Chamberlain P, "*Revamping of a three phase separator based on new internals technology and advanced simulation tools*", Latest Advances In Offshore Processing, Aberdeen 12-13 December 1996.

2 Løken,K, Vangen G, Nordstad K, "*Field Test of internals for oil production*", The 3rd Major International Conference On Current Developments In Production Separation Systems, Aberdeen 23-24 April 1996

3 Nordstad K, Løken K, Vangen G, Urdahl O, "*Light weight separation equipment (LISE) for oil production*", The 4th Annual Event Production Separation System Forum, Oslo 29-30 May 1997.

The application of power fluidics to level control in multiphase separators

G H PRIESTMAN and **J R TIPPETTS**
Chemical & Process Engineering, University of Sheffield, UK
D DICK
BP Exploration, Sunbury, UK

This paper describes recent work on the development of a new no-moving-part fluidic level controller which has the advantages of reliability, no power usage, fast response and solids handling capability and which can be applied to pressurised systems such as oil-gas separators. Details are given of the operating characteristics of the new valve and its application to various multi-phase flow separators. Laboratory tests have been done with both an intensified separation system aimed at solids removal and a more conventional gravity separator operating with air, oil and water. In each case the basic level control function was confirmed and the fast response of the valve was found to be particularly suited to handling mixed phase slugging inflow.

1. INTRODUCTION

Intensified multiphase separators are of current interest for a range of potential applications including pre-separators and inlet conditioners for conventional gravity separators, subsea separators, partial processing and components in multiphase measurement loops. Their function is primarily aimed at separating the liquid and gas phases, but solids handling or separation is sometimes also an important issue. The range of current interest in this area was illustrated in the review of current gas-liquid cylindrical cyclone (GLCC) technology by Kouba and Shoham (1). An example of developments in this area is the work by Kvaerner Process Systems (2) and Statoil with compact vortex based separators. Also offering intensified separation is the proposed Baker Jardine vertical separation and pumping system (Vasps). A further option for intensified separation are designs based on combinations of T-joints, which may have potential as pre-separators (3).

The reduced volume of intensified separators makes them susceptible to changes in the inflow phase fraction, such as occurs with slugging inflow, with significant liquid slugs flooding the device giving liquid overflow, whilst gas slugs lead to gas in the underflow. Furthermore, designing systems that can respond quickly to surges without upset is seen as one of the challenges in reducing the size of vertical production separators. Intensified separators, particularly cyclonic types, can also have a relatively low turn-down ratio and are therefore unable to easily cope with the significant flowrate changes that can occur during the lifetime of an individual well or in combined well flows. Current control systems often rely upon an equilibrium liquid level as a crude measure of the gas-liquid interface, and system response time, power requirement, reliability and the ability to handle solids are all potential problems. Conventional control of the overhead gas flow has been used in a metering application, and internal floating check valves have also been used for overflow control in well testing (1,4).

From the operators viewpoint the operation of a conventional oil-gas gravity separator is a fairly trouble-free affair, given that it is based on sound design. Provided the operator controls the liquid level(s) and the operating pressure at the set points specified, then there should be reasonable confidence that the separator will function satisfactorily. To control an interface level within a separator generally requires a level measurement to be made and this signal to be converted into an instruction to a valve or similar control device to act to maintain the level. But these functions of measurement and control are not 100% reliable and require periodic care, monitoring or maintenance. If a system of level control can be created without an interface sensor and with no moving parts other than the fluids themselves, then there is potential for improved remote, subsea or unattended operation. Ideally the system should also be tolerant to damage or blockage by solids or other contaminants, have a wide operating range, and finally be cheap to manufacture and install.

This paper presents work on a new concept in level control based on a no-moving-part fluidic valve having the advantage of a flow resistance which changes in response to changes in interface level. The concept exploits the classical advantages of fluidic systems of automatic and fast response, no power requirement, reliability and solids handling capability. The system has potential both for extending the operating envelope of intensified separators or for improving level control response and reliability in conventional separators. The basic design and operating characteristics of the valve are presented followed by details of how the new controller can be applied to separator systems. Results are then given of laboratory tests using the valve to control both an intensified solids separator and a more conventional gravity separator operating with air, oil and water.

2. CONTROL VALVE DESIGN AND OPERATION

2.1 Basic Valve Design

The new controller is based on a the turn-up vortex amplifier or TuVA which is shown schematically in Figure 1. The essential features are a vortex chamber with two opposed tangential inlets and one axial outlet. With flow entering through the supply inlet S only and exiting through outlet O only, the through flow is resisted by a fluid vortex within the chamber and the device produces a high resistance to flow. If flow is then introduced through the control inlet, C, the strength of the vortex is weakened and the resistance to flow falls.

In a conventional TuVA (5) the control port diameter is normally smaller than the supply port, such that a small high pressure control flow can be used to modulate a much larger low pressure supply flow. Such a device can be used for filter blockage compensation (6) or jet pump compensation (7). In the proposed level control application the TuVA is

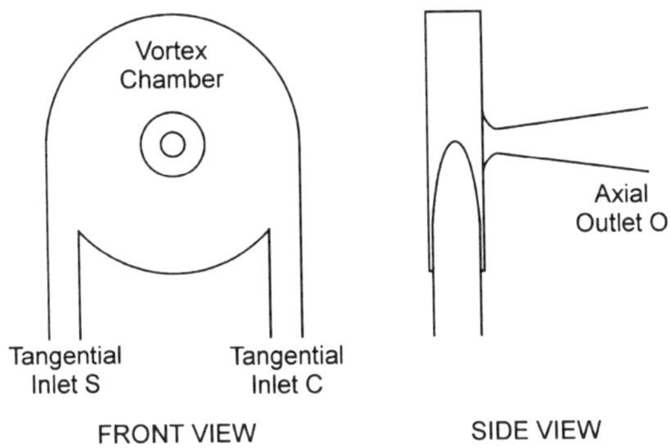

Fig.1 Basic Design of STuVA Control Valve

required to operate under several constraints which require modifications to the standard design. A major constraint is that the maximum available control port pressure is essentially the same as the supply port pressure. Thus to achieve complete cancellation of the vortex, the two supply ports need to be essentially equal in diameter. Thus the requirement is for a Symmetrical Turn-up Vortex Amplifier or STuVA, as shown in Figure 1. This gives it a very high pressure gain and makes it particularly suited to vessel level control.

2.2 Valve Characteristics

There are numerous ways in which the basic operating characteristics of the valve can be represented. The most important features of the characteristics can be quantified by the non-dimensional representation of pressures and flows shown in Figure 2.This plots the variation

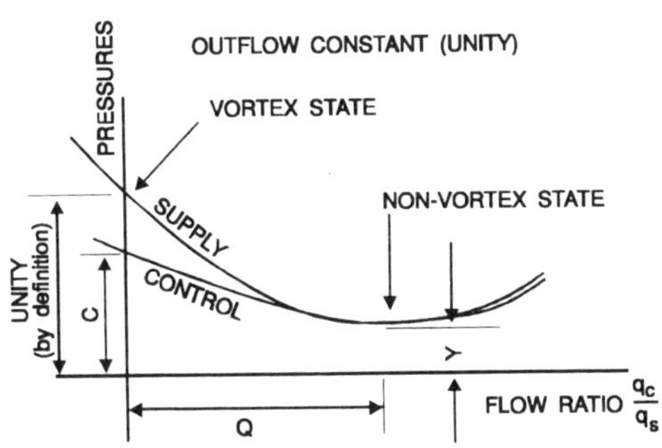

Figure 2 Non-Dimensional Characteristics of STuVA

in the pressures at ports S and C (both measured relative to the outlet pressure O), against the ratio of control to supply port inflows. The plot is based on a constant unit outflow and the supply pressure is defined as unity when the valve is operating in the full vortex state (zero control flow). Thus C is a measure of the control to supply pressure ratio when the control flow is zero and the valve is operating in its maximum resistance state. It can be regarded as the fraction of the supply pressure recovered at the control port when the valve is in the full vortex state. As seen later, for the level controller it is desirable for this to be close to unity. The parameter Y is a measure of the pressure loss through the valve when it is in its minimum resistance state. This should be as small as possible. The flow parameter Q is that producing the minimum resistance state and in a symmetrical device this is expected to be unity.

The actual characteristics of a prototype STuVA are shown in Figure 3 for operation under non-compressible non-cavitating conditions. Values found for the important performance parameters of this basic prototype are a pressure recovery, C, of 0.982 and a minimum pressure loss, Y, of 0.028. Minimum pressure loss occurred at control/supply flow below unity because of a slight difference between the two inlet port diameters. The way in which these characteristics determine the performance of the valve as a level controller is discussed in the next section.

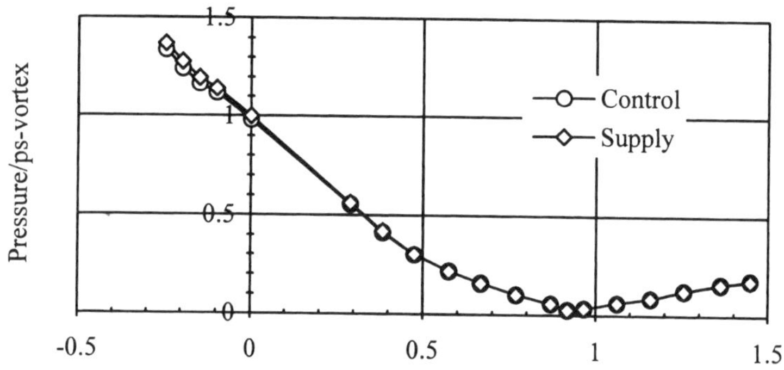

Figure 3. Measured STuVA Operating Characteristics

2.3 Basic Operation of the Level Controller

Figure 4 illustrates the basic application of the STuVA to the control of a gas-liquid interface in a horizontal separator vessel. The supply port is fed from a level well below the surface, whereas the control port is fed from a weir type inlet positioned at the desired control level. In Figure 4a the vessel operating level is seen to be just below the control point and flow enters only through port S. The valve therefore operates with zero control flow and hence a full vortex. Thus when the liquid level is just below the control level there is minimum liquid outflow, Q_{min}. Figure 4b shows operation for a liquid level just above the control point with flow entering continuously through both port S and port C, which are both subject to the same pressure. The valve therefore operates in the non vortex state with maximum liquid outflow, Q_{max}. Thus the level will be maintained close to the control level for all liquid inflows between Q_{min} and Q_{max}. As the flow increases from Q_{min} to Q_{max} the controller will operate in the non-vortex state for an increasing fraction of the operating time. In effect the controller behaves as a no-moving-part variable size outlet from the vessel such that any tendency for the liquid interface level to rise above the control level is opposed by an automatic increase in the effective size of the liquid outlet hole.

Assuming Eulerian similitude, which is reasonable at high Reynolds number, the ratio Q_{max}/Q_{min} can be estimated as $Y^{-0.5}$, giving a flow ratio of approximately 6.0 for the STuVA

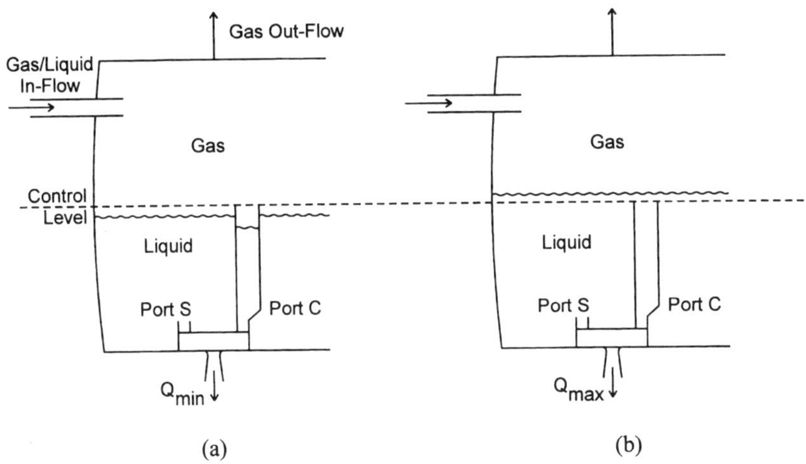

Fig.4 Application of Control Valve to Gas-Liquid Interface Level Control in Separator Vessel

characteristics illustrated in Figure 3. In addition to flow ratio, the pressure recovered at port C during the vortex state is critical for operation in pressurised separators. Any loss of pressure results directly in a fall in liquid level in the control pipe feeding the control port. The valve effectively fails if this level drops sufficiently to permit gas to vent through the valve. A simple consideration of the pressure forces leads to the following relationship determining the drop in liquid level in the control pipe below the vessel interface level, the vessel pressure being measured relative to the valve outlet pressure.

$$\text{Head drop in control pipe} = (\text{vessel operating pressure} + \text{vessel liquid head}) \times (1\text{-}C) \qquad (1)$$

As an example, for a vessel operating at 10 barg draining to atmosphere, a pressure recovery of 99% gives approximately a 1m level drop in the pipe. A major advantage of the STuVA for this application is that a potentially very high pressure recovery close to unity is possible. The prototype valve characteristics shown in Figure 3 was not optimised for pressure recovery and still gave a value of over 98%. This high pressure recovery differentiates the valve from a conventional weir (which effectively has no pressure recovery) and enables it to be applied to pressurised vessels. Clearly in designing the valve it is important to consider both the flow ratio Q_{max}/Q_{min} and the pressure recovery.

3. APPLICATION OF THE STuVA TO SEPARATION SYSTEMS

3.1 Conventional Gravity Separators

Figure 5 illustrates the principle of the application of the STuVA to control the various fluid outflows from a vessel separating a mixed inflow of gas, oil and water. Figure 5a shows how two valves can be used to control the separated oil and water outflows. Note how the valves can be conveniently located within the pressure vessel so as to minimise the number of vessel penetrations and reduce the pressure differential across the main body of the valve. Note also how circular shrouds can be used to set the level from which liquid is supplied into the control pipe. Thus a shroud on the water outlet valve prevents any inflow of oil by ensuring water feed from well below the oil-water interface. The oil valve shroud ensures oil is fed from below the oil-gas interface to avoid inflow of any foam.

During operation the simple balance of hydrostatic forces means that the oil-water interface level, L4, is determined by a combination of the water control pipe inlet level, L2, the oil control pipe inlet level, L1, and the ratio of liquid densities. Thus the thickness of the oil layer is largely determined by the simple relationship:

$$\mathrm{L1} - \mathrm{L4} = \frac{\mathrm{L1} - \mathrm{L2}}{1 - \dfrac{\rho_{\mathrm{OIL}}}{\rho_{\mathrm{WATER}}}} \tag{2}$$

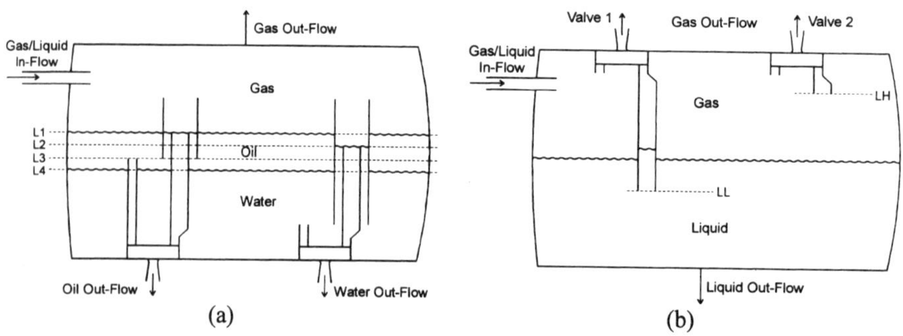

Fig.5 Application of Control Valves to Conventional Gravity Separator

In a conventional oil-gas separator the operating pressure influences the oil-gas equilibrium so the requirement on the gas outlet is often primarily for pressure regulation rather than level control. The use of the STuVA to control the gas outflow is therefore less obvious than for the liquid outflow. Figure 5b, however, illustrates two ways in which the STuVA could be used to control the gas outflow in response to level change. Valve 1 is configured such that the valve normally operates in the minimum outflow state, unless the liquid level falls below a pre-set low level, LL, in which case the valve operates in the maximum outflow state, reducing the gas pressure and tending to reduce liquid outflow. An alternative arrangement, valve 2, is configured to normally operate in the maximum outflow state, unless the liquid level rises above the pre-set high liquid level, LH, in which case the valve switches to the minimum outflow state, increasing the gas pressure and tending to increase the liquid outflow. In either case the STuVA may be used only to avoid extreme level variation, with a more conventional regulator normally controlling the outflow. If the gas STuVA were the only level controller the interface would, of course, tend to locate at the control port inlet level.

3.2 Intensified Separators

Although the application of the STuVA to control the gas outflow from conventional constant pressure separators is perhaps limited, its use with intensified pre-separators may be more useful. Figure 6 illustrates the external mounting of STuVAs to control both the gas phase and a liquid phase in an intensified vertical separator such as could be used as a pre-separator or a solids knock-out pot. In this case more significant liquid level changes are likely in response to slugging inflow. The two valves both tend to operate to control the level. Liquid rising above B2 induces extra liquid outflow by switching the bottom valve to its high outflow non-vortex state. Liquid rising above T2 reduces gas outflow by switching the top valve to its minimum outflow, full vortex state, thus tending to increase the gas pressure and induce extra liquid outflow.

Further applications of the STuVA can be envisaged. These may exploit not just the capability to control level over the flow ratio Q_{max}/Q_{min} at a given operating pressure, but also that of maintaining the level over a narrower outflow range but with a significantly changing operating pressure. Figure 7 illustrates application of the STuVA to control level within a vertical Caisson separator. In this case the oil outflow is induced by a conventional valve restricting the gas outflow. The fluidic valve controls the liquid outflow to prevent excessive liquid level changes with fluctuations either with the liquid inflow or the gas pressure.

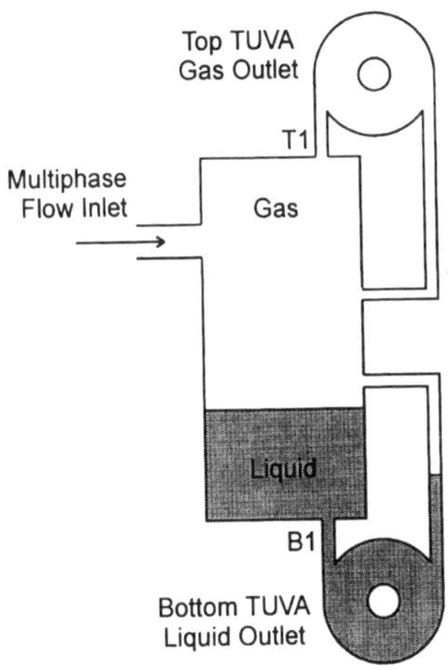

Fig. 6 Application of STuVAs to Intensified Gas-Liquid Separator

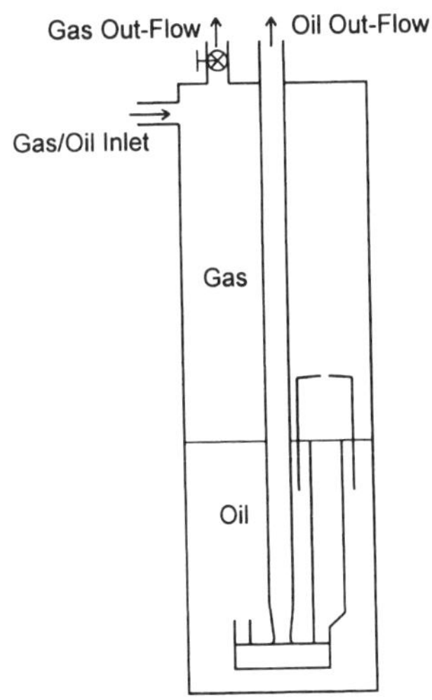

Fig. 7 Application of STuVA to Vertical Caisson Separator

4. SEPARATOR SYSTEM TESTS

Two series of tests were done to investigate the operation of the new control valve in model separator systems. In the first series of tests a single STuVA was used to control the liquid outflow from a vertical intensified separator subject to slugging air-water-solids inflow. The second series of tests used a model gravity separator operating with various combinations of air, oil and water and two STuVAs.

4.1 Intensified Separator

The basic experimental arrangement was as shown in Figure 8. Thus, of the two valve arrangement shown in Figure 6, only the lower liquid controlling valve was being used. Multi-phase slugging inflow was supplied through a 21m long, 75mm internal diameter horizontal pipe. Tests were done using air and water with injected proppant. The air and water feeds were each regulated and measured independently.

The main objective of the system was to control the gas-liquid interface within the vertical separator vessel for a range of liquid flows. In addition it was intended to minimise level variations associated with the arrival of liquid as slugs, which were of relatively large volume given the diameter of the vessel compared to the feed pipe size. Having established a fairly quiescent settling zone at the bottom of the vessel would then give the opportunity for some solids to settle out.

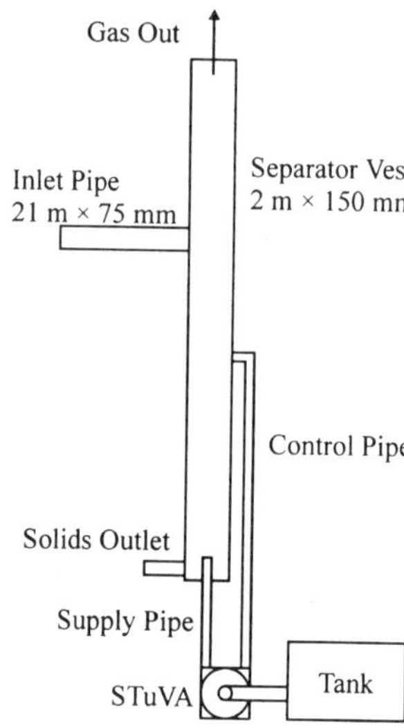

Figure 8 Basic Arrangement of Intensified Separator Rig

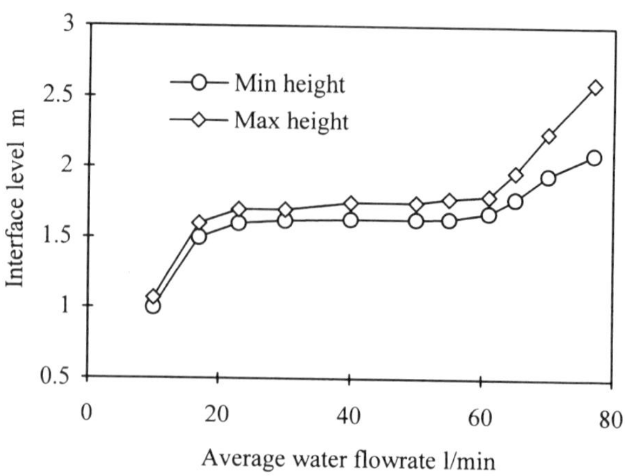

Figure 9 Level Response of Intensified Separator with Slugging Air-water inflow

Figure 9 shows how the gas-liquid interface level varied with changing liquid flowrate. A maximum and a minimum level is shown for each average flowrate, indicating the level variation associated with the slugging nature of the inflow. There is clearly seen to be a significant range of liquid flowrates for which the STuVA is controlling the level and system modifications can enable this range to be extended. Furthermore the level variation associated with the arrival of slugs is seen to be relatively small, confirming the fast response time of the controller. Solids which tended to accumulate at the base of the vessel, below the lower STuVA inlet, were periodically removed by opening the solids outlet. The operation of the STuVA was not affected by the throughflow of any unsettled solids.

4.2 Conventional Gravity Separator Tests

Although the intensified separator with its small residence time is a useful test of the response time of the STuVA, in order to better investigate the level controlling capabilities of the valve further tests were done in a model of a more conventional gravity separator. The basic vessel arrangement was in accordance with Figure 5a. Again fluids were fed through the 21m slug line. In this case, however, an oil phase was used along with the air and water, each phase being independently regulated and measured using rotameters. The oil was Shell Cantonex 11 with a relative density of 0.815. The model separator was made from transparent perspex and was 0.89m in diameter and 1.5 m long. Gas outflow from the vessel could be restricted to enable operation at slightly elevated pressures. Two STuVA valves were used, each mounted internally at the base of the tank. The control pipes were arranged to give L1 = 0.42m and L2 = 0.40m, each measured from the base of the vessel.

Initial tests using just air and water confirmed the operation of the STuVA with the vessel level remaining essentially constant for a flow range Q_{max}/Q_{min} of over 5.0. That this is below the ratio of about 6.0 suggested by the STuVA characteristics can be attributed to the required function in the vessel being more demanding than the characterisation test in that the maximum flow point must be obtained when the control pressures are exactly equal. It is likely that this simultaneity condition has not quite been achieved and some adjustment may improve performance. With flow close to Q_{min} the valve operated almost continually in its high resistance vortex mode. Importantly, in this vortex state the liquid level in the control pipe was maintained close to the vessel liquid level (as indicated in Figure 4a), confirming that pressure was being transmitted from the supply port, in accordance with equation 1. As the flow was increased towards Q_{max} the valve spent an increasing proportion of its time in its low resistance non-vortex mode. Eventually, for an inflow of greater than Q_{max}, the liquid level started to steadily rise in the vessel.

Further confirmation of the important pressure recovery or pressurisation of the control port by the supply port was obtained by operating the separator at a slightly elevated pressure equivalent to a head of about 0.3 m of water above ambient. Despite the increase in gas pressure, no obvious change occurred in the control pipe liquid level. Unfortunately the test rig did not permit higher gas pressures to be used, but further pressurisation tests are planned in an alternative rig.

The final series of tests were done with a combined inflow of air, oil and water. Separate STuVAs controlled the oil and water outflows respectively and a circular shroud prevented any inflow of oil into the water control pipe, as shown in Figure 5a. The two valve arrangement was found to successfully control the liquid levels within the vessel as both the oil and water flowrates were varied between Q_{max} and Q_{min}. The tests also confirmed that the oil-water interface level was largely determined by the simple force balance represented by equation 2, the oil layer typically being maintained at a depth of about 0.11m.

5. CONCLUSION

A new no-moving-part fluidic level controller has been developed based on a special symmetrical design of turn-up vortex amplifier or STuVA. Essentially the controller maintains a fixed gas-liquid interface level for a range of liquid flowrates and can be applied to pressurised vessels. Two parameters in the valve characteristics are identified as particularly important in determining the level control function, a pressure recovery which should be close to unity and a pressure ratio which determines the operating flow range of the valve. Experimental operating characteristics of a prototype valve have been obtained and are presented.

The way in which the STuVA can be applied to control the level in various gas-liquid separation systems has been discussed, including use in conventional gravity separators and smaller intensified systems. Separate control of oil and water outflow from a separator is also considered, including control of the oil layer depth.

Experiments have been done to confirm the operation of the new level controller. A series of tests in an intensified vertical separator with slugging air-water inflow confirmed the basic level control function together with the necessary fast response to handle liquid slugs and a solids handling capability. Further tests done using a model conventional gravity separator demonstrated good level control over a significant flow range together with successful operation for a gas pressure elevated slightly above ambient. Tests included operation with air, oil and water with separate valves controlling the oil and water outflows. Each phase was successfully controlled, as was the depth of the oil layer.

The work to date using basic prototype valves has proven the new level control concept and given encouraging performance. Thus we have a no-sensor and no-moving-part level control system, with tolerance to damage and blockage virtually the same as for a piece of pipework and which should not be expensive to manufacture and install, although has not been investigated in depth as yet. The effective operating range is still a limitation to its wide applicability but further development work is being done aimed at extending the flow range and improving and verifying its performance at significantly elevated vessel pressures.

6. ACKNOWLEDGEMENT

Work reported here has been done as part of the Marinetech Research Programme on the Design and Instrumentation for Process Separation Systems (DIPSS) which is supported by the following organisations: BP Exploration, Amoco (UK) Exploration, Health & Safety Executive, Merpro, Norsk Hydro, Petrobras, Shell and Texaco. Additional support from BP Exploration is also gratefully acknowledged.

7. REFERENCES

1. Kouba, G.E. and Shoham, O., 1996, "A Review of Gas-Liquid Cylindrical Cyclone (GLCC) Technology", Int. Conf. on Developments in Production Separation Systems, Aberdeen.

2. Christiansen, B., 1995, "Compact Centrifugal Separator", Offshore Multiphase and Separations Conference, IChemE, BP London.

3. Azzopardi, B.J., 1993, "T Junctions as Phase Separators for Gas Liquid Flows: Possibilities and Problems", Trans.I.Chem.E., Part A, Vol.71, pp.273-281.
4. Kouba, G.E., Shoham, O, and Shirazi, S, 1995, "Design and Performance of Gas-Liquid Cylindrical Cyclone Separators", 7th Int. Conf. on Multiphase Flow, BHR, Cannes.
5. Mayer, E.A., 1973, British Patent 1339642, London.
6. Oruh, S.N. & Tippetts, J.R., 1978, Proc. 3rd Colloq. Pneumatics-Hydraulics, Sci. Soc. Mech. Eng., Hungary, pp225-238.
7. Tippetts, J.R., 1981, British Patent 1595452, London.

Boosting and Metering

Rotating sealing technology for twin screw pumps used in multiphase production

N NECKER, G KLIER, D MÜLLER-LINK, and **J-U BRANDT**
Feodor Burgmann, Germany

Synopsis:

Since the early days of Multiphase Pumping several concepts have been developed and intensively tested. One of the most promising ones is the Twin-Screw-Pump. As this type of pump could already handle gas void fraction (GVF) of up to 70%, only improvements to an existing design were required. Our paper describes these developments especially with regard to GVF's of up to 100%. As the pumps are no longer running under steady state operating conditions such as standard liquid pumps in general or compressors, a major challenge to be overcome was the sealing technology for the rotating parts. With several examples on hand the most used rotating sealing solutions for Multiphase Pumping are described.

Among the technologies being developed for oil and gas production in the last 10-15 years Multiphase Pumping can be seen as one of the most promising. The significant reduction of capital investment either on- or offshore together with an increased production rate has led to a growing demand on systems to be installed. The following incentives for using Multiphase Pumps have been reported (SPE 27995):

- reduced equipment capital cost and complexity
- elimination of separate oil/water and gas lines
- profitable production of marginal fields
- increased production rates
- reduced weight and space requirements (for offshore applications)
- mitigation of gas hydrate problems
- prolonged reservoir life
- reduced wellhead back pressure.

The biggest challenge to be overcome by a piece of rotating equipment in multiphase service is the variation from 100% liquid to 100% gas at inlet of the fluid to be moved.

In this respect only some minor improvements had to be made to the well-known Twin-Screw-Pump, as this type of pump could handle a GVF (Gas Void Fraction) of up to 70% already. But a certain amount of liquid is needed to seal the gaps between the screws and the liner. Within the patented BORNEMANN design this is achieved by separating liquid at the discharge side of the pump and feeding some of it back to suction internally. Depending on the pressure ratio and temperature level a dry-running phase of approx. one hour can be achieved.
By this the early production gas phase is moved away from the wellhead and transported to treatment facilities downstream of the pump.

The Multiphase Pump is part of a system starting in the reservoir and ending in some system where the phase finally needs to be separated. Subsequently the selection is mainly governed by the general operating data such as:

- average hydrocarbon production (oil+gas) plus water cut
- inlet pressure (desired)
- discharge pressure (required)
- inlet and (sometimes) discharge temperature

The capacity of the pump is determined by the liquid volume plus the gas volume at inlet conditions. The latter one being a function of inlet presssure and temperature only, while the compressability factor may be neglected in this case. The discharge pressure equals the pipeline pressure losses plus the necessary inlet pressure to the treatment facilities downstream of the pump. Additionally the pump design is governed by reservoir and hydrocarbon characteristics such as :

- sand production
- crude oil viscosity and specific gravity
- water salinity
- H_2S and CO_2 content.

The know-how in the pump design may be divided into the pump generally and the rotating sealing elements specifically.

With several examples on hand this paper describes the main requirements of sealing unprocessed Multiphase Well Stream Fluid and gives a general survey of the rotating sealing solutions which are mostly used for Positive Displacement Screw Pumps.

1. PROBLEM TASK

Multiphase operation means continuously changing operating conditions for the pumps and all their components. Unlike sealing systems which are exclusively and specifically applied in liquid or gaseous media mechanical seals for multiphase pumps are operated under rapidly changing conditions and in continuously changing medium phases. Due to the rapid change of the medium from gaseous to liquid and vice versa the mechanical seal has to cope with the following demands:

- viscosity variations
- temperature variations
- erosion (sand particles in the medium)
- high and low operating pressures
- insufficient lubrication (dry running) in the gaseous phase

The chemical analysis of the medium influences the seal additionally.

2. CHEMICAL PARAMETERS

The chemical analysis of the medium differs, depending on the oil field and/or the location. This means that the crude oil can contain

- wax
- salts
- hydrogen sulphides (H_2S) or carbon dioxide (CO_2)
- crack products (hydrocarbons)
- sand particles

The constituents such as wax, salts and crack products (hydrocarbons) deposit on the atmosphere-side of a single mechanical seal, which affects the operating performance of the seal and results in abrasion.

Crude oil constituents such as calcium carbonate or sodium chloride for example tend to precipitate and/or crystallize in the sealing gap. This results in an increased wear of the sliding faces. Other constituents such as hydrogen sulphides of various concentrations lead to a corrosion of metal parts. However, due to the limited wall thickness the strength of the seal components is considerably affected by corrosion.

3. PHYSICAL PARAMETERS

During operation the pressure to be sealed is decreased to atmosphere pressure in the sealing gap. The friction of the sliding faces generates a temperature increase in the sealing gap. The pressure decrease and the simultaneous temperature increase in the sealing gap result in a vaporization of the easily boiling oil constituents. This leads to an insufficient lubrication of the mechanical seals. Moreover, the gas constituents in the medium in multiphase applications affect the seal with regard to dry-run which may result in a destruction of the sliding faces. Gas phases over an extended period of time additionally increase the temperature. This means that the evaporization process in the sealing gap is accelerated. The above physical parameters cause dry running of the mechanical seal.

4. PUMP CHARACTERISTICS

The Twin Rotor Screw Pump consists of two contrarotating screws with their shaft centers located so close to each other that there is only little space available for the installation of two mechanical seals. The requirements can be fulfilled by installing a radially compact seal into the restricted pump envelope.

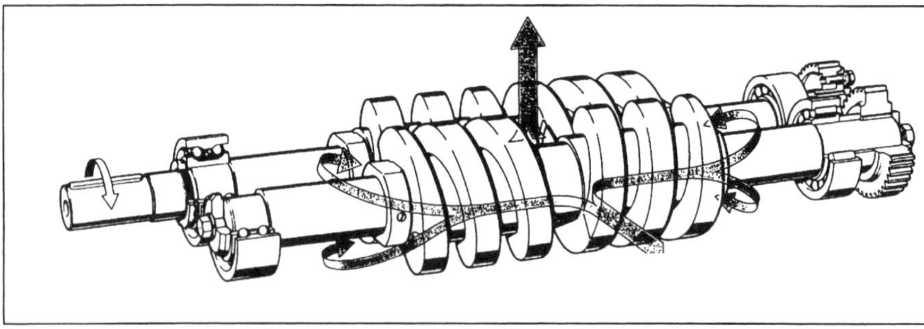

5. PROBLEM SOLUTION

Dependent on the multiphase mixture constituents and the pump design a single seal without additional auxilliary measures can serve the purpose. A permanent liquid flooded seal area, for example, allows the use of a single seal arrangement without any auxilliary systems also in a high pressure application.

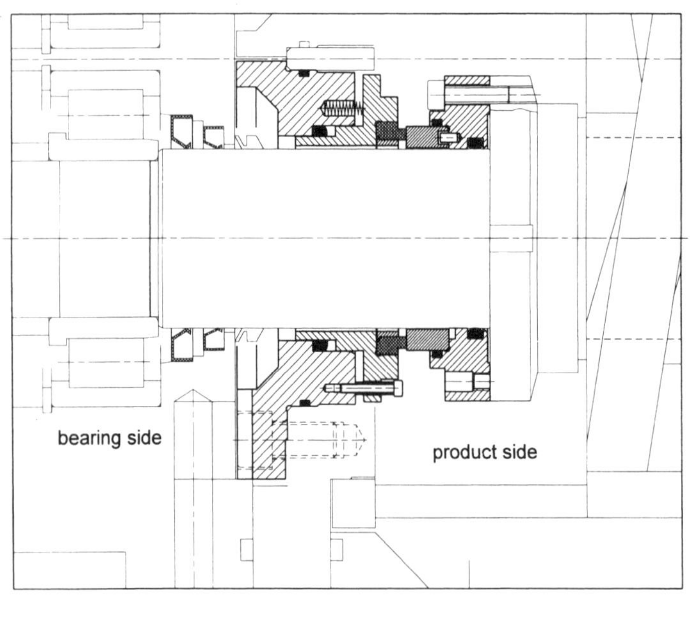

Example:

Single mechanical seal type SHV, installed in a Bornemann pump on an offshore production platform of the Amoco Trinidad Oil Company. This High Pressure Seal has been working for more than 6.000 hours under the following operating conditions: Pressure: 70 bar; Speed: 1.200 rpm; Temperature: 40-70°C; Gas percentage: 78%.

Single mechanical seals can only be used when deposits, precipitation, or crystallization of dissolved medium constituents in the area of the mechanical seal are avoided.

In case of insufficient lubrication and/or dry-run the use of suitable sliding materials can extend the service life of the mechanical seal. If there are no special erosion protection requirements antimony-impregnated carbon graphite versus pressureless sintered silicon carbide is used. Additional measures such as cooling of the sliding faces in combination with a non-pressurized supply fluid stabilize the lubricating film in the sealing gap and reduce the dry-run risk.

This can be prevented by taking the following measures.

5.1 Injection of clean fluid into the seal chamber from an external source (e.g. API 610, plan 32)

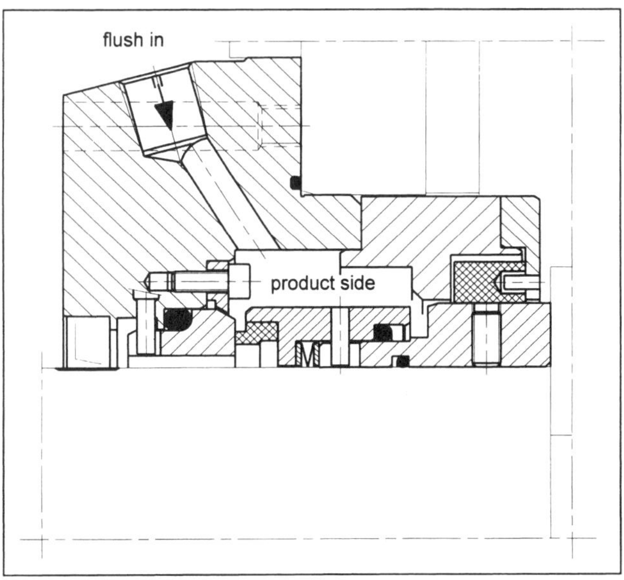

Example:

Single balanced mechanical seal type HJ92, in combination with a floating throttle bushing installed offshore at Chevron (USA) in a Leistritz Twin Screw Pump under the following operating conditions: Pressure: 3-5 bar; Speed: 2.100 rpm; Temperature: 20-50°C; Gas percentage: 95 %

**5.2 Non-pressurized supply fluid (quench)
(e.g. API 610, plan 52 or 62)**

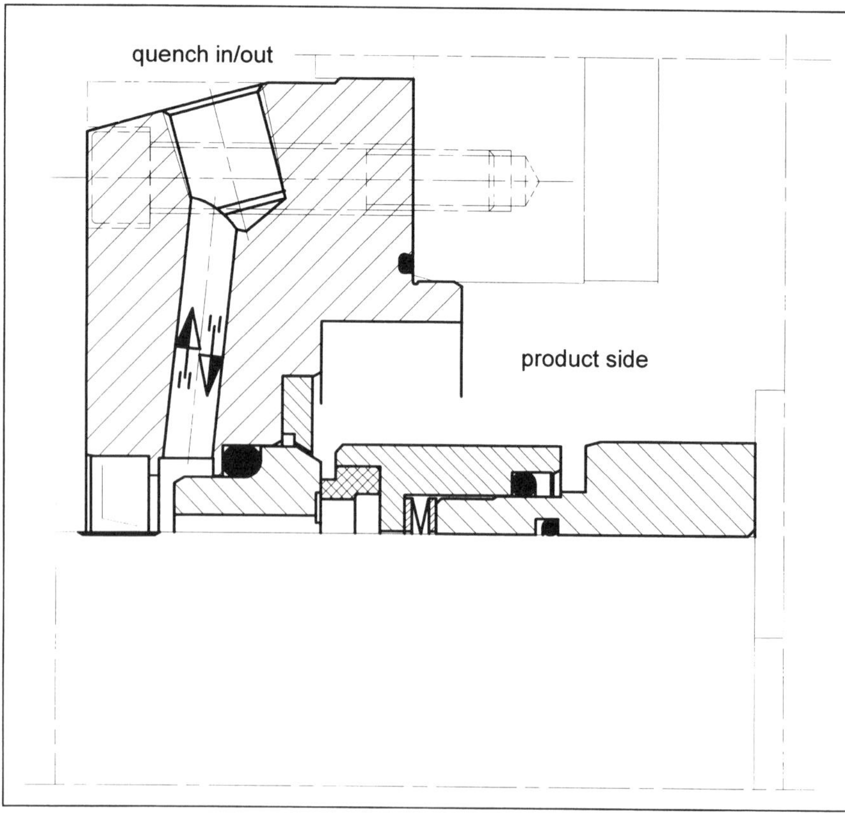

Example:

Single balanced mechanical seal type HJ92, installed onshore at Corpoven Venezuela in a Bornemann pump under the following operating conditions: Pressure 2-4 bar; Speed: 1.800 rpm, Temperature: 32-50°C; Gas percentage: 88%; Viscosity: 300-1.900 Cst
At the inner diameter of the stationary seal face the seal is equipped with hydrodynamic grooves to bring the supply fluid between the faces in order to overcome the gas phase for a substantial period of time.

5.3 Use of a dual mechanical seal with pressurized supply fluid (e.g. API 610, plan 53)

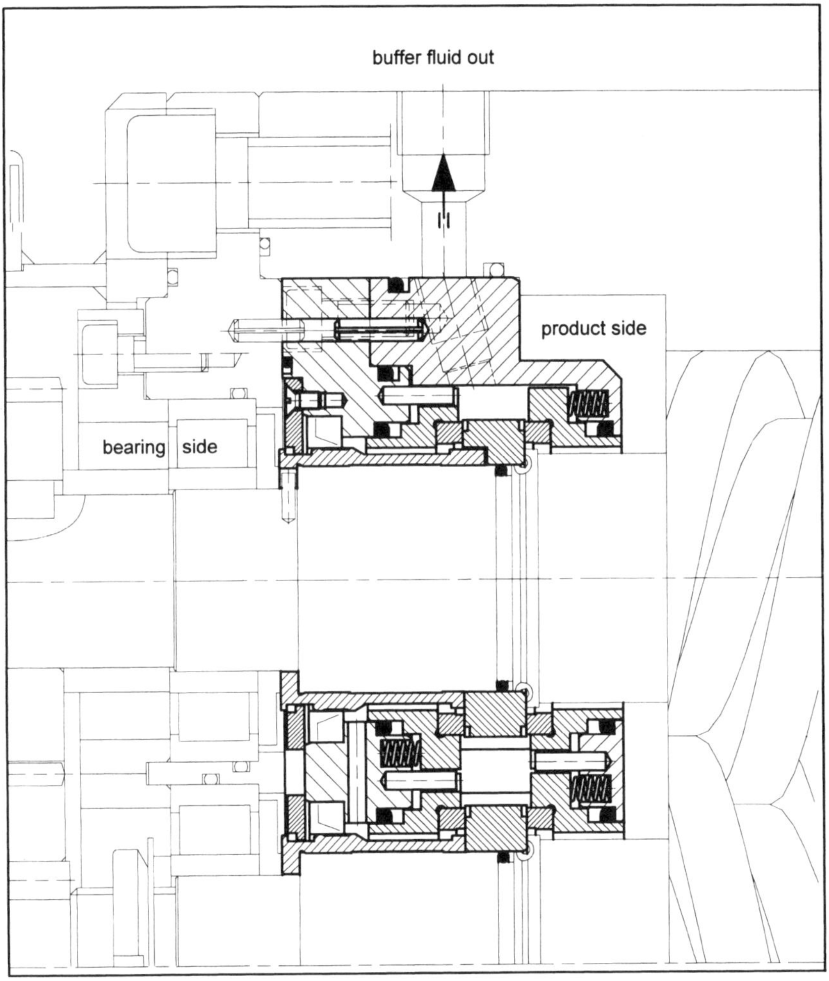

As mentioned before a positive Displacement Screw Pump has 4 shaft ends, i.e. 4 dual mechanical seals are required. With eight seal faces operating it is extremely important to reduce the face loading as well as the power consumption and the starting torque.
A stable double mechanical seal like the above shown seal type SH-D, currently installed onshore at a hydrate test loop of Marathon Oil (USA) could be achieved after a few adjustments of the sliding faces. The seal is operated under the following conditions: Pressure: 2-150 bar; Speed: 1.750 rpm; Temperature: 70 °C; Gas percentage: up to 95%

Corrosion can be avoided by choosing suitable materials. When laying out the mechanical seal the sliding and constructional materials have to be adapted to the constituents of the multiphase medium. This critical material selection should be done in close cooperation with the experienced seal manufacturer.

If the multiphase medium contains abrasive particles such as sand pressureless sintered silicon carbide versus pressureless sintered silicon carbide is used. If there is an increased dry-run risk additional measures such as cooling of the sliding faces by using a dual seal in combination with a pressurized supply system should be taken. In this case the liquid lubrication of the dual mechanical seal is guaranteed in any operating condition.

In order to reduce the risk of dry running and the amount of sliding faces, which consequently means a reduced power consumption as well as starting torque and leakage value, the following solution has been created:

A pressurized dual seal arrangement which is basically a single seal with only one additional seal outboard of the bearings. The interspace in between is being pressurized to the seal chamber pressure plus 10%. This is quite a new design but offers the above mentioned advantages. The following sketch shows the so called "5th seal solution".

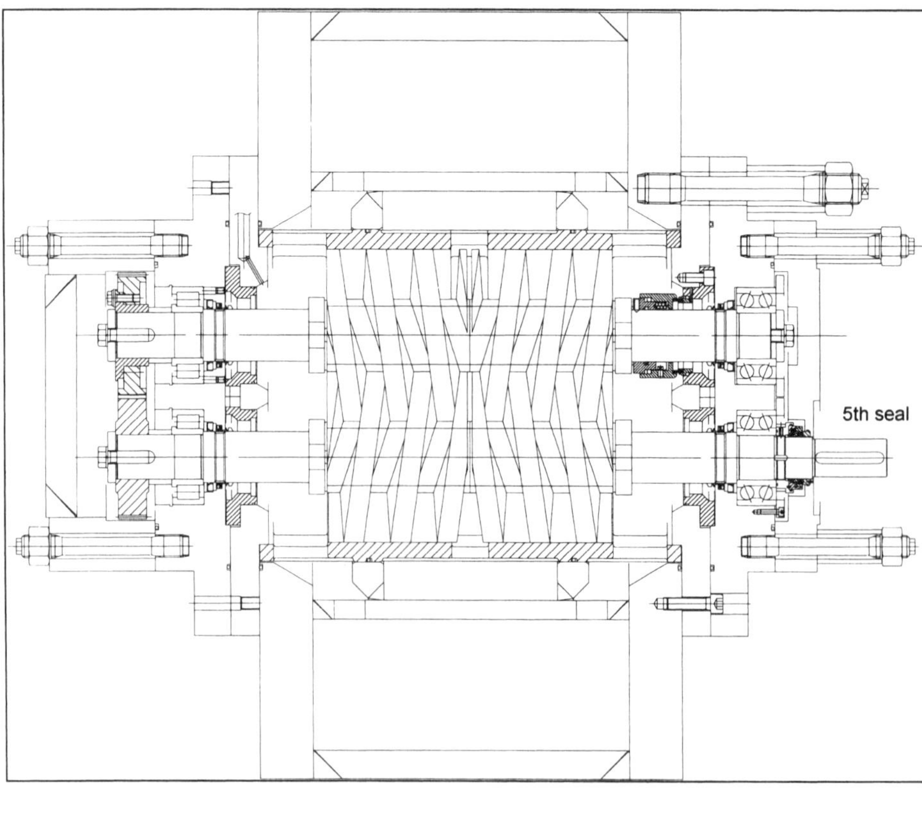

6. CONCLUSION

Twin Screw Pumps have proven their ability to pump multiphase mixtures over a wide range of conditions in several applications all over the world.

The mechanical seal types and sliding materials available today are capable to meet any technical demand defined by multiphase pump services. In order to assimilate the sealing systems with the specific operating conditions the close co-operation between seal manufacturer and end-user has been a basic precondition. Due to constant development there will be further improvement of sealing technology in future. As an example gas-lubricated mechanical seals in multiphase applications resemble the ultimate state of technology and are already realizable today.

A review of the application of jet pump technology in oil and gas industry to increase production

M M SARSHAR and **N A BEG**
CALTEC, BHR Group Limited, Cranfield, UK

ABSTRACT

Production from many fields is restricted because of poor reservoir and production pressure. In many cases, additional boosting is required to maintain an economical level of production. Jet pumps are one of the most cost-effective and reliable equipment to achieve this objective. A high pressure fluid is needed as the motive flow to drive and boost production from low pressure wells or to boost the pressure of processed fluids for transportation purposes. Jet pumps can be used successfully downhole, offshore, subsea and onshore to increase production from oil or gas wells. Their use in oil production, which involves multiphase flow, brings about the challenge of handling two-phase flow (gas and liquid) by the jet pump. The system developed and patented by CALTEC under the trade name of "WELLCOM system" solves this problem as described in this paper.

This paper also reviews the applications of jet pumps and some of the development work carried out in recent years to improve their performance and to extend their range of applications.

Key words: *Jet pumps, ejectors, WELLCOM, well commingling, eductors, multiphase pumping*

1. INTRODUCTION

Lack of sufficient pressure is often the main cause of restricting production from many oil and gas fields. This potential (pressure) energy is required to compensate for losses of pressure along the flow path of hydrocarbon fluids from the reservoir to the wellbore, and losses along the wellbore and pipelines leading to the processing facilities. The practical ways to overcome this problem are the optimisation of the design for the wellbore and pipelines through which the produced fluids pass, and the introduction of a boosting system.

There are a number of boosting systems which help to increase production from oil and gas fields. Each system has its particular range of applications and limitations in terms of both technical viability and economics. In this paper, the application of jet pump technology is reviewed as a cost-effective and reliable way to boost production from many marginal fields. The use of jet pumps relates to downhole, subsea, offshore (topsides) and onshore applications.

2. PRINCIPLE OF OPERATION

The general configuration of a jet pump or ejector is shown in Fig. 1. A high pressure (HP) fluid (gas or liquid) passes through a nozzle as the motive flow. As the high pressure fluid enters the nozzle, part of its potential energy (pressure) is converted into kinetic energy (velocity). This energy conversion leads to a drop in pressure downstream of the nozzle where the low pressure (LP), suction fluids, enter through a chamber. The sudden drop of pressure in this chamber allows the low pressure fluids to be drawn in without being exposed to the full pressure of the motive stream or the downstream pipeline pressure.

The HP and LP fluids then enter a mixing tube, where momentum and energy transfer between the HP and LP fluids takes place. The fluids then pass through a diffuser where the diameter of the pipe increases gradually and the velocity of the mixture reduces. The reduction in the mixture velocity leads to the conversion of part of the kinetic energy to pressure and further pressure recovery takes place. The HP and LP fluids leave the jet pump at an intermediate pressure between the HP and LP pressures.

The jet pump therefore offers mainly two benefits:

1. It protects the LP flow from being exposed to the HP and the downstream (pipeline) pressure.

2. It boosts the pressure of the LP flow to that at the discharge of the jet pump. This is in general the required pipeline pressure which in a conventional manifolding system the LP flow would have been exposed to.

The use of a jet pump therefore allows the LP fluids to operate at a lower pressure in comparison with the conventional manifolding system where the HP pressure is choked down and the LP pressure is restricted by the pipeline pressure at the outlet of the manifold. The use of jet pumps will therefore lead to a boost in production and an increase in total recovery from the field.

The performance of jet pumps and their design varies significantly depending on whether the HP or LP fluids are gas, oil, water or a mixture of these phases. This important issue is discussed in some detail in the following sections.

The systems which involve the use of specially designed jet pumps and the equipment supplied by CALTEC are known by the trade names of WELLCOM-Gas and WELLCOM-Oil for gas or oil production applications respectively.

3. APPLICATIONS

The applications of jet pumps in the oil and gas industry can be addressed with reference to their location within the production system which could be: downhole, subsea, offshore (topsides) and onshore. In each case the application may relate to gas or oil production which affects both the design and the performance of the system.

3.1 Downhole Applications

The majority of the applications of downhole jet pumps is in oil production applications where both the motive and suction flow are a liquid phase. The motive flow can be stabilised oil, water or oil with additives such as Naphta to improve properties such as the viscosity of the produced oil. A good example of this application is the vega field in Sicily (Italy) where the produced oil was highly viscous (15.50 API) (Ref.9). Fig. 2 shows the effect of the added Naphta in concentrations of 25% to 50%, depending on water cut. This Fig. shows that the flowing tubing pressure increased significantly as a result of adding virgin Naphta under different values of water cut ranging between 20% to 60%. Other examples are the wells in BP's Wych Farm in the UK.

Downhole jet pumps are highly reliable with practically no maintenance requirement. This added advantage makes them more attractive in comparison with alternative downhole boosting systems such as electrical submersible pumps (ESPs). ESPs are of course more efficient and can provide a higher level of boost in comparison with downhole jet pumps, but both their capital and operational costs are significantly higher. The selection is therefore field specific and involves assessing the capital cost, the cost of power generation for ESPs or the supply of the motive flow for the jet pump, the maintenance costs and the added production achieved by each system. The efficiency of the liquid-liquid jet pump is affected if free gas is introduced into the suction flow. This effect is discussed in more detail in section 4.3, WELLCOM system. Presence of gas also affects the efficiency of ESPs and could lead to depriming or damage to seals and bearings of the pump. It is for this reason that in such applications centrifugal separators, driven by the pumps shaft, are used to remove the bulk of the free gas to bypass the pump. Even if the pump can handle free gas, its performance is affected because of the energy used to boost the pressure of the additional gas phase.

3.2 Downhole Gas-Liquid Jet Pumps

Use of high pressure motive gas is an alternative option to boost the production of both oil and gas wells, using downhole jet pumps. Use of HP gas as the motive flow has, however, some limitations worth noting. In both oil and gas wells, which produce a mixture of gas and liquid phase, the superficial velocities of gas and liquid phases could be low enough to cause erratic slug and churn flow, which will lead to excessive loss of pressure through the wellbore and will restrict production. In some applications, the produced liquids drop out and accumulate downhole. This will, in turn, lead to the build-up of excess hydrostatic head and will result in killing the well. Gradual accumulation of the liquid phase (oil or water) in the vicinity of the wellbore will worsen this situation further.

A successful method of alleviating this problem has been to increase the velocity of gas through the wellbore so that liquid dropout is prevented and the high velocity gas sweeps the produced liquid phase to the surface. The increase in the velocity of gas is, in general, achieved by selecting a smaller tubing size. Alternatively, gas can be injected downhole to kick the well off and to maintain higher mixture velocities along the wellbore. This method is similar to the conventional gas lift operation in many oil producing wells. A major disadvantage of this system is that the higher velocity of the mixture through the wellbore will result in additional frictional losses and, while the liquid drop out is minimised or prevented by this method, the production will still be restricted because of the significant loss of pressure through wellbore.

Use of downhole jet pumps in such applications could be beneficial but with some limitations. Jet pumps with HP gas as the motive flow and a mixture of gas and liquid (well products) on the suction side do not function efficiently. This is particularly true when the flowrate of the liquid phase is beyond a few percent of the actual volumetric flowrate of produced gas. This limitation can be explained by considering that the energy and momentum of HP gas is low because of the low density of gas in comparison with the average density of the LP oil (or water) and the gas mixture as the suction flow. The liquid phase density being many times greater than that of gas, means that presence of a liquid phase, even in a small percentage by volume, has a dramatic effect on the total density and mass of the LP flow in comparison with the mass of HP gas. For this reason the gas-liquid or gas-multiphase jet pump will operate with a very low level of efficiency and the boost in pressure achieved downstream of the jet pump will be in most cases insignificant.

A further important factor to note in this application is the risk of choking in the mixing tube if the designers of the jet pump do not take correctly the account of this important effect.

The velocity of sound is high in the medium involving pure gas or liquid phase. Typical velocity of sound through air is 343 m/sec and 1483 m/sec in water at normal pressure. These values drop dramatically in gas-liquid mixture as shown in Fig. 3. The steep slope at either side of the graph shows that changes by only a few percent in volume of the liquid present in gas phase will reduce the velocity of sound significantly. With air-water mixture of 50/50, the velocity of sound could drop to 25 m/sec which may, in the case of jet pumps, result in choking the flow. This phenomenon dictates that the designer of the gas-gas jet pumps should check the choking effect when a liquid phase (oil or water) also flows through the jet pump.

CALTEC has developed a novel downhole system known as WELLBOOST which uses jet pump technology and is suitable for applications when gas lifting is not effective in low pressure oil or gas wells and presence of the liquid phase restricts production. This system is undergoing its initial tests in CALTEC laboratories.

3.3 Subsea Applications

The subsea application of jet pumps relates primarily to manifolding cases where products from a number of wells are manifolded for transportation by a single pipeline. The production pressure of the lines from various wells could be different. This often arises because of production from satellite fields with different production characteristics, production from different zones or the fragmentation of the reservoir. In such applications, the energy from HP wells can be used to boost the production from LP wells. The boost in the pressure of the combined products helps to transport the produced fluids efficiently by a pipeline to a processing platform - see Fig. 4. The jet pump, in this case, helps to increase production from LP wells and in some applications, it may eliminate the need for a separate LP pipeline.

In the case of oil production, if high pressure wells do not exist, motive flow can be supplied from platform as shown in Fig. 5. This boosting method is particularly ideal for deepwater applications where the hydrostatic head of the produced fluids within the riser imposes a significant back-pressure on producing wells. The jet pump in this case is the reliable subsea component while the booster pump, which may need frequent maintenance, is located on the platform. The line providing the motive flow can be dosed with chemicals and inhibitors if required. The chemicals are used to prevent formation of hydrates or wax or, in applications involving the production of heavy viscous oil, to reduce the viscosity of the produced oil in order to minimise transportation problems. The viability and the economics of this option depends on the complexity and the cost of the piping system required to manifold the HP and LP products within a subsea template. The cost of the subsea jet pump is, in fact, an insignificant proportion of the manifolding cost in most applications.

3.4 Applications on Platforms

In both oil and gas production applications, the jet pump can be used to boost production from low pressure wells using energy from high pressure wells. In gas production applications, if high pressure wells do not exist, high pressure gas from the outlet of compressors can be used as the motive flow. In some applications the pressure of the low pressure gas produced from the second stage or the third stage separators can be boosted in the same manner using HP gas from a compressor or gas from HP wells - see Fig. 6 . This method could often lead to the elimination of low pressure compressors or flaring of gas with substantial benefits.

Fig. 7 shows an application of jet pumps when low pressure oil wells are involved. The production pressure of wells is, in this case, dictated by the pressure of the separator which is required to operate at a level demanded by the gas export line. The booster pump provides the motive flow, using oil from the outlet of the separator. Similar applications can also relate to onshore production.

4. PERFORMANCE OF JET PUMPS

4.1 Gas-Gas Jet Pump

The performance of jet pumps in gas production application is dictated mainly by the following factors: HP/LP pressure ratio, LP/HP flow ratio and presence of liquid in the motive or suction gas. Fig. 8 shows the typical performance of jet pumps under different HP/LP pressure ratios.

For each production condition an optimum design is desired for the key features such as the size of the nozzle and the mixing tube. If significant changes occur during the production life of the field to both HP and LP flow conditions, changes to the above said features may become necessary in order to maximise the efficiency and the performance of the system. It is for this reason that the jet pumps supplied by CALTEC have interchangeable parts which enable both the nozzle and the mixing tube to be changed if required. Change-out operation for such units is relatively easy and can be achieved within a short period.

Presence of liquids (oil or water) in the suction flow affects the performance of the gas-gas jet pump if the liquid content exceeds a few percent by volume. Presence of the liquid phase could also lead to choking as discussed earlier. A software has been developed which predicts these effects on the performance of the jet pump. If the quantity of the liquid phase is excessive, separation of the liquid phase may become necessary if optimum performance is desired. A similar effect on performance may be experienced if some liquid is present in the HP motive flow. Presence of a liquid phase in either HP or LP flow under transient conditions such as start-up is in general tolerated by the jet pump and, although it still affects its performance and efficiency temporarily, the fluids pass through the system without much concern and the system performance stabilises rapidly.

Noise

Gas-Gas jet pumps may generate noise which could be above the limit set by regulatory authorities or by local demands. The noise level generally set for manned platforms is 85 dB. This limit may vary depending on the hours of exposure for personnel or the location of the jet pump in relation to the surrounding areas, particularly in onshore applications.

The level of noise emitted depends primarily on the pressure ratio and flow ratio of HP and LP gas and the wall thickness of the jet pump, the LP and the discharge lines. The noise level is generally specified at 1.0m away from the pipeline or the jet pump.

Silencers may be required downstream of the jet pump on the LP and the discharge line in order to prevent noise travelling through these lines and extend the area subjected to an unacceptable level of noise. Such noise suppression devices are inline units installed close to the jet pump on the LP and discharge lines. The noise emitted directly through the body of the jet pump may be below that experienced in the LP and the discharge line. This is because the wall thickness of jet pump components is usually above that of the LP and the discharge lines.

4.2 Liquid-Liquid Jet Pumps

The performance of liquid-liquid jet pumps is relatively well understood. Fig. 9 shows the typical performance and the efficiency (η) of the jet pump handling a liquid phase. The term pressure ratio 'N' in this case is a good indication of the jet pump performance. The slope of this line varies significantly and depends primarily on the ratio of the area of the nozzle to mixing tube. Presence of gas in either HP or LP products affects the performance of the jet pump and its design as discussed below.

4.3 Liquid-Multiphase Jet Pumps (WELLCOM-Oil)

Production of oil involves a mixture of gas and liquid phases; the liquid phase may be oil or a mixture of oil and water. The jet pump does not function well if the motive flow is in two phase (gas and liquid) form. For this reason, if the HP fluids are supplied from high pressure wells, separation of gas from the liquid phase is necessary in order to achieve best performance from the jet pump. The system for oil production application was developed and patented by CALTEC in 1994, and extensive laboratory and field trials demonstrated its viability and performance clearly. The system, known as WELLCOM-Oil, consists of three key components as shown in Fig. 10. A compact inline separator, known as WELLSEP, a specially designed jet pump and a commingling spool. The separator has the duty to separate the bulk of the HP liquid phase as the motive flow for the jet pump. The separated gas phase bypasses the jet pump and is combined with the flow from the outlet of the jet pump, using a commingling spool. All components are compact and simple, with no moving parts. The separator is a very compact unit and, in comparison with conventional gravity separators, it is tiny. It requires no active pressure or level control, ideal for subsea and offshore applications.

Figs. 11 and 12 show the typical performance of the WELLCOM system. It can be seen that in addition to parameters such as HP/LP pressure ratio and LP/HP flow ratio, the gas volume fraction (GVF) of the LP flow plays an important role in the performance of the system. The significant effect of the GVF of the LP flow on the overall performance of the system is expected as an increase in the volume of the LP gas results in more HP energy being dissipated to compress the LP gas and to increase the pressure of the LP gas and liquid mixture. Fig. 13 shows the results of field trials of a full-scale WELLCOM unit during 1996-97 in Agip's Trecate field in Italy. The shaded area shows the level of boost achieved (discharge - LP pressure) which varies depending on the LP flowrate. Fig. 14 shows the typical effect of the GVF of the LP flow on the performance of the system. These graphs show that for oil production applications, the WELLCOM system can be used effectively to boost production from LP oil wells. The performance of the system becomes, however, less attractive when the GVF of the LP flow is high as shown in Fig. 14. In such cases, the Dual WELLCOM system described below will be significantly more efficient.

4.4 The Dual WELLCOM System

The Dual WELLCOM system shown in Fig 15 consists of two compact separators and two jet pumps. In this case, gas and liquid phases are separated from both HP and LP fluid mixtures. A gas-gas jet pump is used to boost the pressure of the LP gas, using the energy from HP gas. In a similar manner, a liquid-liquid jet pump is used to boost the pressure of the LP liquid phase. Table 1 shows the comparison of the two systems for a typical scenario involving a high GVF value for the LP flow.

4.5 Liquid-Gas Jet Pumps

Compression of gas using high pressure liquid as the motive flow is feasible, but the range of its operation is limited to specific volumetric flow ratios of gas to the liquid phase. The design of the jet pump in this application requires particular attention in order to achieve efficient mixing of the liquid and gas phases. Work by CALTEC and others (Ref. 2) has shown that the normal design of the mixing tube may result in poor mixing of the two phases and very low efficiency of the jet pump.

Following an extensive development work, CALTEC has developed a novel feature for the jet pump in this application which maximises the efficiency and gas handling capacity of the jet pump. Fig. 16 shows the efficiency of the improved jet pump under a range of LP to HP volumetric flow ratios. This graph shows the efficiency drops below 20% for actual gas-liquid flow ratios in excess of 3.5. An application of this system is shown in Fig. 17 where the boosting system enables an increase in production from low pressure oil wells using conventional single phase (liquid) booster pumps. This system is known as "WELLCOM Boost", and is a competitive and simpler alternative to multiphase pumping systems for a specific range of GVF values.

4.6 Gas-Liquid Jet Pumps

If high pressure gas is the motive flow and liquid (oil or water) is the suction fluid, the jet pump can handle a small volume of the liquid phase before its efficiency drop to an unacceptably low level. The main reason for this limitation is the major difference between the density of the motive gas and that of the LP liquid phase. Gas-liquid jet pump application is therefore limited to cases when the introduction of a small volume of liquid phase is involved (e.g. chemical injection) or removal of liquids from sumps at relatively small flowrates and without the need to boost the pressure of the LP liquid phase. In such applications, most of the HP pressure is used to lift the liquid phase and, as a result, the discharge pressure will be very close to that of the LP liquid phase. Choking is also an important issue to check as discussed in section 3.2.

5. DESIGN AND PREDICTION OF JET PUMP PERFORMANCE

Jet pumps, despite their simplicity and reliability, require correct design of their key dimensions and components in order to achieve optimum efficiency. This issue becomes particularly important in the oil and gas industry when multiphase flow is involved and a reasonable level of boost in the LP pressure is desired downstream of the jet pump.

CALTEC has developed and validated a comprehensive range of proprietary software which can be used for both optimum design and for predicting the performance of the jet pump under various operating conditions. In applications where the operating conditions for both HP and LP flow may change, the design can be modified to optimise production under such conditions. Introduction of exchangeable nozzle and mixing tubes for the jet pumps supplied by CALTEC is a cost-effective way to maintain optimum production when the flow conditions change. In applications where the production conditions change significantly on a daily or seasonal basis, because of the variations in demand, more than one jet pump may be used to cope with such major changes, in order to maintain optimum performance.

6. MONITORING AND CONTROL

Jet pumps do not require active control and are self-regulating. In applications where the LP line is not protected against exposure by the pressure and the flow from the HP line, installing a non-return valve on the LP line upstream of the jet pump is recommended. If, for operation reasons, the jet pump needs to be bypassed on occasions, isolation valves such as double block valves or equivalent may be used. This will allow the change-out operation to be made safely while the production continues by bypassing the jet pump.

Installation of pressure and temperature transmitters for the HP, LP and discharge lines is recommended. These instruments are not required for operation purposes but are a good means of monitoring the performance of the system if required.

7. HYDRATE FORMATION

Formation of hydrates relates primarily to applications where both free gas and water are present in the mixture. If the HP flow is gas, the temperature of the HP gas drops as a result of Joule-Thomson effect downstream of the nozzle. This gas is then combined immediately with the LP gas as it passes through the mixing tube and the diffuser where further compression of gas takes place and some heat is also generated, because of frictional losses. The process can be treated close to the ideal isothermal condition. The temperature downstream of the jet pump can be estimated and is primarily governed by the Joule Thomson effect, the volumetric flowrate and temperature of LP gas and the level of pressure recovery at the outlet of the jet pump.

Hydrates do not form within the jet pump even if the operating pressures and temperatures are within the hydrate formation band. This is because hydrates require some incubation period to form and high velocities of the mixture through the jet pump prevent the lattice of hydrate molecules to agglomerate. Formation of hydrates upstream or downstream of the jet pump along the pipeline do, however, require checking based on the operating pressures and temperatures. Conventional prevention methods, such as injecting glycol or methanal, may be used if formation of hydrates is predicted within the said lines during the operation or during shut-down when the temperature may fall.

8. ECONOMIC EVALUATION

The economic evaluation of the jet pump can be made by comparing the associated capital and operating expenditure against savings, gains or the added revenue as a result of increasing production.

The capital expenditure generally consists of primarily four factors:

(a) Capital cost of the jet pump system
(b) Cost of the associated valves, instrumentation, etc
(c) Installation costs
(d) Additional cost of modifications to the piping system (HP - LP and discharge lines) to accommodate the jet pump system.

The capital cost of the jet pump system varies depending on size, pressure rating and the material specified or needed for the system. It is difficult to quote an accurate range because of the vast variation caused by the said factors. Factors such as presence of H_2S, CO_2 and sand may demand the selection of high grade materials, or a specific design and material for components such as the nozzle and the mixing tube which are subjected to high velocities of the mixture.

An important factor which affects the total capital cost is the modification to the piping system needed for installing the jet pump system. The associated cost may be minor in many applications, but it may be substantial in applications where major re-routing and modifications to the piping system is desired.
The savings and gains may include one or more of the following:

- Added revenue from increased production
- Extension of field life and increase in total recovery
- Prevention of flaring or wastage of LP gas
- Eliminating the need to install additional compressors or pumps
- Eliminating the need to install additional pipelines (as often needed for LP production)
- Optimisation of operation and added flexibility, particularly in gas production application when demand for gas supply changes significantly on seasonal or daily basis
- In pipeline and gas gathering systems, optimisation of the operation and minimising the use of compressors. In some applications, the number of compressors used within the gathering and transportation system may be reduced with considerable saving achieved
- In gas storage applications, increased early season deliverability, retaining more late-season heartland deliverability and lower N.E.G.

In each application, one or more of the above may have specific monetary benefits which can be defined in present value cash terms. The operating cost of the jet pump is next-to-none unless the HP flow is supplied using compressors or booster pumps (oil production case).

The simplicity and the reliability of the jet pump system, together with its low cost in comparison with other alternative boosting systems, will result in most cases in achieving payback periods ranging between a few weeks to a few months.

It is worth noting that in practically all applications, including the jet pumps used for the Hewett field and seven applications in Italy (onshore and offshore), the payback period for the system has been a matter of only a few weeks from the added revenue.

9. CONCLUSIONS

Experience so far has shown that jet pumps are a cost-effective way to optimise production and boost production from low pressure wells. Jet pumps have a wide range of applications in both oil and gas production. Their simplicity, low-cost and reliability make them good candidates for many applications and more economical, in comparison with other alternative boosting systems which involve compressors or separation systems and booster pumps.

Good knowledge of the performance and design details are required in order to supply jet pumps with optimum design. It is for this reason that CALTEC has, over the past several years, successfully carried out extensive development work, testing and field validation work in order to optimise the design and develop a range of software capable of designing and predicting the performance of jet pumps in each application.

The field experience in practically all recent field application examples has confirmed the economic attractions and the reliability of jet pumps in downhole, offshore or onshore applications.

10. REFERENCES

1. M M Sarshar, N A Beg (CALTEC, UK)
"A Cost-effective Way to Boost Production from Low Pressure Fields, Using Jet Pump Technology". ETCE-4736, Energy Sources Technology Conference & Exhibition, ETCE/ASME, USA, 1998

2. M M Sarshar, N A Beg, W L Loh (CALTEC, UK), M Villa, F Paone (Agip SpA)
"Field Testing and Applications of WELLCOM System Using Jet Pump Technology" 1997. Multiphase 97, 8th International Conference on Multiphase Production, Cannes, France, 18 - 20 June 1997.

3. I S Yuden (JKX Oil & Gas, Guildford, UK), Y A Sazanov, V N Yeliseev, B A Malov (Orenburgneft JSC, Moscow)
"Jet Pump Feeds Corrosion Inhibitor in Russian Waterflood". Oil and Gas Journal, Jan 27, 1997

4. M Villa, G DeGhetto, F Paone (Agip SpA), G Giacchetta, M Bevilacqua (Univ. Of Ancona)
"Ejectors for Boosting Low Pressure Oil Wells: Field Application", (OTC #8504) The Offshore Technology Conference, Houston, USA, 1997

5. M Villa, G DeGhetto, F Paone (Agip SpA)
"Use of Ejectors for Boosting Low Pressure Oil and Gas Wells". The Offshore Mediterranean Conference & Exhibition in Ravenna, Italy, March 21, 1997

6. M M Sarshar, W L Loh (CALTEC, UK)
"Applications of WELLCOM System to Enhance Production". The 7th International Conference on Multiphase Production, organised and sponsored by BHR Group Limited, Cannes, France, June 7-9, 1995

7. A J Green, A T Reade (BHR Group), Kevin Ashton (Phillips Petroleum Co UK Ltd)
"Gas Production Improvements Using Ejectors". SPE #26684, The Offshore European Conference, Aberdeen, September 7-10, 1993

8. M M Sarshar (CALTEC, UK)
"An Alternative Cost-effective Option to Boost Production". (C508/028/96), 6th European Conference on Fluid Machinery for the Oil, Petrochemical and related Industries, IMechE, 13-14 May 1996

9. N Conti, P Giunta, R Pozzi
"Enhanced Oil Production by Diluent Lift". Offshore Mediterranean Conference & Exhibition, Ravenna, Italy, March 1993.

Nomenclature:

$$N = \frac{P_3 - P_2}{P_1 - P_3}$$

$$M = \frac{LP\ Flowrate}{HP\ Flowrate}$$

$$GVF = \frac{Qg_2}{Qg_2 + Ql_2}$$

Liquid-Liquid jet pump efficiency:

$$\eta = M \times N$$

Liquid-Gas jet pump efficiency:

$$\eta = \frac{P_2 Q_{g2} \ln \frac{P_3}{P_2}}{Q_{l1} (P_1 - P_3)}$$

Liquid-Multiphase jet pump efficiency:

$$\eta = \frac{P_2 Q_{g2} \ln \frac{P_3}{P_2} + Ql_2 (P_3 - P_2)}{Q_{l1} (P_1 - P_3)}$$

GVF	Gas Volume Fraction
M	LP to HP flow ratio
N	Pressure ratio
P_1	HP pressure at the inlet of the jet pump
P_2	LP pressure at the inlet of the jet pump
P_3	Discharge pressure at the outlet of the jet pump
η	Efficiency of the jet pump
Ql_1	Volumetric flowrate of liquid phase at HP flow
Ql_2	Volumetric flowrate of liquid phase at LP flow
Qg_2	Volumetric flowrate of LP gas @ P_2 and given temp

TABLE 1: Comparing the performance of single and dual WELLCOM systems

Case	High Pressure (HP)				Low Pressure (LP)				Discharge Pressure
	Pressure barg	Liquid Flowrate m^3/h	Gas Flowrate Sm^3/d	GVF	Pressure barg	Liquid Flowrate m^3/h	Gas Flowrate Sm^3/d	GVF	barg
A	70	100	423069	75%	30	40	221662	90%	34
B-1	70	0	423069	100%	30	0	221662	100%	40
B-2	70	100	0	0%	30	40	0	0%	45

Case A: Single WELLCOM system **Case B-1 & B-2:** Dual WELLCOM system, gas & liquid jet pumps

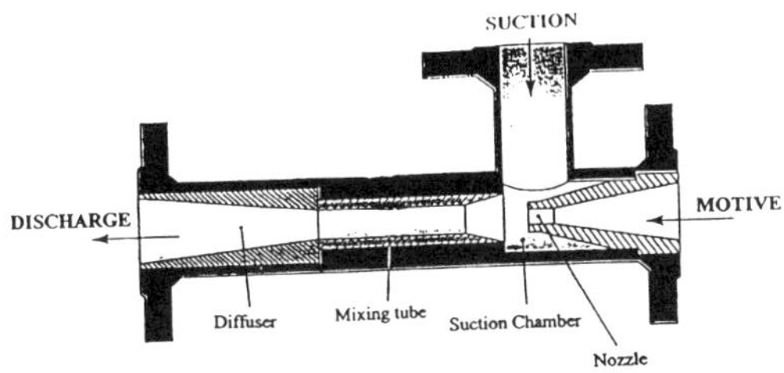

FIG 1 GENERAL CONFIGURATION OF A JET PUMP

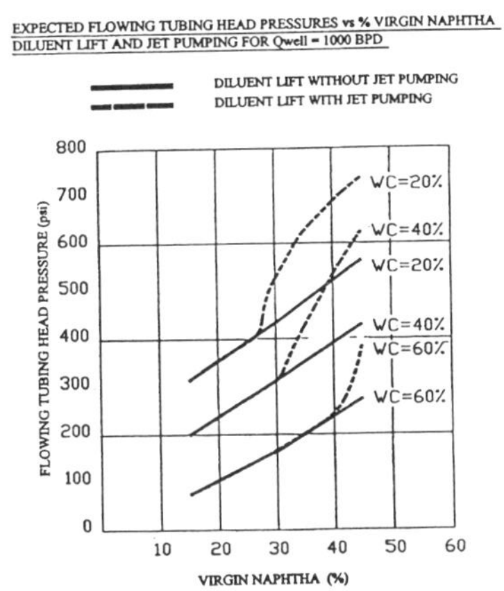

FIG 2 EXAMPLE OF THE PRODUCTION PRESSURE INCREASE USING A JET PUMP DOWNHOLE

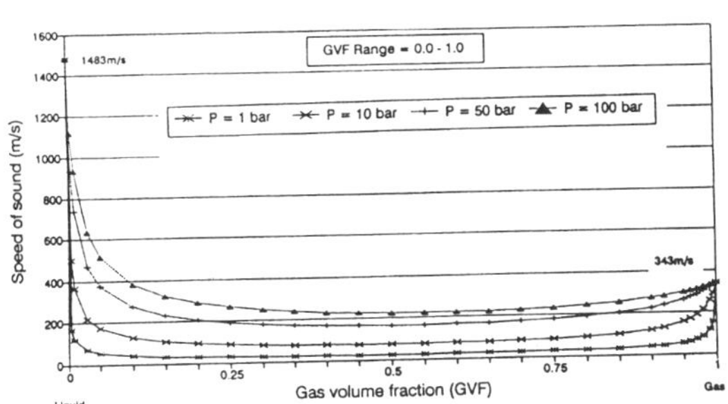

FIG 3 SPEED OF SOUND IN GAS-LIQUID MULTIPHASE FLOW

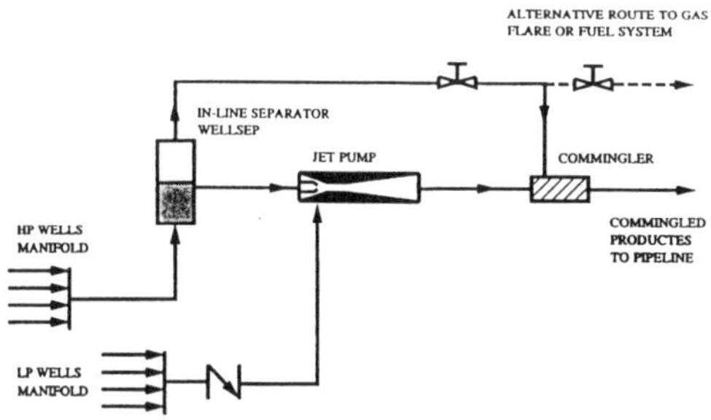

FIG 4 SUBSEA MANIFOLDING USING WELLCOM SYSTEM

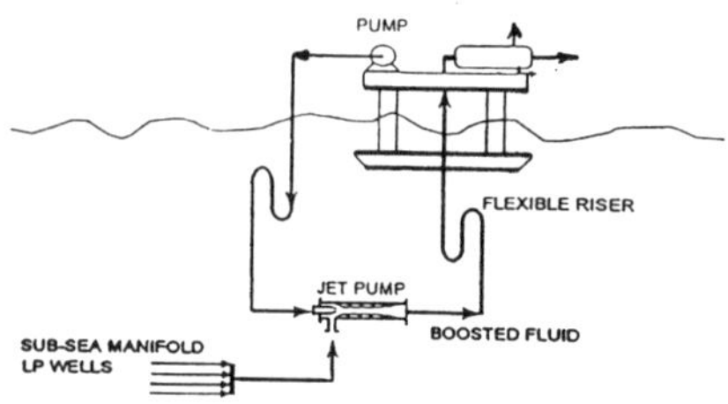

FIG 5 USE OF SUBSEA JET PUMP TO BOOST OIL PRODUCTION

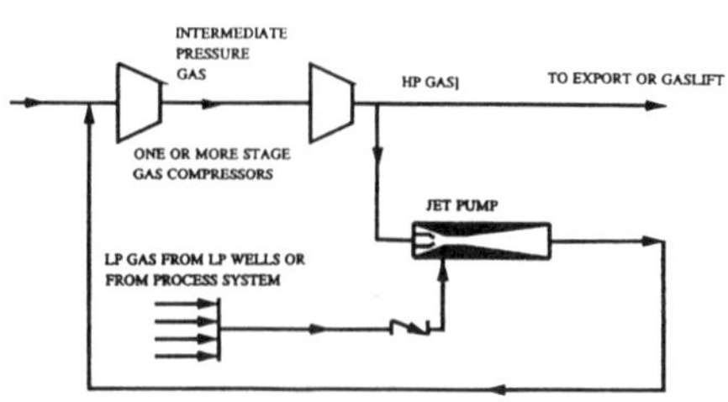

FIG 6 USE OF JET PUMPS TO INCREASE THE PRESSURE OF LP GAS

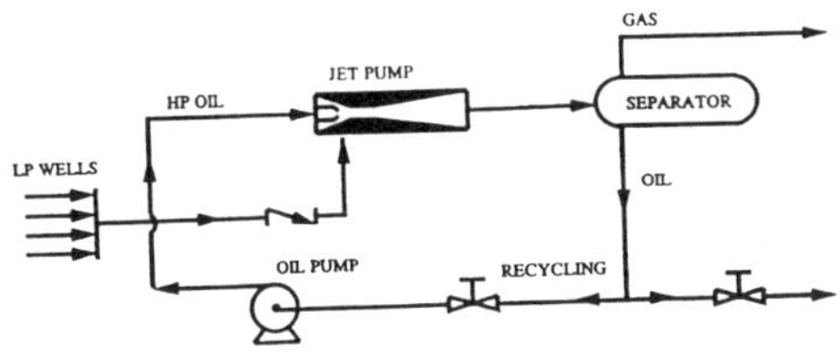

FIG 7 USE OF A JET PUMP AND BOOSTER PUMP TO INCREASE OIL PRODUCTION (WELLCOM BOOST)

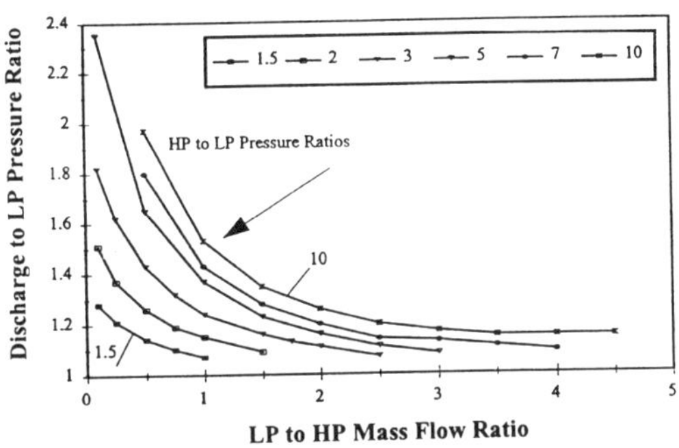

FIG 8 PERFORMANCE CURVES OF OPTIMUM GAS JET PUMPS UNDER VARIOUS HP TO LP PRESSURE RATIOS

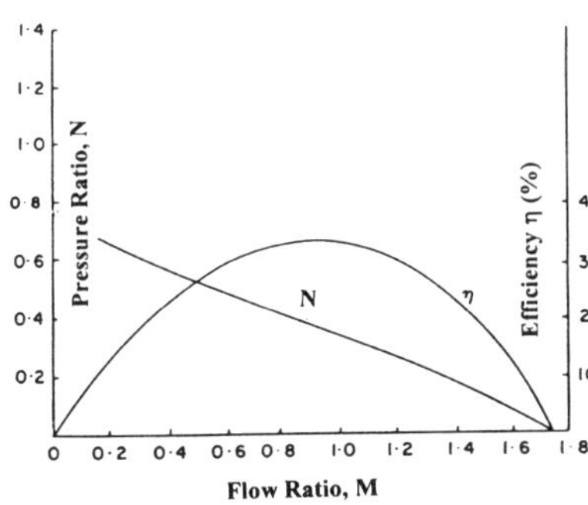

FIG 9 JET PUMP PERFORMANCE IN LIQUID-LIQUID OPERATION

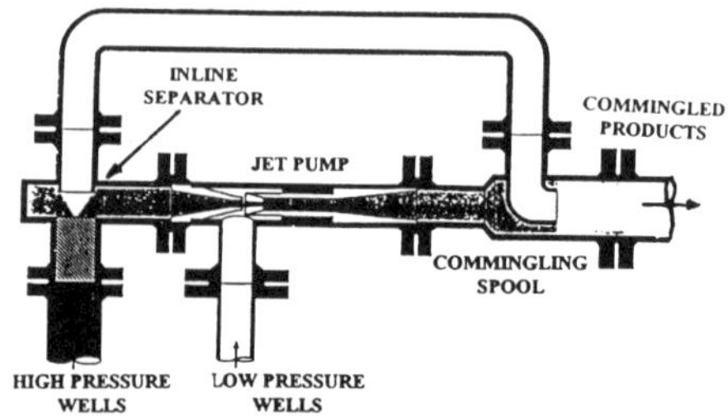

FIG 10 GENERAL CONFIGURATION OF THE WELLCOM SYSTEM FOR OIL PRODUCTION

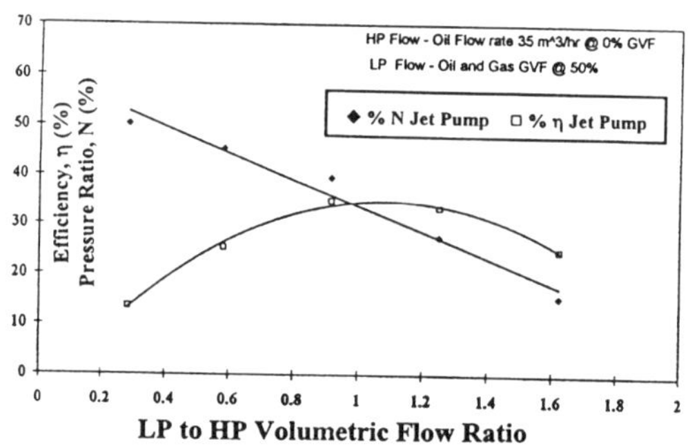

FIG 11 TYPICAL PERFORMANCE CURVES OF A LIQUID-MULTIPHASE JET PUMP

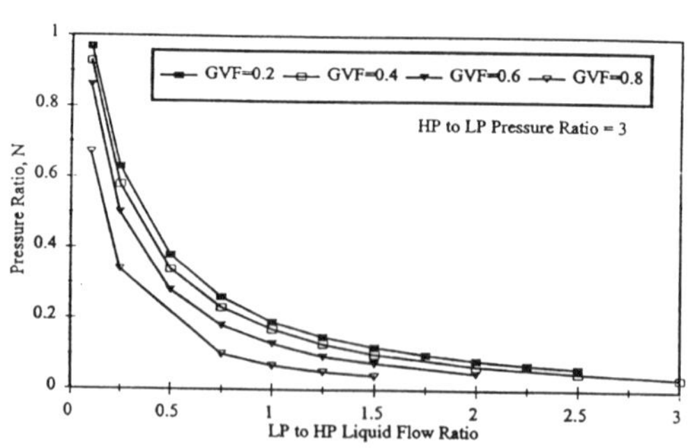

FIG 12 TYPICAL PERFORMANCE CURVES FOR AN OPTIMISED WELLCOM-OIL SYSTEM WITH A RANGE OF GVF FOR LP FLOW

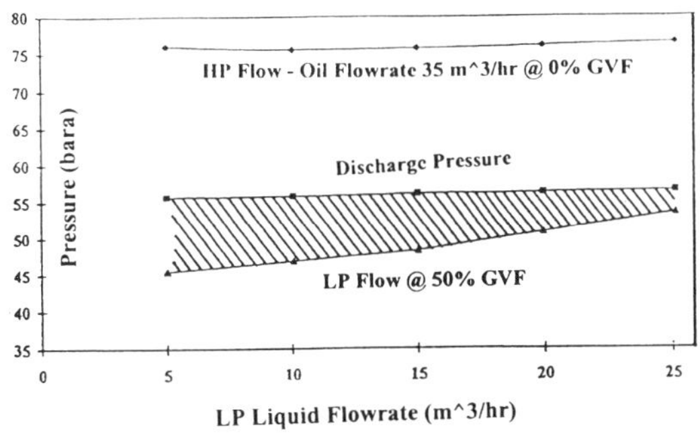

FIG 13 TYPICAL PRESSURE BOOST ACHIEVED BY WELLCOM-OIL SYSTEM IN TRECATE FIELD

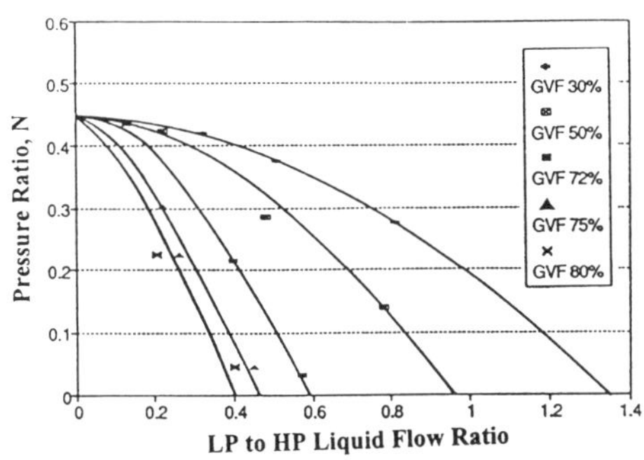

FIG 14 OIL-MULTIPHASE JET PUMP PERFORMANCE AT DIFFERENT LP GAS VOLUME FRACTIONS

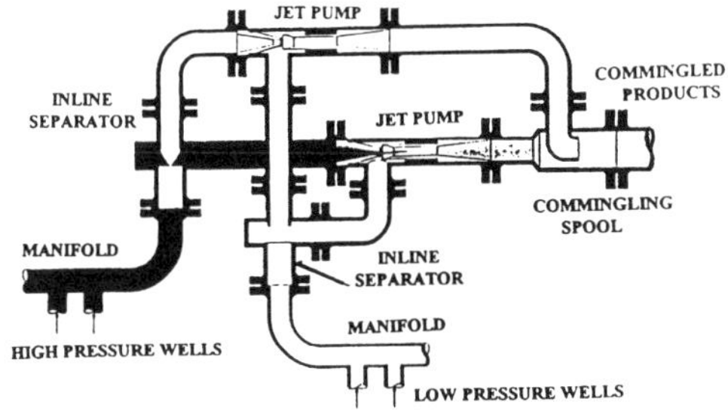

FIG 15 DUAL WELLCOM SYSTEM FOR OIL PRODUCTION APPLICATIONS

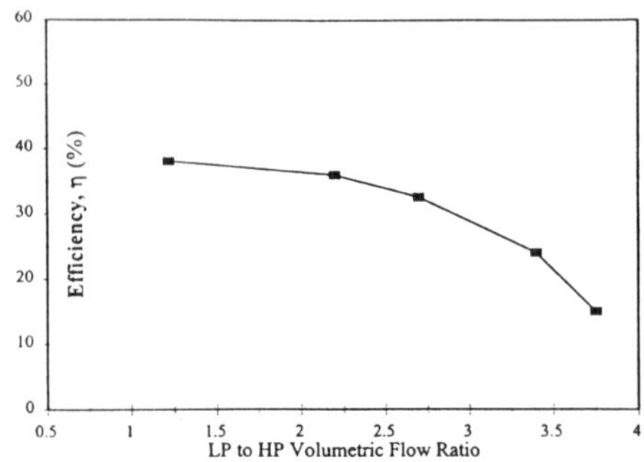

FIG 16 TYPICAL EFFICIENCY CURVE FOR LIQUID-GAS JET PUMP

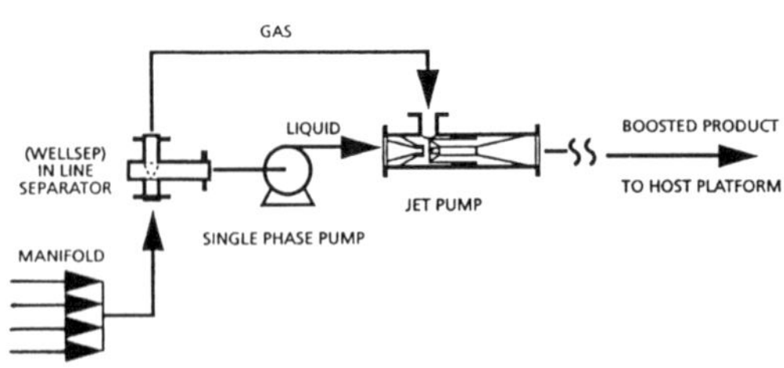

FIG 17 WELLCOM-BOOST SYSTEM

Sealing technology – the key to reliable multi-phase pumps

R MAURISCHAT
Leistritz AG, Germany

SYNOPSIS

This paper shows the different mechanical seal arrangements used in twin screw pumps for multiphase applications. For each seal arrangement benefits and problems are discussed. As reference actual multiphase pump applications are mentioned. It shows that not only one seal arrangement can be used for all multiphase pump applications. To ensure a reliable and cost effective operation of multiphase pumps the right seal solution has to be selected. New sealing solutions can only be developed in close cooperation with the operators.

1. INTRODUCTION

Multiphase pumping (MPP) technology has proven beneficial in numerous applications all over the world. This technology is an acceptable alternative to the conventional way of separating multiphase fluid from a well into gas and liquid phases, then handling each separately.

There are two major types of multiphase pumps available, the screw pump and the helico-axial pump. The twin screw pump has the most applications, operating with lower flow rates and higher differential pressures. The helico-axial pump has fewer applications, operating with high flow rates and low pressure ratio.

The LEISTRITZ MPP is based on our standard screw pump design which is used in numerous successful applications all over the world. The MPP has two dual flow rotors, meaning the incoming flow is split upon entering the pump suction chamber and directed into the two inlet sides of the rotors (Sketch I). A small tolerance is maintained between the rotors and the casing by externally lubricated timing gears and bearings. Due to the use of external timing gears and bearings, four mechanical seals are required in this type of MPP.

In the following the sealing technology of screw pumps in MPP applications will be discussed in detail.

Sketch I Twin Screw Multiphase Pump

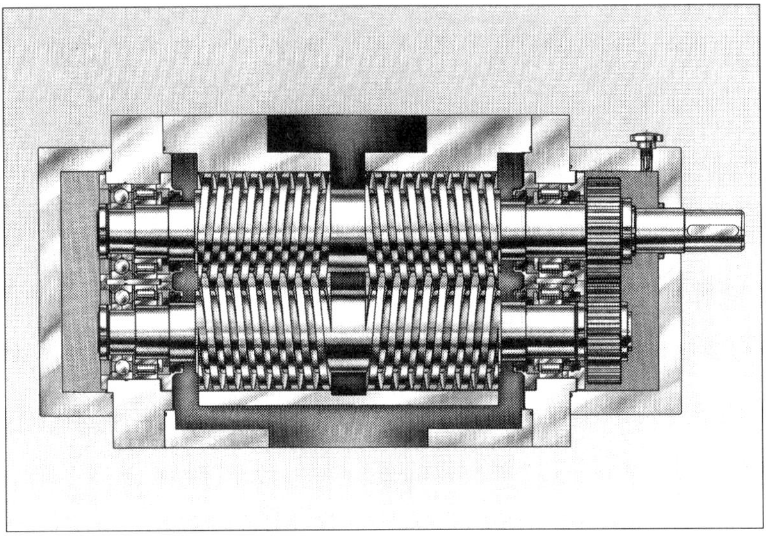

2. HISTORY OF MECHANICAL SEAL APPLICATIONS IN MULTIPHASE PUMPS

It's common knowledge that standard twin screw pumps can handle gas fractions (GVF) at inlet conditions up to 60% without design modifications. Consequently, the first step towards improving the GVF capabilities was to determine the limits of this existing design.

During the period beginning 1985 to 1990 several tests were conducted with off-the-shelf type screw pumps. The pumps were equipped with single mechanical seals (Sketch II).
Early in the testing the limits of this seal configuration were obvious and for GVF higher than 70% no acceptable seal life time could be achieved.

Applying this new found knowledge, the first prototypes of real MPP were equipped with double mechanical seals (Sketch III) and an external seal flush system. The seal flush was necessary to maintain lubrication of the seal faces at all times. Thus providing safe pump operation independent of GVF, viscosity or solid content of the pumped fluid.

The use of double mechanical seals provides an acceptable seal life, solving the seals problem through a relatively conservative approach.

3. SINGLE MECHANICAL SEAL

Laboratory tests on multiphase pumps equipped with single seals were extremely encouraging, leading us to believe these seals can handle GVF's up to 90%. Unfortunately, single seals were not as reliable in the field and we found this seal arrangement limited to 70% GVF's. The limiting factor of single mechanical seals is the multiphase product itself. Wells do not produce a homogeneous mixture of gas and liquid, they produce gas and liquid slugs. This means the pumps are exposed to slug flow and during gas slugs the seal runs dry, which then causes thermal damage to the seal.

Listed below are examples of results and causes of single mechanical seal failures:

- The seal needs sufficient liquid between the seal faces to build up a hydrodynamic film. If the pump operates under high GVF conditions, the seal faces may not have enough liquid available and prematurely wear.

- A lack of liquid also creates excessive heat in the seal faces, which could lead to cracks in the seal faces and ultimately leaks.

- During high GVF operation the pump shaft rotation adds to the problem of insufficient liquid in the seal chamber. The centrifugal forces caused by the shaft's rotation drive liquid away from the shaft and form a gas ring around the seal faces. Thus preventing liquid from reaching the seal faces.

- During operation under any GVF condition, solids within the pumped multiphase fluid will act like a grinding stone between the seal faces and decrease the service life of the seal.

Sketch II Single Mechanical Seal

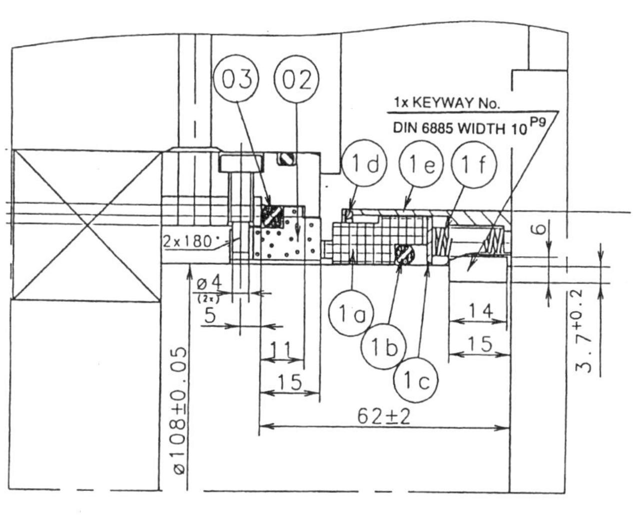

Fortunately some of the above single mechanical seal problems can be eliminated by using different seal face material combinations. Carbon seal faces are capable of self lubrication, which decreases the seal face temperature due to its inherent thermal conductivity capabilities. Carbon is typically used for multiphase flow with high GVF's and a low solid content.

If the solid content is high, a hard face combination like Silicon Carbide vs. Tungsten Carbide is used. However, this combination is not tolerant of dry running periods and the faces can be permanently damaged after a short dry running period.

Currently there are MPP's with single mechanical seals in operation. Typically a seal change is required every 3 to 6 month, but the users like single mechanical seals because flush systems are not necessary and the seals are relatively inexpensive.

4. DOUBLE MECHANICAL SEAL

A double mechanical seal arrangement is the most conservative approach from a reliability point of view. A typical double seal is shown in sketch III.

The seal faces are here lubricated by a clean external flush fluid. The flush is injected into the seal chamber at a pressure higher than the pump's inlet pressure. This arrangement ensures the flush will lubricate the seal faces and forces the flush to leak to the product side. Thus avoiding contamination of the seal flush or seal faces with crude oil or gas.

By using an external seal flush, the mechanical seals can handle high solid contents and GVF's up to 100%.

Sketch III Double Mechanical Seal

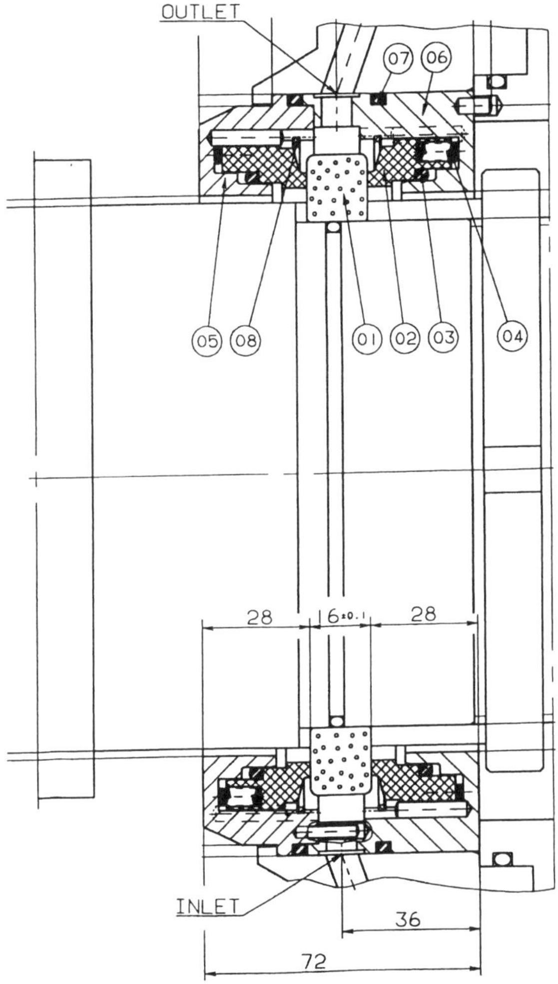

Chevron in Canada operated a MPP under severe slug conditions (gas slug duration of 1.5 to 2 hours) with a double mechanical seal arrangement. This pump has operated for 14 months with no downtime caused by the seals.

During a scheduled maintenance, the seal manufacturer inspected the seals and found no significant damage on the seal faces. The seal flush eliminated premature wear due to thermal stress or abrasives in the product (reference SPE 3659 report). After the maintenance the pump was put back into service.

A reliable seal flush system is crucial to the successful operation of double mechanical seals. A typical P&ID for the seal flush is shown in sketch IV. The seal flush system normally consists of a tank, electrically driven flush pump, pressure control valve, accumulator, filter, heat exchanger and interconnecting piping and instrumentation.

Sketch IV Seal Flush System

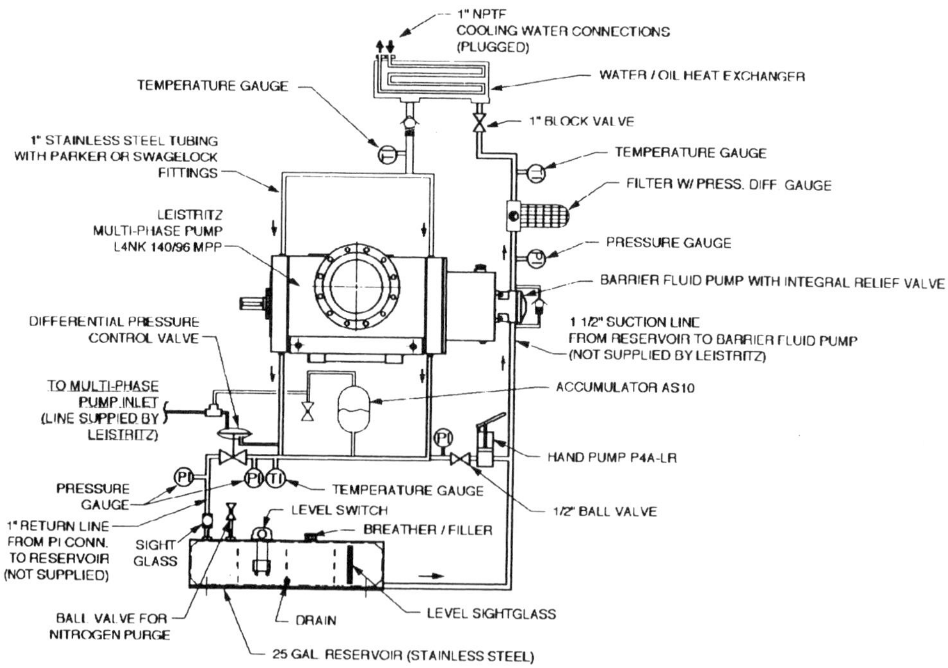

There are two negative aspects to the seal flush system, but they can be minimized by using the correct approach. The first prohibitive factor is the number of components required to effectively operate the system. This additional equipment increases the maintenance requirements, which may lead to decreased system reliability. However proper system monitoring and routine maintenance will greatly reduce unexpected down time.

The second potential obstacle is the operating personnel. They may not be accustomed to operating a seal flush system. Upon installation of this equipment, all responsible personnel should attend an orientation class for training relevant to safe and proper MPP and seal flush system operation.

Nevertheless there are MPP applications such as multiphase fluids with extremely high GVF, high H_2S, solids or explosive product contents, which require double mechanical seals and seal flush systems. These applications exist in everyday situations and we are constantly re-evaluating our system to ensure the overall reliability and effectiveness.

5. SINGLE MECHANICAL SEAL WITH THROTTLE BUSHING

The use of a single mechanical seal with throttle bushing was developed in close cooperation with Chevron. This seal arrangement was derived form API 610 Plan 11, which is typically used in rotodynamic pumps. This seal arrangement consists of a single mechanical seal (described in section 3) and a close clearance bushing. Reference the cross section illustrated in Sketch V.

Lubrication of the seal faces is provided by fluid from the MPP's discharge and injected into the seal chamber. Fluid is collected in a liquid reservoir/knockout specially designed to accommodate the volume of liquid needed to sustain the longest gas slugs.

Sketch V Single Mechanical Seal with Throttle Bushing

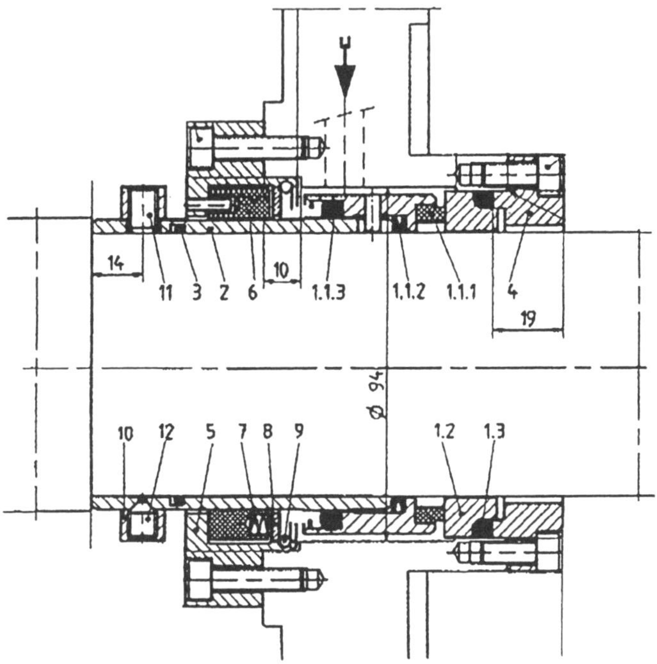

The single mechanical seal with throttle bushing has proven extremely reliable in field applications. Chevron has operated a MPP in the Gulf of Mexico since February 1996 with this seal arrangement. It was operating approximately 18 months without any seal problems. After this period the pump was shut down for maintenance, due to a bearing failure. During the shut down, Chevron inspected the seals and found little evidence of wear on the seal faces.

In February 1997, Texaco, Houston received a MPP with single mechanical seals and throttle bushing arrangement. The MPP continues to operate trouble free for 14 months now and has enjoyed an availability of 99%.

In March 1997, Chevron converted another MPP with double mechanical seals to single seals with throttle bushing arrangement. This pump continues to operate without any problems.

The data collected from the above installations proves the single seal with throttle bushing solution is a very promising alternative for future applications.

6. OTHER SEAL ALTERNATIVES

Fortunately, seal manufactures are cooperating with the pump manufactures to explore new sealing methods. One possibility is to machine grooves (hydro pads) in the seal faces, which brings fluid between the faces. These hydro pads allow liquid between the seal faces against the pressure gradient between atmospheric pressure and pump inlet pressure. The seal flush is on the atmospheric side of the seal and originates from an external seal pot. This solution seems to work successfully in some applications.

A more promising alternative seems to be the upstream pumping solution. This method replaces the hydro pads with profiles similar to those found on gas seals. This shape builds up a defined pressure between the seal faces, which should be higher than the multiphase fluid pressure at the other side of the mechanical seal.

Both of the above mentioned solutions rely on a hydrodynamic pressure build up across the seal faces. This means the seals have an optimum design point, determined by the pump size, speed, and viscosity of the flush fluid. If the MPP's operating conditions change, the seal faces are subject to premature wear or excessive leakage of the flush fluid into the multiphase fluid.

For some applications gas seals will also become a viable alternative to a liquid seal flush system. Unfortunately the relatively high price of these seals prohibits them form becoming an attractive alternative at this point.

7. CONCLUSION

It was shown that there is more than one seal arrangement available for MPP's. The proper seal arrangement must be determined on a case by case basis to ensure reliable and cost effective MPP operation. Clearly the future in sealing technology is simple seal arrangements and auxiliary systems.

In recent years, the most successful way of simplifying sealing systems was through close cooperation between MPP users and manufactures. If we can maintain these relationships, we will discover more alternatives that lead to reliable sealing solutions.

Nasties

Integration of fluid flow effects within a risk-based pipeline integrity management (PIM) process

D RICHARDSON and **C McGOVERN**
NeoCorr Engineering Limited, Calgary, Canada

SYNOPSIS

A quality Pipeline Integrity Management (PIM) process is a risk based, systematic process that identifies the controls required to maintain integrity. Through the application of a risk based PIM process, limited funds and resources can be deployed in the areas of highest risk to reduce the likelihood of internal corrosion failures. The process can easily be applied to any pipeline system from the very simple to the very complex. The failure modes and effects analysis (FMEA) within the PIM process clearly identifies the pipeline function, operating envelopes, deterioration mechanisms, associated risks, and maintenance requirements. Pipeline maintenance controls identified within the PIM process are easily justified, as they are the result of a structured risk based analysis.

At the heart of a successful PIM program is accurate characterization of the internal corrosion hazard profile. The Paper demonstrates how two-phase flow modelling data is utilized to identify the formation of detrimental fluid flow patterns contributing to corrosive environments in sour gas pipelines, including a description of the protective scale disruption processes that underlies the technical assessment of risk within the FMEA model.

Three sour gas production pipeline case studies are presented involving different situations where flow is important for corrosion considerations.

Key Words: Pipeline Integrity Management (PIM), failure modes and effects analysis (FMEA), Pipeline Maintenance Plan, detrimental flow patterns, internal pitting corrosion, corrosion mitigation, corrosion inspection, corrosion monitoring

1. INTRODUCTION

The producing and transportation segments of the oil and gas industry are facing increasing failure rates in pipelines, at a time when public and regulatory pressure for better control is increasing. As can be seen from the Alberta Energy Utility Board illustration (Figure 1) there are a myriad of threats to pipeline integrity. The three most prevalent threats include: internal corrosion, external corrosion and third party damage. Pipeline failures caused by internal and external corrosion are causing producing and transmission companies to expend substantial funds to solve their particular problems. It is certain that these expenditures are rarely, if ever, optimized, in terms of both dollar value and desired technical outcome, because of short-term, reactive response to pipeline failures. A quality Pipeline Integrity Management (PIM) process provides a proactive approach, based on sound technical foundations. The corner stone of this technical foundation is the inherently structured failure modes and effects analysis, which identifies integrity-related threats and details mitigation controls within a formal Pipeline Maintenance Plan.

Within the PIM process, a formal Pipeline Maintenance Plan document is prepared as a tool for communicating the outcome of the risk evaluation and provides the basis for scheduling appropriate monitoring, inspection, and mitigation controls to address the hazard. The Paper provides a brief outline of a quality Pipeline Integrity Management (PIM) process for control of internal pipeline corrosion as the foundation of a formal Pipeline Maintenance Plan. Three case studies are presented involving different situations where flow has played an important role in explaining failure events, and providing the design basis for the mitigation program for a new pipeline.

2. FAILURE MODES AND EFFECTS ANALYSIS WITHIN A PIM PROCESS

The success of any Pipeline Integrity Management (PIM) process is directly related to: 1) the understanding the operating company has of its pipeline system with respect to deterioration mechanisms; and, 2) the company's ability to implement the appropriate mitigation, inspection and monitoring programs. The failure modes and effects analysis (FMEA) within the PIM process identifies the internal corrosion risk associated with pipeline operations and forms the basis for scheduling mitigation, monitoring and inspection activities.

The failure modes and effects analysis (FMEA) is divided into five discrete activities:

2.1 FMEA -- Functional Block Analysis:

Block diagrams are used to illustrate pipeline systems to allow a systematic functional block analysis (Figure 2). The block diagrams are used to show primary and secondary function, detail any inter-relationships and identify the impacts of loss of function. The drawings are intentionally kept simple in nature and are broken down into as many sub-systems as required. The results from this analysis provide a clear understanding of the pipeline function and effect of loss of function to the overall system.

2.2 FMEA -- Operating Envelopes

The range of anticipated operating conditions for the pipelines defines the operating envelope limits for the subsequent deterioration mechanism analysis (Figure 3). Any future operation that occurs outside the defined limits may trigger a need to re-assess the pipeline risk assessment model to assure the appropriate maintenance activity follows the excursion.

2.3 FMEA -- Deterioration Mechanisms:

All known sources of information are reviewed to compose a detailed list of potential damage mechanisms that could reduce pipeline integrity (Figure 4). Each pipeline is assessed against the list to determine which mechanisms are likely and activities are detailed in the Pipeline Maintenance Plan to address each individual threat. It is normal for a Pipeline Maintenance Plan for production pipelines to focus on internal, bottom-of-the-pipe corrosion as the main deterioration mechanism. Kang, Wilkens and Jepson (1996) [3] describe the effects of pipe inclination on flow regimes and individual flow characteristics and the subsequent effect on corrosion.

2.4 FMEA -- Risk and Hazard Analysis

The hazard / risk analysis is a critical step in the PIM process and ensures that the evaluator understands the actual hazards and risks and that sufficient activity is directed to the areas of highest risk. The analysis should be performed at two levels: 1) a high-level risk-based analysis to identify the highest priority pipelines within an operating system; and, 2) a mechanism-specific analysis on the high priority pipelines revealed by the high-level analysis. The mechanism specific analysis is a more detailed review of the actual hazards utilizing flow modelling and provides the technical basis from which mitigation, monitoring, and inspection controls are scheduled. Ho-Chung-Qui and Williamson (1987) [4] provide details of sour gas failure mechanisms and appropriate mitigation practices.

By fully understanding the actual risk with respect to Public Safety, Corporate Image, Cash Flow, and the Environment, the Corrosion Specialist can prioritize the pipelines. Richardson (1995) [5] explains how a risk-based approach ensures field implementation of mitigation, monitoring and inspection controls and costs are directed towards pipelines that represent the greatest exposure to the operating company. Figure 5 illustrates the results of a risk assessment for 42 individual pipelines, and graphically represents their relative ranking according to the likelihood of a failure, and the consequences associated with an accident.

2.5 FMEA -- Development of a Pipeline Maintenance Plan:

The results from the deterioration mechanism analysis are reviewed against the pipeline operating conditions and a simple cause and effect analysis is performed. The failure modes, failure effects, safeguards and recommended actions are documented. The failure effects are evaluated for consequence and probability and plotted on the risk summary matrix. The mitigation controls are also documented within the formal Pipeline Maintenance Plan.

The formal Pipeline Maintenance Plan is the summary of the entire FMEA process. The activities within the Plan reflect the actual risk of loss of integrity. For the plan to be successful the identified activities must follow the Quality Maintenance Model of "Plan, Schedule, Execute, Analyze and Improve" (Figure 6). The American Petroleum Institute (1990) [6] published an initial Recommended Practice document to address a process for hazard assessment as the basis for a reliability-centered maintenance program.

3. INCORPORATION OF FLUID FLOW EFFECTS WITHIN FMEA

This section has been included in this Paper because of the strong relationship between corrosion initiation and multiphase fluid flow in both oil and gas pipelines. An understanding of this relationship is essential when assessing which areas are susceptible to corrosion – it provides information about "where" corrosion will occur, and strengthens the quality of the Pipeline Maintenance Plan. Bich and Goerz (1995) [2] identified a need to better understand the contribution of flow regimes to assist with the explanation of unusually high pitting corrosion rates for sour gas pipelines containing high levels of CO2.

3.1 Definition of a Corrosive Environment

Typically, the environment inside a gas/hydrocarbon-condensate pipeline is characterized by the presence of the following constituents:

a) a gas phase containing inert hydrocarbon species and varying concentrations of the corrodents H_2S and CO_2;

b) a hydrocarbon-condensate; and

c) varying amounts of liquid water (both formation water and water of condensation), usually containing salts.

The presence of H_2S and/or CO_2 in combination with saline water composes an environment causing various forms of corrosive attack of low-alloy carbon steel line pipe. H_2S and CO_2 are not corrosive in a dry form, however when dissolved in water they are very corrosive. Another significant source of corrodents is the dissolved H_2S and CO_2 in liquid hydrocarbon-condensate. While the condensate is not an electrolyte and not corrosive itself, it can dissolve far more acid gases than water and can act as an acid gas reservoir to replenish the aqueous phase acids as they are consumed.

A corrosive environment can be defined as an environment that will cause corrosion attack to an exposed metal surface. In a sour multiphase system there is usually a continual film of hydrocarbon-condensate (if present) and/or protective iron sulphide scale covering the entire metal surface. Therefore, for pitting corrosion to occur, protective film/scale disruption is required to expose the metal surface to the corrosive environment and the regeneration of new protective film/scale must be hindered.

3.2 Two-Phase Fluid Flow Patterns

As liquids and gases are transported down the pipeline the phases form distinct patterns. The patterns most relevant to corrosion in wet, sour multi-phase systems are; stratified, wavy, intermittent (slug) and annular mist. They are dependent on many factors, the most prominent being angle of inclination of the pipeline. The fluid flow pattern is critical when considering the overall corrosivity of a pipeline system. Certain flow patterns (i.e. stratified and intermittent flow) have the potential to cause disruption to protective films and the initiation of corrosion, while other flow patterns (i.e. annular-mist) do not contribute to the initiation of corrosion.

The characteristic flow patterns for horizontal and near-horizontal pipe are described:

STRATIFIED

This flow pattern generally forms at low gas and liquid velocities where liquid flows along the bottom of the pipe and gas flows along the top. As the liquid travels along the pipeline, the water and hydrocarbon-condensate phases separate due to gravity effects.

If solids greater than 50 microns in cross section are present they may settle at points along the pipeline depending on the size and density of the solid, the viscosity, density, and rate of travel of the stratified liquid phase. The prediction of whether solids will settle is dependent on factors such as the vertical velocity component of the turbulent eddies. The horizontal bulk velocity and Stokes law may be used to tell how far a solid will travel before it settles. The negative affect of settled solids is that they can produce conditions that result in scale disruption and corrosion.

WAVY (flat terrain / low liquid fraction)

This flow pattern is similar to stratified flow except as the gas rate increases relative to the liquid the gas disturbs the smooth liquid interface and waves are formed. Under flat and low liquid fraction conditions, the wavy flow regime results in a mixing of the water and condensate phases and moves solids along with the liquid phase.

WAVY (upward inclines / high liquid

In areas of the pipeline that are upwardly inclined, have high liquid fraction, or where hydrocarbon-condensate ratios are very high, there is a potential for preferential accumulation of the water phase. Under these *fraction)* conditions the hydrocarbon-condensate phase will slip over the water phase resulting in a liquid phase with two different flow rates. Even though the bulk liquid phase appears to be flowing, in wavy flow, the water phase may be moving at a reduced velocity. In this case solids greater than 50 microns in cross section may settle in areas of stratified water flow.

If the wavy front simultaneously contacts the entire pipe surface and there is a thick enough liquid film to sustain "slug" growth, intermittent flow can occur. If the liquid film is not thick enough the gas "blows through" and the "slug" collapses.

INTERMITTENT (SLUG)

This flow pattern forms at higher gas and liquid velocities and results in a liquid/gas interface (slug) being formed when a liquid wavy simultaneously contacts the entire pipe cross section. The nose of the slug has an axial (forward) velocity much higher than the normal fluid velocity. The slug front is a highly turbulent, mixing zone where gas bubbles are entrained and contact the bottom of the pipe with a very high velocity that causes the bubble to implode. This implosion effect is analogous to cavitation in pump impellers and can cause disruption to hydrocarbon-condensate films and protective iron sulphide scales. Due to the high gas and liquid velocities intermittent flow regimes will move solids with the liquid phase.

According to Kang, Wilkens and Jepson (1996) [3], the area most susceptible to intermittent flow is the transition area from a downhill or horizontal profile to an upward inclination. The frequency of intermittent flow diminishes as the fluids travel further downstream from the transition to an upward inclination. In the transition areas, the frequency of intermittent flow is much greater than of areas away from the transition point.

ANNULAR DISPERSED

This flow pattern forms at very high gas and liquid velocities where the liquid phase is transported as both minute droplets entrained in the gas phase and as a liquid film around the pipe wall. The gas phase travels down the center of the pipe at higher velocity than the liquid on the pipe wall. The solids are easily moved with the liquid phase. The effective liquid fraction is typically less than three percent of the cross sectional area of the pipe.

3.3 Protective Film Disruption Processes

Pitting corrosion is a highly localized deterioration mechanism occurring on the pipe surface with little or no general corrosion. Pitting typically occurs as a process of local anodic dissolution where metal loss is accelerated by the presence of a small anode and a large cathode. For pitting to occur the protective scale or film must be removed or displaced to expose bare metal to a corrosive environment. Bich and Goerz (1996) [2] conclude it is important to understand how, and under what operating conditions, a protective film or scale will be disrupted from the pipe wall by considering flow regimes and their individual characteristics.

Before the pipe wall will become susceptible to corrosion pitting, several disruption processes must take place, they include (in sequential order):

1. The displacement or removal of hydrocarbon-condensate (if present) from the pipe surface;
2. The degradation, removal or displacement of protective iron sulphide scale from the pipe surface; and,
3. The local resistance of the bare metal to replenish the disrupted hydrocarbon-condensate film or iron sulphide scale.

Once formed, most iron sulphides are tenacious protective scales that under most conditions, self-repair when damaged. Therefore, for localized pitting corrosion to initiate, the protective iron sulphide scale must be disrupted and not allowed to self-repair. During normal pipeline operation there are several environmentally related (i.e. not physically disrupted - pigging) mechanisms of scale disruption. The two most prevalent environmental disrupters include:

- settled solids onto the pipe surface, sulphur, formation particles and corrosion products, and,
- physical stripping in areas of high frequency intermittent (slug) flow.

In areas of the pipeline that experience stratified flow patterns it is possible for sulphur and solids to settle onto the pipe wall according to Ho-Hung-Qui and Williamson (1987) [4]. In these areas the scale is likely disrupted due to the setup of differential corrosion cells beneath the settled sulphur and solids. Elemental sulphur can also break down the protective iron sulphide scale and enhance the cathodic reactions by directly consuming electrons via solid state reactions or as a polysulphide. Based on correlating field observations, the settling of solids greater than 50 microns in cross section is not likely during wavy, intermittent, or, annular dispersed flow due to inherently higher liquid velocities. However, any flow interruption will settle solids and create an environment for scale disruption as described above.

In areas of the pipeline that experience intermittent (slug) flow patterns, physical-stripping action causes disruption to the protective iron sulphide scale. As described, the turbulent nature of the liquid/gas front causes a high-shear cavitation-like action on the metal surface at the location of imploding gas bubbles. This physical process is sufficient to disrupt the protective scale and will also prevent the self-repair of any scale disrupted. The observations of pipe removed from areas of high intermittent flow confirm large areas of scale disruption. In areas of scale disruption, the presence or absence of corrosion pits and the rate of corrosion attack is a function of the corrosivity of the fluid according to deWaard and Lotz (1993) [1]. Stratified, wavy and annular dispersed flow does not have a turbulent gas/liquid front and therefore physical stripping may not be possible in these flow regimes.

Following the physical removal of the protective scale in the areas experiencing intermittent flow, it is very unlikely the process of self-repair can compete with the aggressive nature of the fluid flow. The bare metal surface is therefore very susceptible to corrosion attack.

4. CASE STUDIES DEMONSTRATING EFFECTS OF FLOW ON CORROSION

It is highly relevant to the evaluation of the corrosion susceptibility profile for a wet, sour gas production pipeline to predict the flow regimes and their effect on the corrosion risk assessment, even at the design stage. The importance of including two phase flow analysis within the corrosion assessment is demonstrated for three situations: 1) at the design stage for a new pipeline; 2) as an explanation of a corrosion failure attributed to detrimental slug flow regimes; and, 3) as an explanation of a high corrosion rate failure attributed to the formation of stagnant water traps.

4.1 Case Study 1: Design Stage for New Pipeline

4.1.1 Operating Conditions

H2S mol %	CO2 mol %	Chloride mg/l	Temperature C
17.6	5.5	25,682	55

Pressure KPa	Line Size mm	Length m	Topography
6,400	202.7	25,670	Rugged

4.1.2 Hazard Assessment – Fluid Corrosivity

Based on deWaard and Lotz (1993) [1] corrosion rate model, the anticipated maximum corrosion rate could be 10.2 mm/year – unmitigated pipeline penetration is less than one year. It is anticipated the predicted corrosion rate may increase due to high chloride levels in the produced water.

4.1.3 Hazard Assessment – Fluid Flow Conditions

The pipeline is expected to operate with severe, detrimental slug flow regimes on uphill sections immediately upon start-up of the pipeline.

Gas Flow Rate e3m3/day	Required Gas Flow to Eliminate Detrimental Flow Conditions e3m3/d	Dominant Flow Regime
57	400	Slug

Figure 7 is a corrosion susceptibility graph for the pipeline at an anticipated start-up flow condition of 57 e3m3/day of gas. The graph demonstrates several locations where detrimental slug flow will contribute to high-rate corrosion attack.

4.1.4 Pipeline Performance History

There has been no history for this new pipeline, but similar nearby pipelines producing to the same processing plant have experienced high corrosion rate failures in a period consistent with the predictions of deWaard and Lotz (1993) [1].

4.1.5 Observations & Conclusions

Based on the highly corrosive fluids and the presence of detrimental flow conditions at anticipated start-up conditions, the operator pre-conditioned the pipeline with a batch inhibitor between pigs. The ongoing mitigation controls will rely on continuous injection of inhibitor supplemented with monthly batch inhibitor application between pigs. Real time corrosion monitoring equipment has been installed at two locations (at each extremity of the pipeline) as confirmation of mitigation effectiveness at regions of detrimental flow.

4.2 Case Study 2: Corrosion Failure Attributed to Slug Flow

4.2.1 Operating Conditions

H2S mol %	CO2 mol %	Chloride mg/l	Temperature C
19.8	5.8	13,000	25

Pressure KPa	Line Size mm	Length m	Topography
4,040	114.3	2,920	Rolling / Foothills

4.2.2 Hazard Assessment – Fluid Corrosivity

Based on deWaard and Lotz (1993) [3] corrosion rate model, the anticipated maximum corrosion rate could be 1.8 mm/year – unmitigated pipeline penetration in approximately three years.

4.2.3 Hazard Assessment – Fluid Flow Conditions

The pipeline operates consistently with detrimental slug flow regimes on uphill sections of the pipeline with water fractions averaging 33 percent in the regions of slug flow. The total water hold-up capacity in the pipeline is approximately 15.0 m3 – the daily water production is approximately 1.0 m3/day.

Gas Flow Rate e3m3/day	Required Gas Flow to Eliminate Detrimental Flow Conditions e3m3/d	Dominant Flow Regime
21	60	Slug

Figure 8 is a corrosion susceptibility graph for the pipeline at normal operating conditions demonstrating the locations of high corrosion susceptibility, and failure location.

4.2.4 Pipeline Performance History

Following is a chronological sequence of events for the pipeline:

- The pipeline was inspected with an in-line inspection tool after three years operation. Isolated pitting (25% wall loss) was identified in uphill sections between 450 and 500 metres.
- Weekly pigging was initiated as a supplement to continuous injection program.
- A permanent inspection bellhole was installed at 490 metres.
- Corrosion probes and quarterly ultrasonic measurements on a single pit provided confirmation that the mitigation controls were effective – the existing pit did not grow.
- Corrosion failure occurred at 1,430 metres, 40 months after completing the in-line inspection. There was no corrosion indication at failure site during the initial inspection.

– Subsequent in-line inspection (post failure) confirmed the mitigation controls were only effective in regions of slug flow throughout initial 1,000-metre section of pipeline.

4.2.5 Observations & Conclusions

– Thorough laboratory analysis of the pipeline indicated an absence of corrosion inhibitor on pipe surface at the failure location.
– Comparison of daily water production with the calculated hold-up volume indicates the weekly pigging was preventing inhibited produced water from travelling the entire length of the pipeline. The pigging program was removing water too frequently from the pipeline.
– Low daily water volumes were not sufficient to carry inhibitor along entire pipeline length during the seven-day interval between pigging. The slug regions near the end of the pipeline were not receiving protection from the inhibitor.
– Future mitigation controls must be scheduled with an understanding of insitu water volumes compared with daily water production volumes.

4.3 Case Study 3: High-Rate Corrosion Failure Attributed to Stagnant Water Traps

4.3.1 Operating Conditions

H2S mol %	CO2 mol %	Chloride mg/l	Temperature C
18.2	5.1	150,000	35

Pressure KPa	Line Size mm	Length m	Topography
6,300	219.1	15,350	Severe Valleys

4.3.2 Hazard Assessment – Fluid Corrosivity

Based on deWaard and Lotz (1993) [1] corrosion rate model, the anticipated maximum corrosion rate could be 3.5 mm/year – unmitigated pipeline penetration in approximately one year.

4.3.3 Hazard Assessment – Fluid Flow Conditions

The pipeline operates consistently with detrimental stratified flow regime in uphill regions averaging approximately 45% liquid fraction. The total steady-state water holding capacity in the pipeline under present operating conditions is 95 m3 – total daily water production is approximately 1.5 m3/day.

Gas Flow Rate e3m3/day	Required Gas Flow to Eliminate Detrimental Flow Conditions e3m3/d	Dominant Flow Regime
220	650	Stratified

Figure 9 is a corrosion susceptibility graph for the pipeline at normal operating conditions illustrating locations of extreme stagnant water trap formation, and failure location.

4.3.4 Pipeline Performance History

Following is a chronological sequence of events for the pipeline:

- The pipeline was commissioned based on a plan to mitigate corrosion with continuous inhibition.
- A permanent bellhole inspection location was installed during construction at approximately 13,500 metres (near end of pipeline) and real time monitoring equipment was installed to provide a measure of mitigation effectiveness.
- After 84 days of operation, the pipeline failed from internal corrosion at approximately the 1,500 metre interval.
- The corrosion monitoring equipment at the permanent bellhole did not provide an indication of high-rate corrosion activity.
- Subsequent in-line inspection post-failure indicated severe pitting corrosion in uphill sections of the pipeline.
- The pitting severity was the highest in the initial regions of the pipeline, diminishing along the route. There was no corrosion activity after approximately 7,000 metres.

4.3.5 Observations & Conclusions

- Upstream well production delivered a substantial amount of well stimulation fluids and associated solids into the pipeline.
- The steady-state water capacity in the pipeline was not satisfied during the 84 days of operation. All of the water that was produced into the pipeline remained insitu.
- Stagnant water traps that formed along the pipeline permitted corrosive solids to settle, causing disruption of the iron sulphide scales and rapid pitting rates.
- A greater awareness of the fluid composition and the water transit profile has resulted in re-commissioning the pipeline with a more appropriate pigging and batching mitigation program.
- A new permanent bellhole location for verification of the mitigation program has been installed in the early section of the pipeline.

– A two phase flow study was successful at explaining the corrosion damage profile in the pipeline and provides substantial technical enhancement for the design of future mitigation controls.

5. SUMMARY

A quality Pipeline Integrity Management (PIM) process based on a failure modes and effects analysis (FMEA) generates a formal Pipeline Maintenance Plan that provides operators with an enhanced capability to schedule activities appropriate to corrosion susceptibility profile of a pipeline.

A formal Pipeline Maintenance Plan allows the operator to identify changes to the corrosion risk as they occur and an opportunity to immediately modify the mitigation controls appropriate to the changes in the risk. Adherence to the Plan provides a means for scheduling an adequate response to changes in product flow rates and fluid compositions. A technical basis incorporating two-phase flow data within the Plan justifies pigging and batch inhibition programs as a supplement to the continuously injected inhibitor programs to address the elevated threat associated with high-frequency slug flow regimes and / or the risk associated with solid-phase settling in stagnant water traps. Finally, the Plan provides the basis for scheduling appropriate monitoring and inspection activities to provide the necessary verification of the Plan's effectiveness.

6. ACKNOWLEDGEMENTS

Use of flow regime transition models are highly relevant (even at the design stage) to predict the flow regimes and their effect on the corrosion risk assessment. The pipeline simulation data used within the case study examples was derived using PIPEFLO pipeline simulation software provided under license from Neotechnology Consultants Ltd., Calgary, Alberta. The steady state prediction method OLGA-S was utilized within PIPEFLO to complete the two phase modelling of wet sour gas pipelines. OLGA-S is made available through a license agreement between Neotechnology Consultants Ltd. and ScandPower A/S, Kjeller, Norway.

7. REFERENCES

1. C. de Waard, U. Lotz, "Prediction of CO2 Corrosion of Carbon Steel", NACE International, CORROSION/93, No. 69.
2. N. Bich, K. Goerz, "Caroline Pipeline Failure: Findings on Corrosion Mechanisms in Wet Sour Gas Systems Containing Significant CO2", NACE International, CORROSION/96, No. 26.
3. C. Kang, R. Wilkens, W.P. Jepson, "The Effect of Slug Frequency on Corrosion in High Pressure, Inclined Pipelines", NACE International, CORROSION/96, No. 20.
4. D.F. Ho-Chung-Qui, A.I. Williamson, "Corrosion Experience and Inhibition Practices in Wet Sour Gas Gathering Systems," NACE International, CORROSION/87, No. 46.
5. D.J. Richardson, "Managing Internal Pipeline Corrosion Based on Assessment of Risk," NACE International, 1995 Canadian Region Western Conference, Regina, SK.
6. "Management of Process Hazards," API Recommended Practice No. 750, American Petroleum Institute, January, 1990.

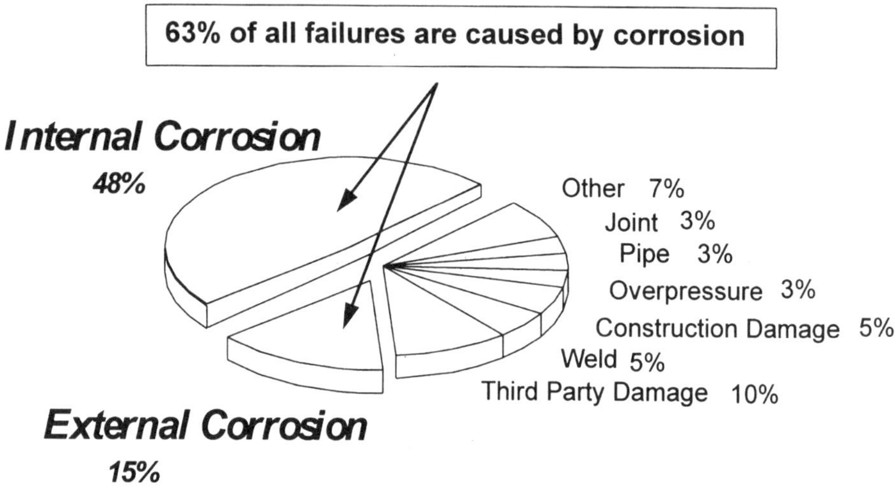

Figure 1 -- AEUB Pipeline Failure Statistics by Failure Type

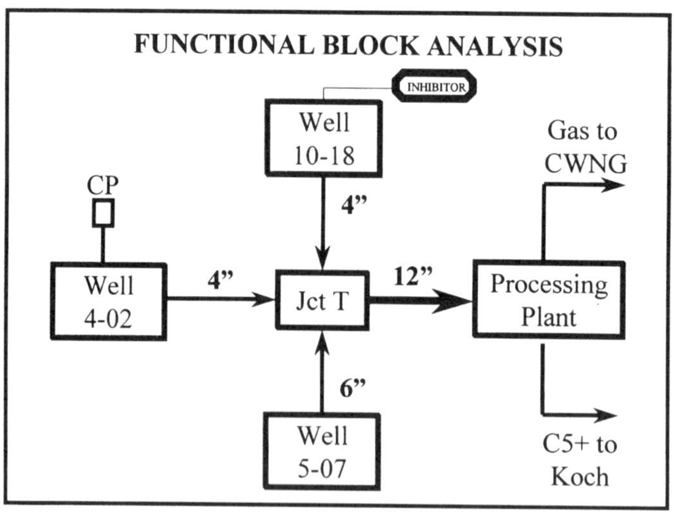

Figure 2 -- Functional Block Analysis -- Simple Block Flow Diagram

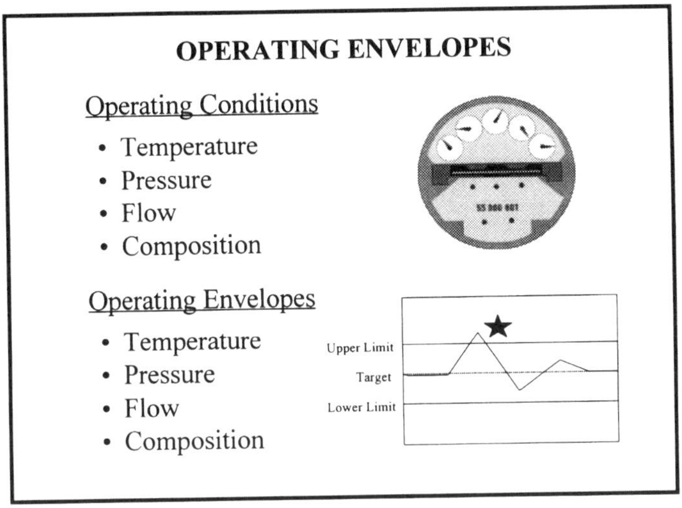

Figure 3 -- Operating Conditions versus Operating Envelopes

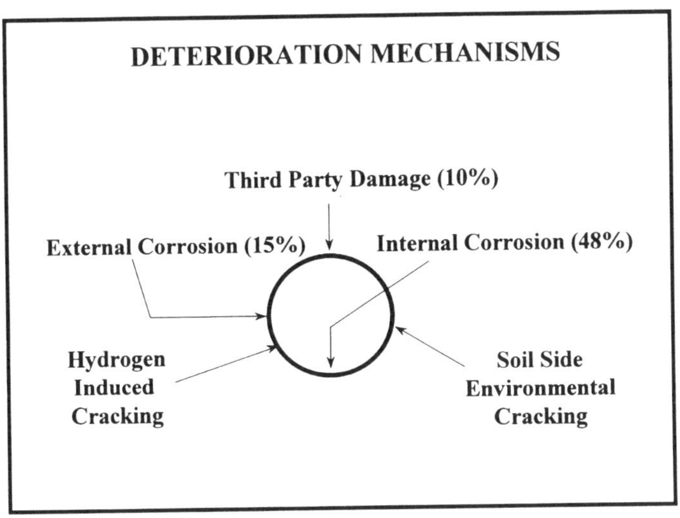

Figure 4 -- Deterioration Mechanisms Effecting Pipelines

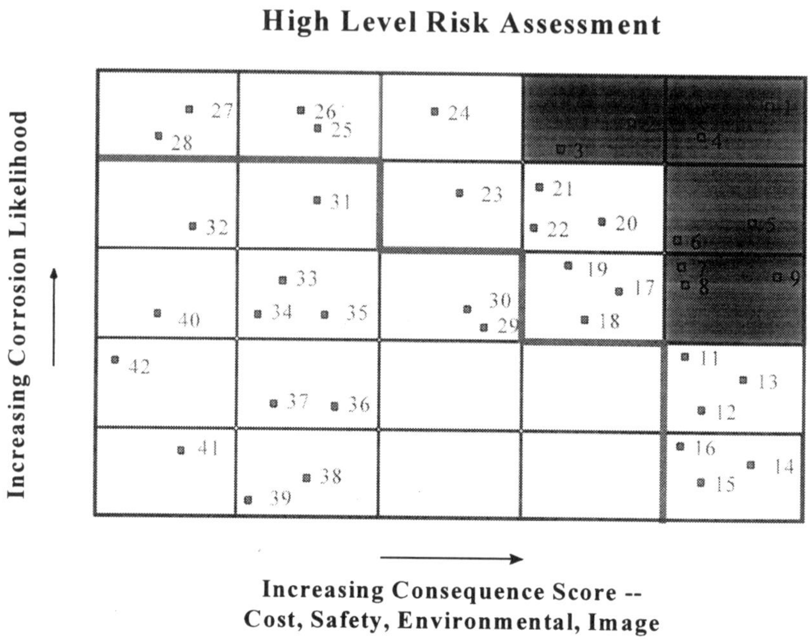

Figure 5 -- Risk Matrix Presentation of Pipeline Hazard Evaluation

Figure 6 -- Implementation of a Pipeline Maintenance Plan

Figure 7
Internal Corrosion Susceptibility Profile
Sour Gas Production Pipeline
Case 1: Design Stage

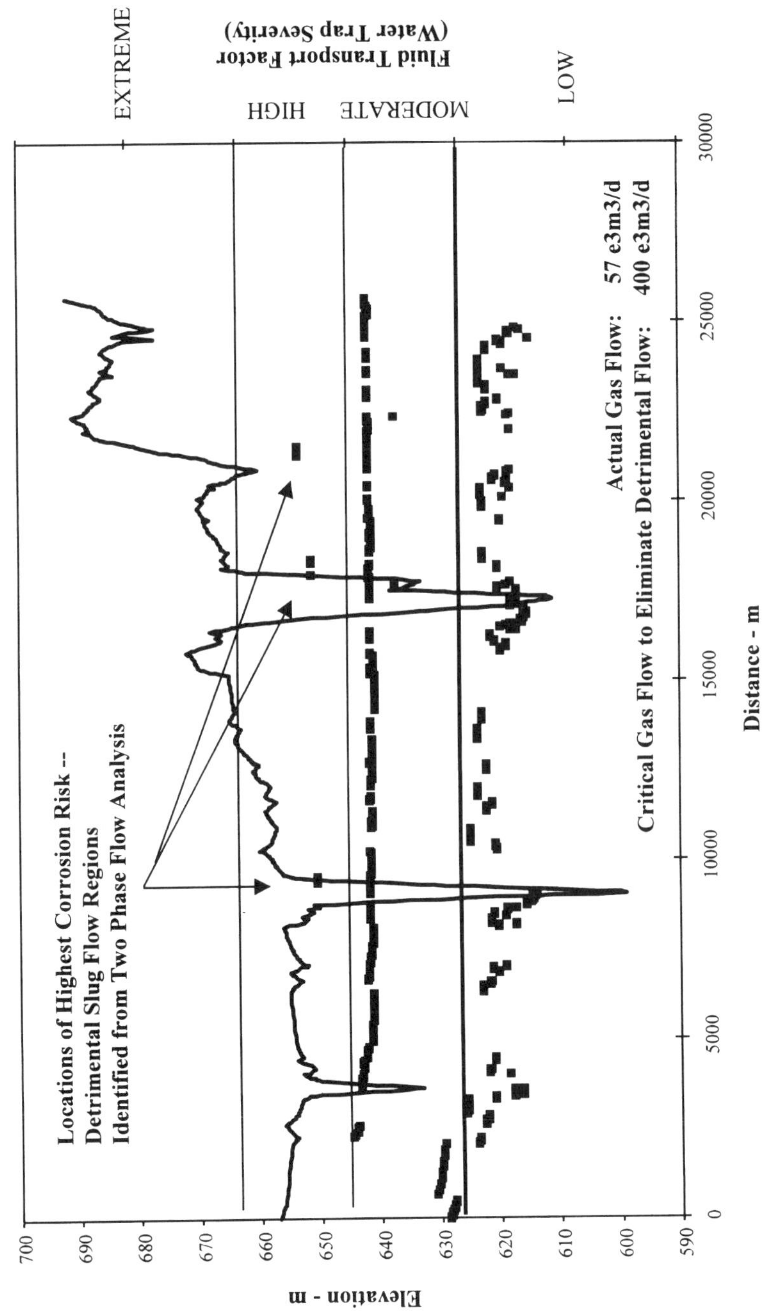

Figure 8
Internal Corrosion Susceptibility Profile
Sour Gas Production Pipeline
Case 2: Failure Attributed to Slug Flow Regimes

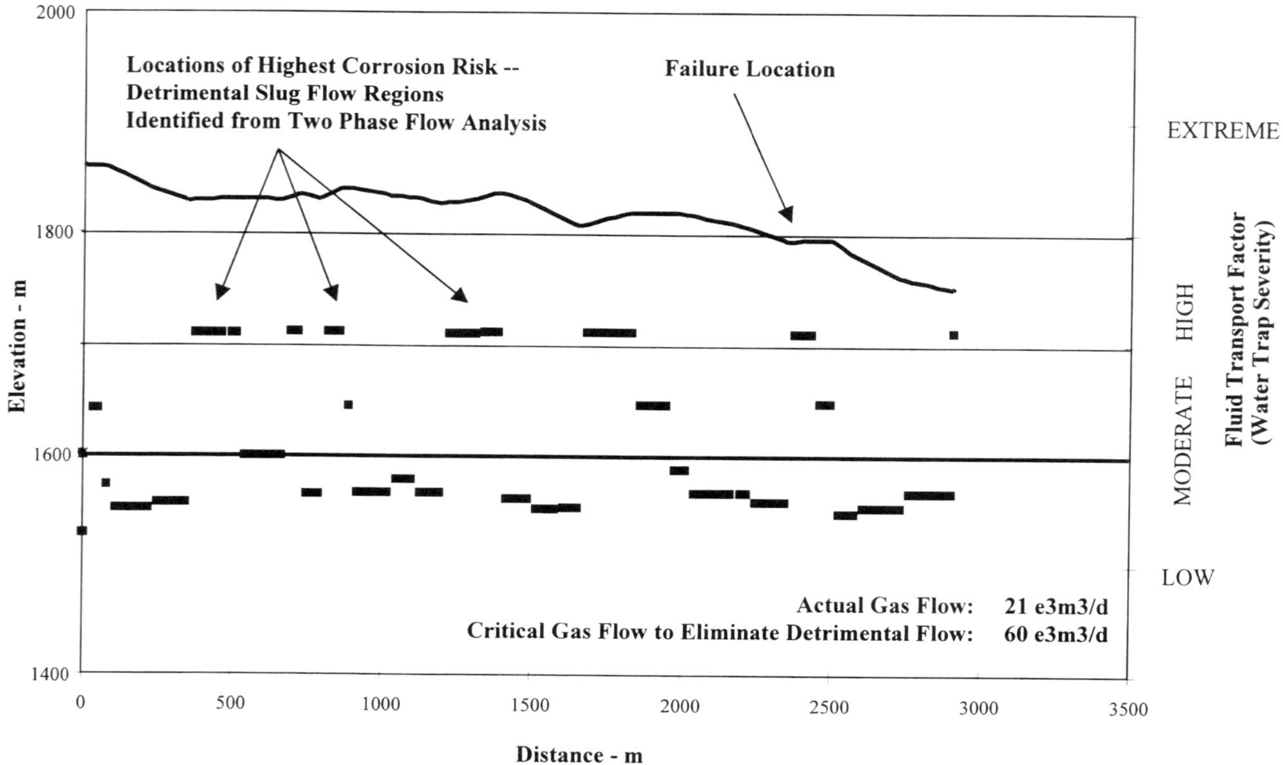

Figure 9
Internal Corrosion Susceptibility Profile
Sour Gas Production Pipeline
Case 3: Failure Attributed to Severe Stagnant Water Traps & Settling Solids

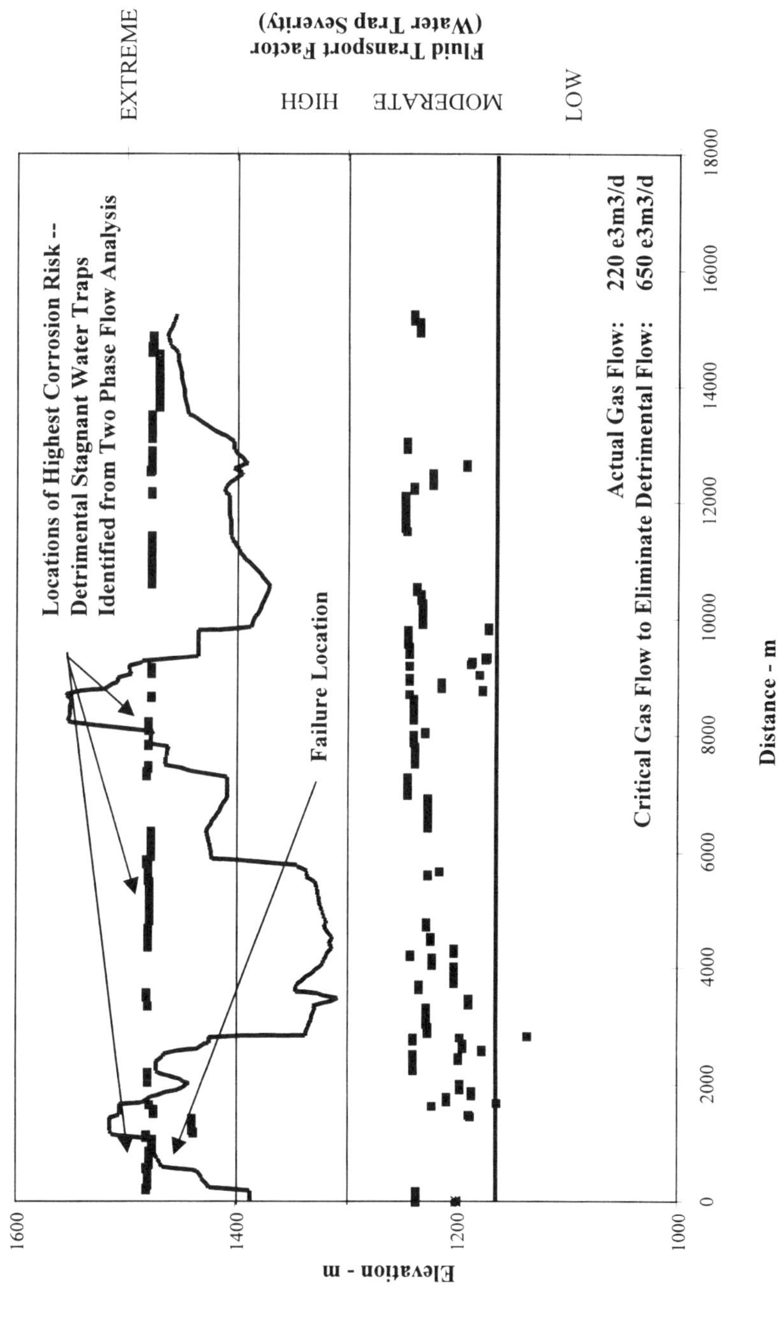

Wax deposition in offshore pipeline systems

O B RYGG
Well Flow Dynamics AS
A K RYDAHL
Calsep A/S
H P RØNNINGSEN
Statoil A/S, Norway

ABSTRACT

This paper describes a model developed to predict wax deposition in wells and pipelines. The simulation model uses a standard steady state multi-phase point model to predict pressure drop and liquid hold-up along the pipeline. By integration in time, the effect of wax deposition on pressure drop and temperature is calculated. A multi-component wax model is used to continuously estimate the wax precipitation along the pipeline and the viscosity of the current composition. The wax deposition is estimated from the diffusion of wax from the bulk towards the wall as a result of temperature gradients as well as shear dispersion effects. Varying inner pipe wall friction due to wax deposition is also included.

The model has been applied to a number of single phase and multi-phase wells and pipeline systems. Two such cases are discussed in this paper. Wax build-up, temperature and pressure drops over time have been simulated in each case.

INTRODUCTION

The increasing use of multi-phase pipelines for transportation of unprocessed well streams offshore has revealed several challenges with respect to fluids effects, such as hydrates, asphalthenes and wax.

The formation and deposition of wax have mostly been focused around single phase transportation of paraffinic oils. Recent experiences have proved that also in unstabilized oil and condensate systems wax deposition may be a significant problem. Consequently there is a need for a modeling tool incorporating both the latest developments in thermodynamic wax modeling and multi-phase effects.

The formation of wax in oils for various pressures, temperatures and fluid compositions has been studied in several papers for both single phase and multi-phase systems[1,2], and is therefore only briefly covered here.

WAX DEPOSITION MODEL

Following is a description of the various elements of the developed multi-phase flow wax deposition model.

Model structure

The wax deposition model consists of a number of sub-models each taking care of one particular technical aspect. For instance, the flow model is responsible for the pressure drop calculation, the thermodynamic wax model determines the number and properties of the existing phases and the wax deposition model predicts the amount of wax that deposits in a section of the pipeline.

Since the wax deposition build-up usually is slow compared to flow disturbances in a pipeline, a semi-stationary model is suitable to predict wax deposition over longer periods. Semi-stationary means that changes in boundary conditions like flow rates, pressures and temperatures are taken into consideration as well as the variation in inner pipe diameter due to wax deposition.

The model is based on a compositional concept, and consequently all components are treated individually in each sub-model. A component mass balance keeps track of all components, and the maximum possible amount of wax deposited is equal to the absolute amount of wax in the composition.

In **Figure 1** an illustration of the simulator structure is shown. To perform a simulation it is necessary to discretize the pipeline into a number of sections. Within each section the pressure and composition are assumed to be constant. There is a temperature gradient from the centre of the pipe to the pipe wall, due to the surrounding temperature being lower than that of the transported fluid. The equilibrium phases and properties are for each section determined by the thermodynamic wax model. The viscosity is calculated by the viscosity model, and the flow model calculates the resulting pressure drop and flow regime based on this information. Given the flow regime, the wax deposition is determined by the deposition model. An energy balance determines the bulk and wall temperatures. When all sections have been simulated, the pressure drops are summed to give the total pressure drop. Depending on the boundary conditions, iterations are performed to match input specifications. The modular structure of the program means that each model can be modified, improved or even exchanged if better models become available.

Wax deposited in one period will affect the results in the following time periods. When wax deposits in one section, the inner diameter of that section decreases which again leads to higher pressure drop in the pipeline. Also the energy balance will be influenced by the wax deposition, as the wax serves to insulate the pipeline and thus results in a temperature increase over time for a waxy section.

Multi-phase flow models

The modular structure of the code makes it easy to switch between different multi-phase point models and correlations. In the present simulations the OLGAS point model has been used (Bendiksen et al.[3]). When the flow model predicts single phase flow, a standard Moody correlation is used.

Thermodynamic wax model

The thermodynamic wax model (TWM) by Rønningsen et al.[2] was applied in the program. The TWM is based on compositional information and is able to predict wax appearance temperatures and pressures with good accuracy. Amounts of wax precipitated is predicted as well. The quality of the TWM is substantiated through the large amount of experimental wax data that was used to develop the model. More than 80 different compositions were used, of which 65 were live oils.

The TWM's primary advantages to previous presented models are:

- satisfactory description of the effect of pressure on wax precipitation
- satisfactory description of the effect of dissolved gas on wax precipitation

These two effects are crucial when multi-phase systems are considered.

Wax deposition model

Two different wax deposition mechanisms are included in the model. The wax can deposit on the wall by diffusion through the laminar film layer or by shear dispersion effects near the wall.

<u>Molecular diffusion:</u>

It is assumed that the temperature gradient makes dissolved wax diffuse from the bulk towards the wall. Wax already precipitated in the wax phase does not contribute to the diffusion but may redissolve to compensate for the wax deposited on the wall.

The volume rate of wax deposited by molecular diffusion for a given wax forming component *i* is calculated from the relation:

$$\text{Vol}_{\text{wax}}^{\text{diff}} = \sum_{i=1}^{\text{NWAX}} \frac{D_i \left(c_i^b - c_i^w\right) S_{\text{wet}} \, MW_i}{\delta \, \rho_i} 2\pi \, r L \tag{1}$$

where c_i^b, c_i^w are the molar concentrations of the wax component *i* dissolved in the oil phase in the bulk and at the wall respectively, S_{wet} is the fraction of the wetted perimeter, NWAX the number of wax components, MW_i and ρ_i the molecular weight and density of wax component i, r the current inner pipe radius and L the length of the pipe section. The thickness of the laminar sub-layer, δ (see **Figure 2** and **Figure 3**), is calculated from the Blasius[4] expression, and the diffusion coefficient, D_i, of the wax forming component i is calculated from a correlation by Hayduk and Minhas[5].

Shear dispersion:

The volume rate of already precipitated wax deposited from shear dispersion, can be estimated from the following correlation proposed by Burger et al.[6]:

$$\text{Vol}_{\text{wax}}^{\text{shear}} = \frac{k^{*} C_{\text{wall}} \dot{\gamma} A}{\rho_{\text{wax}}} \tag{2}$$

where k^* is a shear deposition rate constant, C_{wall} is the volume fraction of precipitated wax in the oil at the inner wall temperature, $\dot{\gamma}$ is shear rate at the wall, A is the surface area available for deposition and ρ_{wax} is the average density of the wax. The shear dispersion mechanism is often assumed to be negligible (Brown et al.[7] and Hamouda[8]).

Total wax deposition

The total rate of increase in thickness of the wax layer accounting for both diffusion and shear dispersion is found from the expression:

$$l_{\text{wax}} = \frac{\text{Vol}_{\text{wax}}^{\text{diff}} + \text{Vol}_{\text{wax}}^{\text{shear}}}{(1-\phi) 2\pi \, r \, L} \tag{3}$$

where ϕ is the porosity of the deposited wax, i.e. the volume fraction of the immobile deposit consisting of trapped oil and voids. The porosity is usually assumed to be in the range from 0.6 to 0.9. As seen in **Equation 3** the thickness of the wax layer is averaged around the pipe perimeter even if the inner pipe surface is only partially wetted with liquid.

Wetted perimeter:

In multi-phase flow the wetted inner surface area depends on the local flow regime predicted and the liquid hold-up. For stratified smooth and wavy flow the wetted perimeter can be found from the relation:

$$S_{\text{wet}} = 2\varphi \, r \tag{4}$$

where the angle φ can be found from the expression for liquid holdup, HOL, see **Figure 2**:

$$\text{HOL} = \frac{\varphi - \sin\varphi}{2\pi} \tag{5}$$

As a first approximation for annular, bubble and slug flow regime, the whole perimeter is assumed to be wetted.

Temperature model

The temperature of the flowing fluid can be obtained from a steady flow energy equation using the relation:

$$\left(\dot{m}_g Cp_g + \dot{m}_o Cp_o\right)\Delta T + \left(\dot{m}_g \frac{\partial H_g}{\partial p} + \dot{m}_o \frac{\partial H_o}{\partial p}\right)\Delta p + \left(\dot{m}_g + \dot{m}_o\right) g \, \Delta y = -UA\left(T_b - T_\infty\right) \tag{6}$$

where $\dot{m}$ is the mass flow rate, Cp the heat capacity, H the enthalpy, p pressure, y elevation, T_b fluid bulk temperature, T_∞ ambient temperature, U overall heat transfer coefficient and Δ denotes the difference from the previous segment. The kinetic energy flux is small for low velocity flow and is therefore neglected. We assume an average temperature for the gas and liquid flow.

The internal heat convection, h_i, between the fluid and the pipe wall (or the wax layer) is calculated using the Dittus-Boelter[9] equation:

$$Nu = 0.023\,Re^{0.8}\,Pr^{0.3} \tag{7}$$

For two phase flow, mixed Nusselt, Reynolds and Prandtl numbers are defined through superficial velocities, densities, and thermal conductivity for each phase. Heat transport through the pipe walls is calculated using the following relation

$$Q_r = \frac{T_b \quad T_\infty}{R_{tot}} = U_1 A_1 (T_b - T_\infty) \tag{8}$$

where Q_r is radial heat transfer rate, R_{tot} total thermal resistance, U_1 heat transfer coefficient with respect to internal pipe diameter and A_1 inside pipe surface area.

The thermal resistance for each wall layer is defined as:

$$R_{tw} = \frac{\ln\left(r_2/r_1\right)}{2\pi L k} \tag{9}$$

where r_2 is outer radius of the layer, r_1 is internal radius of the layer, L is pipe (section) length and k is thermal conductivity.

Total thermal resistance is obtained by adding the external heat transfer to the sum of the resistances for each wall layer. The calculation of overall thermal resistance is done once for each simulation. The modification of the resistance due to the wax layer is added locally at each simulation step by calculating the thermal resistance in the wax layer from equation (9) and the internal heat transfer between the fluid and the wax. For the thermal conductivity of the wax layer we use the conductivity of an oil with the same composition as the local wax layer itself. The modified overall heat transfer is then used in equation (6) to calculate the local temperature.

To calculate the wax diffusion to the wall, an estimate of the temperature at the wall is needed. Considering the relation for radial heat transfer, the inner wall temperature can be obtained from the following relation:

$$Q_r = \frac{T_b \quad T_\infty}{R_{tot}} = U_1 A_1 (T_b - T_\infty) = h_i A_1 (T_b - T_w) \tag{10}$$

Viscosity correlation

Viscosity is of importance to the accuracy of the flow model pressure drop calculation. An inaccurate viscosity may result in erroneous pressure drops. The viscosity of gas and oil phases is generally well determined using the corresponding states principle (Pedersen and Fredenslund[10]). However, when wax is present as precipitate in the oil phase, this will change the overall viscosity of the combined oil and wax phases.

The viscosity of a given composition primarily depends on temperature, shear rate and pressure. The best situation when evaluating the viscosity of a composition would be to have experimental viscosity data available covering all these parameters. For fields in production, some data may be available, but usually these are limited to viscosity as a function of temperature at ambient pressure.

When no viscosity data are available, a simple correlation is used to estimate the viscosity of the oil and wax mixture given temperature and pressure:

$$\ln(\eta_{oil+wax}) = \ln(\eta_{oil}) + X\,Vol_{wax} \tag{11}$$

where Vol_{wax} is the volume fraction of solid wax in the liquid phase obtained from the thermodynamic model, η_{oil} is the viscosity of the oil without wax, and X is a constant determined from experience with similar compositions. This simple model takes into consideration pressure and temperature but neglects the shear effect thus assuming Newtonian behaviour. Alternatively, a series of the above equations, all valid for a particular shear rate, could be applied. In the present cases, viscosity estimates of the pipeline fluids were available and were used in the calculations (**Table 1**).

APPLICATION OF THE MODEL TO OFFSHORE PIPELINE SYSTEMS

The model has been successfully applied to a number of multi-phase pipeline systems. The following cases were selected to illustrate the applicability and flexibility of the model.

Case A; Flowline from subsea template to platform

A water-free multi-phase oil well stream is transported from a subsea template to a North Sea platform, see **Figure 4**. From the 10+ mole%, molecular weight and density the composition is estimated to C80. The components from C7 and on are lumped into 12 pseudo-components each of which are divided into a wax forming and a non-wax forming component. This division is based on density and molecular weight information. The wax forming components are essentially n-paraffins which have a lower density than other hydrocarbons of the same molecular weight. With a detailed compositional analysis of the heavy end, the wax content in each carbon number fraction could have been assumed equal to the n-paraffin content. Some system data and conditions are given in **Table 1**. The composition of the well stream fluid is given in **Table 2**. The measured wax appearance temperature, or cloud point, of the stabilized oil (i.e. 1 bar) is 34 °C. The calculated cloud point of the wellstream fluid is 30 °C at 30 bar. Since the outlet temperature of the pipeline is seen to be about 13 °C, a large part of the pipeline is evidently operating well below the cloud point. There is some free gas phase in the flowline at the operating conditions. In a 3 month period with relatively stable conditions, commencing about half a year after the flowline started to carry oil, field measurements indicate no significant increase of flowline pressure drop. The line has not been pigged yet, and there are no direct measurements of wax deposition.

Figure 5 shows how the calculated wax deposition is distributed along the flowline as a function of time. Wax starts to deposit 5.5 km downstream of the subsea template when the temperature at the wall falls below the cloud point. After 3 months, a maximum wax layer thickness of about 10 mm is calculated assuming a wax porosity of 0.6. The choice of this parameter is somewhat arbitrary and it can in fact be used as a tuning parameter. Most likely, this will be system and fluid dependent, but in black oil systems it seems reasonable to assume a fairly high porosity. It should be noted that there is not a linear relationship between the increase in porosity and the corresponding increase in thickness of the immobile layer. For instance, increasing the porosity from 80% to 90% will give a doubling of the layer thickness, other things being equal. Burger et al.[6] reported a porosity of about 84% for the Trans Alaska pipeline. The reason for using a somewhat lower porosity in this case is due to an anticipated effect of the two-phase flow regime.

The calculated pressure drop and deposited wax volume in the flowline are shown as a function of time in **Figure 6**. The pipe wall roughness is here assumed not to increase due to the formation of a wax layer. The calculated pressure drop agrees well with field measurements which indicates 11-12 bar pressure drop along the line. The increase in pressure drop with time is seen to be moderate (1.5 bar in 90 days), also in reasonable agreement with field experience.

In general, the effect of the wax layer on the roughness, may be viewed as a second potential tuning parameter. Bern et al.[11] assumed the roughness to increase by an amount equal to the thickness of the wax layer. **Figure 7** shows the effect of increasing the pipe wall roughness. When the roughness is defined as 10% of the local wax layer thickness, i.e. up to about 1 mm, the pressure drop is seen to increase significantly with time. Since no such increase is actually observed, and assuming that wax deposition really takes place (which is very likely when operating so far below the cloud point), it seems reasonable to conclude that the increase in roughness is marginal in this case. It should be noted that there is also some uncertainty in the live oil viscosity which will influence the calculated pressure drop, but that is probably relatively constant and should not introduce any variation of pressure drop with time.

Since the calculated increase of pressure drop is mainly a diameter effect in this case, a lower porosity could be used to further improve the agreement with field observations. Alternatively, an adjusted diffusion coefficient (lower) or introduction of certain shear erosion could be applied. As long as no direct wax thickness data are available, it is impossible to decide which adjustment would be most correct.

In summary, by choosing a reasonable wax porosity and making the assumption that the pipe wall roughness is not increased, the calculations are in good agreement with observations. It should be noted that pressure drop is not necessarily a good measure of wax deposition.

Case B; Flowlines from two fields to a field centre

Fully stabilized oil with less than 0.5% water is transported from two different North Sea fields to a field centre. Field 2 is connected to the main flowline from field 1 via a Y-junction, see **Figure 8**. The pipeline geometries were discretized using 50 sections for the flowline from field 1 and 27 sections for the connection from field 2. With a constant 24000 Sm^3/d of stabilized oil at a temperature of 50 °C coming from field 2, the arrival temperature at the junction is 47 °C, which gives rise to a temperature increase in the last part of the line. The additional flowrate coming from field 2 was simulated as a mass source to the original flowline from field 1. The total flow rate was 34200 Sm^3/d with 10200 Sm^3/d continuosly producing from field 1. The composition of the produced fluid from field 1 is given in **Table 3**. The measured cloud point of the stabilized oil is 35 °C. Some other data and conditions are given in **Table 1**.

The flowline is being pigged at regular 14 days intervals. A fairly small amount of wax (about 10 kg) is normally collected in the pig receiver. However, the removed wax is believed to be largely dissolved and dispersed in the oil ahead of the pig. Inlet and outlet pressure and temperature measurements were done daily between two pigging operations. No significant temperature changes were observed. However, a steady increase of pressure drop was observed (**Figure 10**).

Figure 9 shows the calculated wax deposition profile along the flowline. In this case, a wax layer porosity similar to the value reported by Burger et al.[6] for a stabilized black oil, i.e. 85%, was used. Again, this value has not been verified and is thus somewhat arbitrary. Onset of wax deposition is seen to occur about 6 km downstream of the inlet when the wall temperature falls below the cloud point. A shift to lower deposit thickness is seen after field 2 is coming in.

Figure 10 shows the wax volume and increase in pressure drop in the line compared to the measured field data. Assuming a thin layer of wax is still covering the pipe wall after pigging, the roughness was increased from 0.05 mm to 0.12 mm in order to fit the observed pressure drop just after pigging. As a base case, no further increase of roughness was assumed as a result of additional wax deposition. Compared to the approximately 5 bar increase of pressure drop observed in the field, a somewhat smaller increase is calculated. This could indicate either that a too slow wax deposition is calculated (which could be adjusted by the porosity), or that the effect on roughness is somewhat larger in this case. It is shown in the graph how the pressure drop would increase somewhat faster if the roughness is let to increase by 15% of the local wax layer thickness. In fact, this illustrates the problem that we, at the moment, do not have sufficient data basis to estimate accurately the wax porosity and roughness characteristics in a given case. More data are obviously needed on these parameters in order to further improve modeling reliability.

CONCLUDING REMARKS

A multi-phase dynamic wax deposition simulator has been developed which incorporates latest developments in thermodynamic wax modeling and multi-phase flow modeling. The deposition part is based on traditional diffusion and shear dispersion mechanisms.

By reasonable choices of some adjustable parameters (i.e. wax layer porosity and pipe wall roughness), it has been demonstrated that good agreement is achieved between calculated and observed pressure losses in two pipelines with wax deposition.

More data are needed on wax porosity and roughness effects, as well effects of multi-phase flow regime, in order to further improve reliability of wax deposition predictions.

ACKNOWLEDGMENT

We are grateful to Statoil for providing the field data.

NOMENCLATURE

A	Area (m^2)
C	Volume fraction (-)
C_p	Heat capacity (J/(kg K))
c	Molar concentration (mole/m^3)
D	Diffusion coefficient (m^2/s)
g	Gravitational acceleration (m/s^2)
H	Enthalpy (J/kg)
HOL	Hold-up (-)
k	Thermal conductivity (W/(m K))
k*	Shear deposition rate constant (kg/m^2)
L	Length of section (m)
l	Increase in thickness (m/s)
MW	Molecular weight (kg/mole)
	Mass flow rate (kg/s)
Nu	Nusselt's number (-)
NWAX	Number of wax comp. (-)
p	Pressure (Pa)
Pr	Prandtl's number (-)
Q	Heat transfer rate (J/s)
q	Heat flux (J/s)
R	Thermal resistance (W/K)
Re	Reynold's number (-)
r	Pipeline radius (m)
S	Perimeter (-)
T	Temperature (K)
U	Heat transfer coefficient (W/(m^2K))
Vol	Volume increase (m^3/s) or volume fraction (-) (Equation (11))
X	Viscosity increase constant (-)
y	Elevation (m)

Symbols

δ	Thickness of laminar sub-layer (m)
η	Viscosity (Pa s)
$\dot{\gamma}$	Shear rate (s^{-1})
ϕ	Porosity (-)
φ	Angle (-)
ρ	Density (kg/m^3)

Sub- and superscripts

b	Bulk
conv	Convection
diff	Diffusion
i	Component index
o, g	Oil and gas
r	Radial
shear	Shear dispersion
t	Thermal
tot	Total
w	Wall
wax	Wax
wet	Wetted
1	Internal
2	Outer
∞	Ambient (-)

REFERENCES

1. Pedersen, K.S., "Prediction of Cloud Point Temperatures and Amount of Wax Precipitation", SPE Production and Facilities, Feb. 1995, 46.
2. Rønningsen, H.P., Sømme, B.F. and Pedersen, K.S.:"An improved thermodynamic model for wax precipitation: experimental foundation and application." 8th International Conference on Multiphase '97, Cannes, France, 18-20 June, 1997.
3. Bendiksen, K., Malnes, D., Moe, R. and Nuland, S.: "The Dynamic Two-Fluid Model OLGA: Theory and Application." SPE Production Engineering, May, 1991.
4. Blasius, H., "Das Ähnlichkeitsgesetz bei Reibungsvorgängen in Flüssigkeiten", Forsch. Ver. Deut. Ing. 131, 1913
5. Hayduk, W. and Minhas, B. S., "Correlations for Prediction of Molecular Diffusivities in Liquids". The Canadian Journal of Chemical Engineering, April 1982, Volume 60, 295-299.
6. Burger, E.D., Perkins, T.K. and Striegler, I.H. "Studies of Wax Deposition in the Trans Alaska Pipeline". Journal of Petroleum Technology, June 1981, 1075-1086.
7. Brown, T.S., Niesen, V.G. and Erickson, D.P., "Measurement and Prediction of Kinetics of Paraffin Deposition", SPE 26548, 68th Annual Technical Conference and Exhibition of SPE, Houston, TX, 3-6 October, 1993

8. Hamouda, A., "An approach for simulation of paraffin deposition in pipelines as a function of flow characteristics with a reference to Teeside oil pipeline", SPE 28966 (1995), presented at SPE Int. symposium on Oilfield Chemistry, San Antonio, 14-17 February 1995
9. Dittus, F.W., and L. M. K. Boelter, University of California Publications on Engineering, Vol. 2, p. 443, Berkley, 1930.
10. Pedersen, K.S. and Fredenslund, Aa.: "An improved corresponding states model for the prediction of oil and gas viscosities and thermal conductivities", Chem. Eng. Sci. 42, 1987, pp. 182-186.
11. Bern, P.A., Withers, V., Cairns, J.R., "Wax deposition in crude oil pipelines", Proceedings Eur. Offshore Petr. Conf., London, 1980, 571-578.

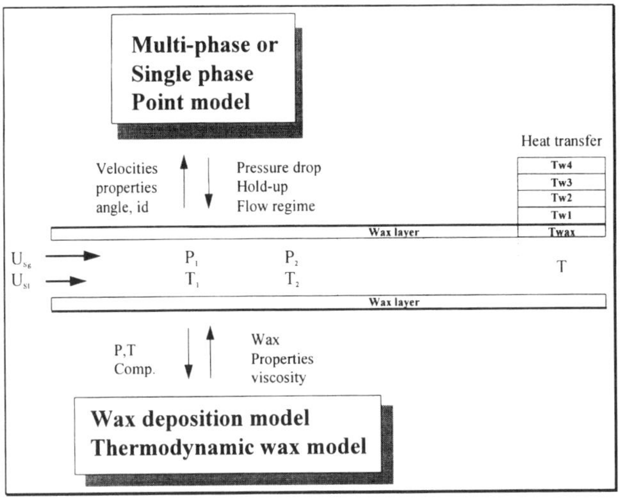

Figure 1 Definition of flow model structure.

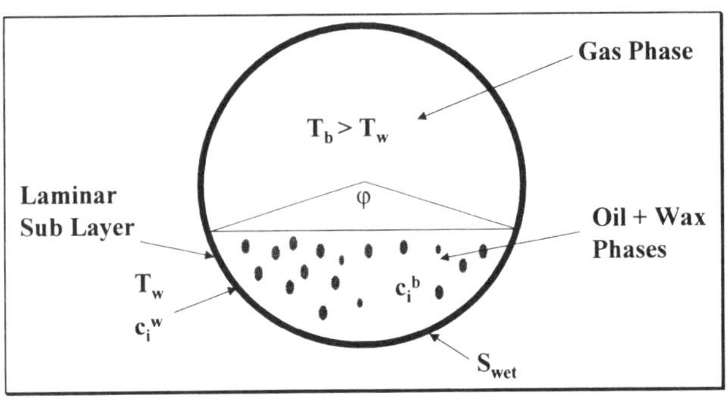

Figure 2 Distribution of wax, oil and gas in a pipeline.

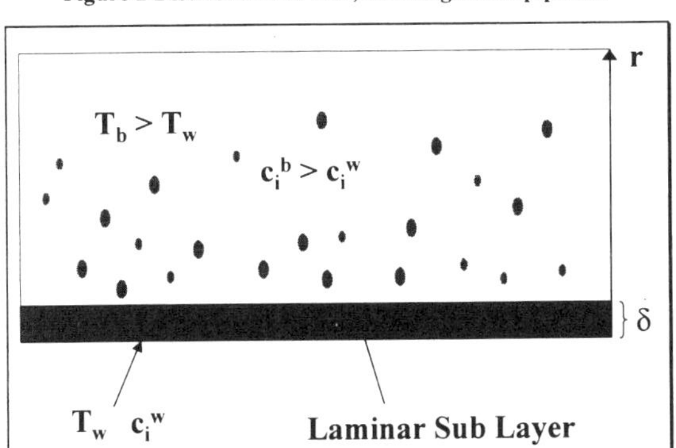

Figure 3 Laminar sub layer near pipe wall.

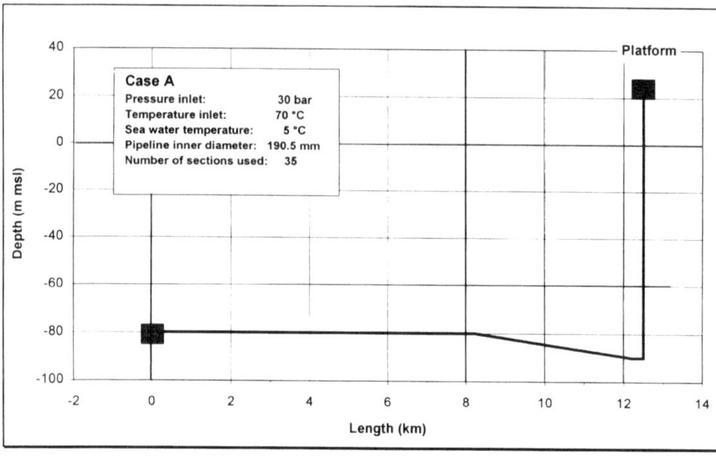

Figure 4 Case A. Pipeline configuration.

	Case A	Case B - Field 1	Case B – Field 2
Pipeline length (km)	12.5 km	38 km	3.6 km
Pipeline inner diameter	190.5 mm	374.6 mm	374.6 mm
Flowrate	1720 Sm³/d	10 175 Sm³/d	24 000 Sm³/d
Inlet temperature	70°C	70°C	50°C
Pressure	30 bar, inlet	35 bar outlet	
Surrounding sea water temp.	5°C	5°C	
Pipeline materials riser	14 mm steel - 1mm paint - 6 mm concrete	15.9 mm steel - 25 mm rubber	
Pipeline materials pipeline	14 mm steel - 1mm paint - 6 mm concrete	11 mm steel - 6 mm asphalt - 40 mm concrete	
Pipeline trenching	Whole pipeline	First 2 km & last 3 km	First 2 km
Viscosity	2 cp @ 70°C 13 cp @ 0°C	1.5 cp @ 40°C, 2 cp @ 30°C, 5.5 cp @ 15°C, 10.3 cp @ 5°C	

Table 1 Case A and B. System data.

Component	Mole %	Mole wgt. (g/mole)	density (g/cm³)
N_2	1.030	28.073	
CO_2	1.200	44.010	
CH_4	16.180	16.043	
C_2H_6	4.550	30.070	
C_3H_8	8.260	44.097	
iC_4H_{10}	1.520	58.124	
NC_4H_{10}	5.370	58.124	
iC_5H_{12}	2.290	72.151	
NC_5H_{12}	3.020	72.151	
C_6	3.980	85.100	
C_7	6.000	93.200	0.732
C_8	5.790	107.10	0.750
C_9	3.860	119.70	0.770
C_{10+}	36.950	286.00	0.884

Table 2 Case A. Fluid composition.

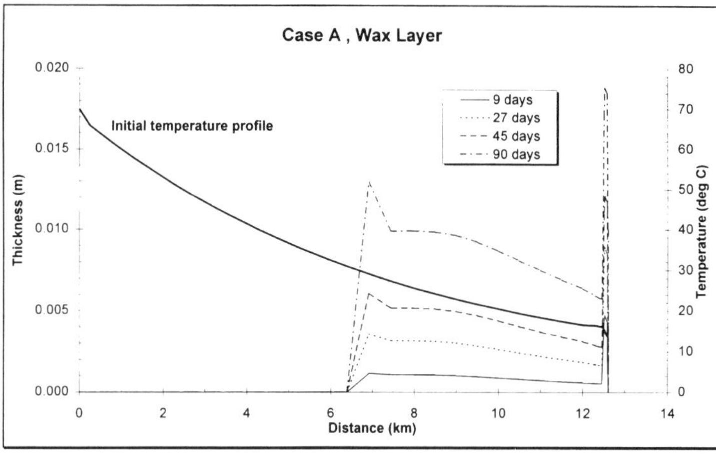

Figure 5 Case A. Wax layer in flowline and initial temperature profile.

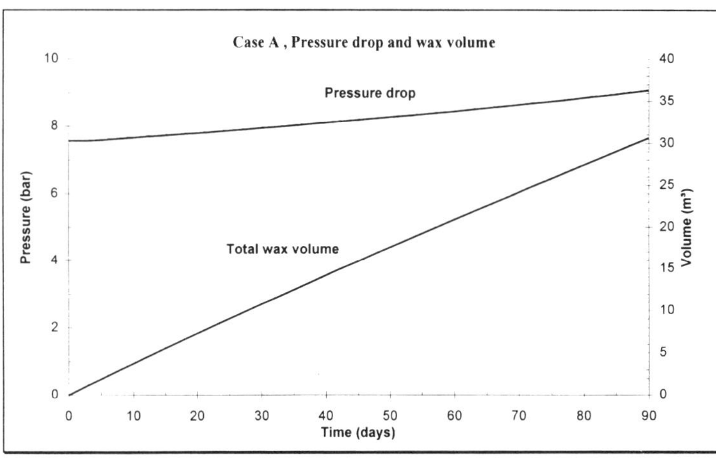

Figure 6 Case A. Pressure drop and total wax volume.

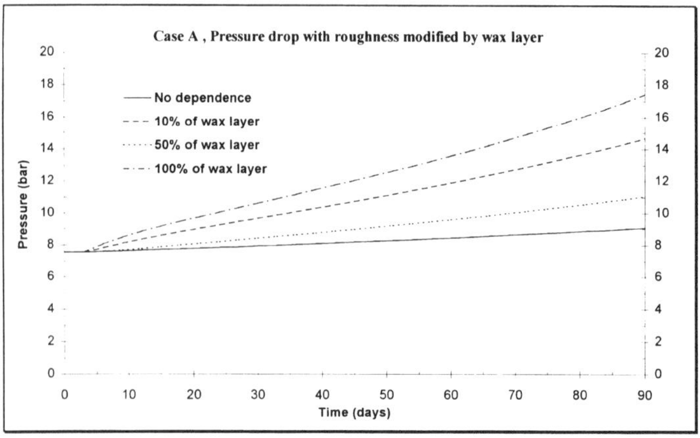

Figure 7 Case A. Total pressure drop with different wax layer roughness.

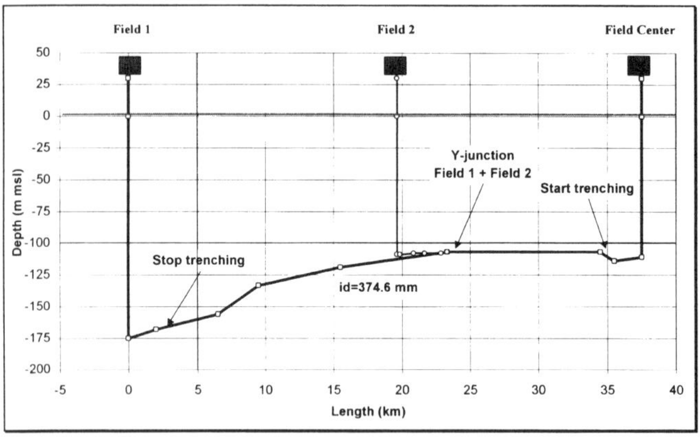

Figure 8 Case B. Pipeline configuration.

Component	mole %	mole wgt. (g/mole)	density (g/cm³)
CH_4	0.021	16.043	
C_2H_6	0.254	30.070	
C_3H_8	1.236	44.097	
iC_4H_{10}	0.588	58.124	
nC_4H_{10}	2.512	58.124	
iC_5H_{12}	1.955	72.151	
nC_5H_{12}	3.485	72.151	
C_6	6.842	85.100	
C_7	12.856	92.200	0.734
C_8	13.985	105.40	0.758
C_9	9.190	119.80	0.777
C_{10+}	47.076	253.50	0.847

Table 3 Case B. Fluid composition.

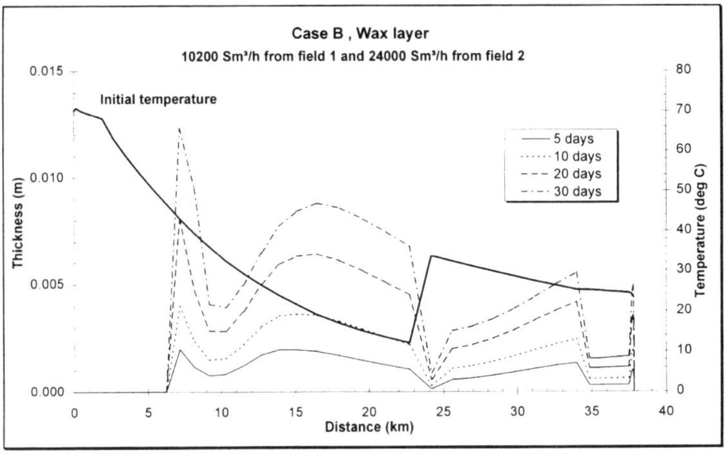

Figure 9 Case B. Wax layer in flowline and initial temperature profile.

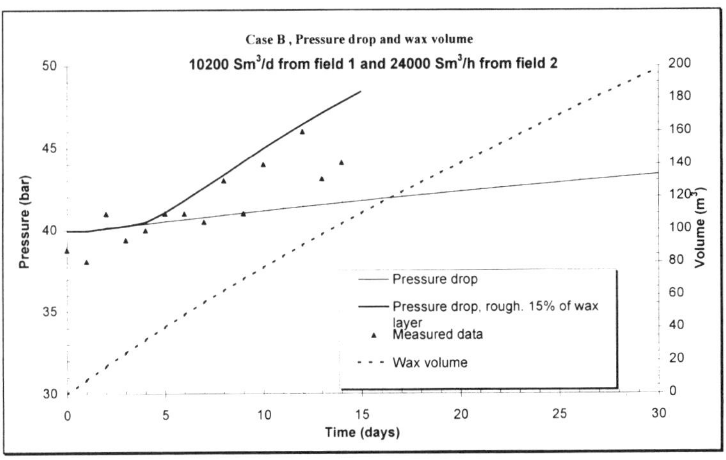

Figure 10 Case B. Pressure drop and total wax volume.

Direct heating of multiphase flowlines for hydrate control on Åsgard

T VANVIK
Statoil A/S, Norway
R MOE
Aker Engineering A/S, Norway

The Åsgard Field Development off shore of Norway will be among the world's most extensive subsea developed oil and gas production systems when put in operation in 1999. The large number of subsea flowlines demands proper hydrate control measures to avoid operational problems on the field. The different methods for hydrate control that will be used on the field are discussed in this paper. The newly developed concept of direct electric heating of the subsea production flowlines is described in more detail, including the capabilities and requirements of the heating system.

1 INTRODUCTION

The Midgard, Smørbukk and Smørbukk Sør (South) fields are located on Haltenbanken in the Norwegian Sea, 300 km off shore of Mid-Norway at a water depth of about 300 m. The three fields together are named Åsgard - home of the Gods in Norse mythology - and are currently being developed jointly by the Norwegian oil companies Statoil and Saga. The field location is shown in Figure 1.

The Midgard field was discovered by Operator Saga in 1981, while the Smørbukk fields were discovered in 1984 and 1985 by Operator Statoil. In the fall of 1994 Statoil and Saga agreed to participate on equal terms in an integrated project team, with Statoil as Operator, in the co-ordinated field development of Midgard, Smørbukk, and Smørbukk Sør. The participating interests of the licensees in the Åsgard Unit are shown in Table 1.

Production from the Åsgard fields will start within the first quarter of 1999, and gas export starts in the 4th quarter of 2000.

Table 1 Licensees in Åsgard Unit

Licensee	Interest
Statoil	60.50
Norsk Agip AS	7.90
Total Norge AS	7.65
Mobil Exploration Norway Inc.	7.35
Neste Petroleum AS	7.00
Saga Petroleum a.s.	7.00
Norsk Hydro Produksjon a.s.	2.60

1.1 Field installations

The field development concept for Åsgard is based on production from two separate floating installations: A Floating Production, Storage and Off loading ship (FPSO) for oil production, and a gas and condensate production centre, a semi submersible floating production unit (SEMI). The FPSO is located in the area between Smørbukk and Smørbukk Sør, while the SEMI is positioned centrally on Smørbukk. The FPSO and SEMI are named Åsgard A and B, respectively. A Floating Storage Unit (FSU), Åsgard C, will be installed and connected to Åsgard B. Rich gas will be exported from Åsgard B to Kårstø (near Stavanger) through a new 700 km long 42 inch pipeline. The maximum annual contract quantity of gas export is 10.8 GSm^3/year at sales specifications (exit Kårstø).

Figure 1 Field location

Åsgard A will produce oil from the Smørbukk and Smørbukk Sør fields, and has an oil production capacity of 32,000 Sm^3/d and a storage tank capacity of 145,000 m^3. The production plant has a gas re-injection capacity of 21 MSm^3/d at a pressure of 425 bara. In addition, 7 MSm³/d of gas may be transferred between Åsgard A and B, in case the FPSO or the SEMI produces excess gas. The oil production starts early in 1999.

Åsgard B starts production nearly two years later, and will start exporting gas in October year 2000. Gas and condensate will be produced from the Smørbukk and Midgard reservoirs and up to 11 MSm³/d of gas can be re-injected in Smørbukk from Åsgard B. The design gas export rate is 38.5 MSm³/d. The condensate processing and export capacity is 15,000 Sm³/d, which will be exported to the FSU; Åsgard C.

Totally 59 subsea completed wells are planned, including 16 gas injectors and one produced water disposal well. The wells are connected to 3 templates at Midgard, and at Smørbukk and Smørbukk Sør there are 11 and 3 templates, respectively. The produced water will be disposed in the Smørbukk reservoir from Åsgard B. All templates will have 4 well slots each. An illustration of the field layout is shown in Figure 2. The gas injection templates are shown in white, while the oil and gas production templates and the produced water injection manifold are shaded. The latter is located close to Åsgard B, on the north side.

About 300 km of flowlines will be installed on the sea bed, to transport the well stream to the floating processing plants. Generally two flowlines are connected to each production template, and the steel pipes are connected to the floating installations by flexible risers. The subsea flowlines vary in length from 3 km to about 50 km. The flowline route length between the Smørbukk templates and Åsgard B is from 6 to 8 km. These 10 inch flowlines will be electrically heated when required.

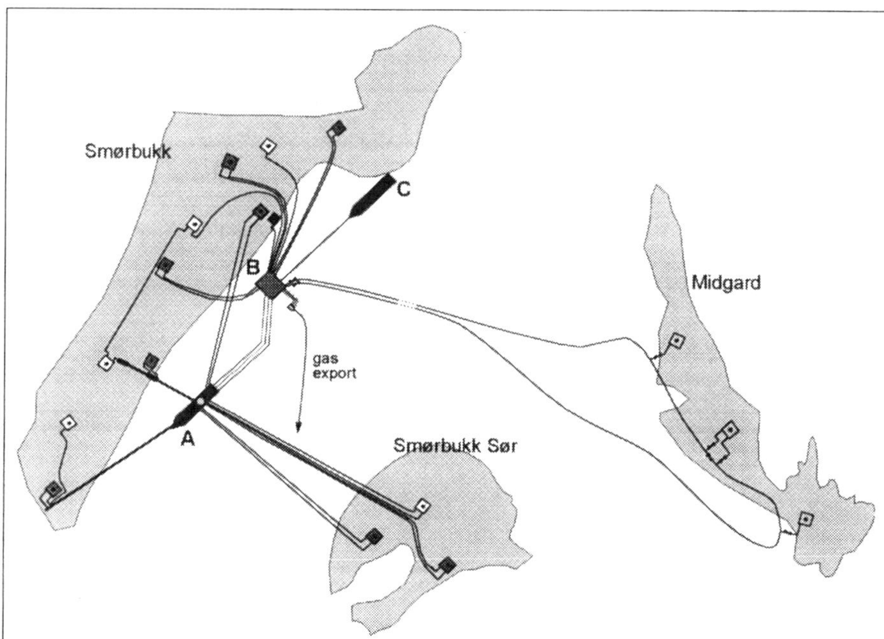

Figure 2 Åsgard fields, flowlines and installations

1.2 Fluid characteristics

The fluids from the three Åsgard fields vary from light gas and condensate (60°API) to oil (42°API). The well stream from Midgard is mostly gas, which presumably has water and hydrocarbon dew point at reservoir conditions - initially 255 bara and 90°C. An initial gas/oil volume ratio (GOR) of 6500 Sm^3/Sm^3 has been forecasted, but this value will increase as the reservoir pressure decreases during production. No formation water will normally be produced from this reservoir.

The southern and western regions of Smørbukk also contain mainly light hydrocarbons, but have more significant volumes of condensate. The initial GOR may vary from 700 to 1500, and water cuts ranging from 10% to 30% are anticipated during the first years of production.

The main oil reserves are situated in Smørbukk Sør and the northern parts of Smørbukk. Product stream GOR is expected to be below 500 during the first production years. Some formation water will also be produced, and the relative amount is generally expected to increase with time. Typically the initial water cut is expected to be less than 30%.

Smørbukk has quite high initial reservoir pressure; up to 505 bara, which generally will decline throughout the field life. The bottom hole temperature has been measured to about 150 - 165°C. Smørbukk Sør has a little less extreme reservoir conditions; maximum 440 bara and 150°C.

2 HYDRATES AND HYDRATE CONTROL METHODS

Gas hydrates may be described as ice- or snow-like crystals that can be formed when light hydrocarbon molecules; methane, ethane, propane and butane, get in contact with water. Other small molecules, like CO_2 and H_2S, may also form hydrates. These small molecules fit into a lattice made up of H_2O, and effectively stabilise the grid. Thus hydrates may, in contrast to ice and snow, be present at room temperatures, and their stability increases with pressure. This is illustrated in Figure 3, which shows typical hydrate formation conditions for Åsgard fluids. Stable hydrates may form to the left of the shown hydrate equilibrium curve, although some super cooling is generally required to initiate the hydrate growth.

At a normal flowline pressure of 150 bara, hydrates may be present at temperatures lower than approximately 22°C, as long as free water is present. Some hydrates may stick to each other, and thus form larger lumps that can block pipes, valves and other equipment. Hydrates will melt if heat is supplied or pressure is reduced sufficiently. The melting time depends on how effectively heat may be transferred to the hydrates.

The hydrocarbons from all Åsgard wells will be transported through subsea flowlines to the floating processing plants. Usually the well stream will contain some amount of formation water. Even if formation water is not produced, some water will condense from the well stream as the temperature drops when heat is lost to the surrounding sea water. The sea water temperature is normally between 6 and 8°C at the sea bottom in this area. The fact that a water phase will always be present in the production flowlines implies a potential for hydrate formation. Given the large number of wells and flowlines on the Åsgard fields, careful planning is vital in order to deal with and avoid hydrate related problems.

Because Åsgard is committed to gas sales contracts, the field regularity is especially important. Smooth operation of the processing plants will strongly depend on the operator's ability to avoid hydrate problems.

Traditional methods for hydrate prevention include use of chemicals, like methanol and different glycols. These chemicals reduce the hydrate formation (and melting) temperature, thus shifting the equilibrium curve shown in Figure 3 to the left. Methanol is a widely used hydrate inhibitor, and it can be used to reduce the formation temperature by 20°C at a concentration of 40% (by weight) in the free water.

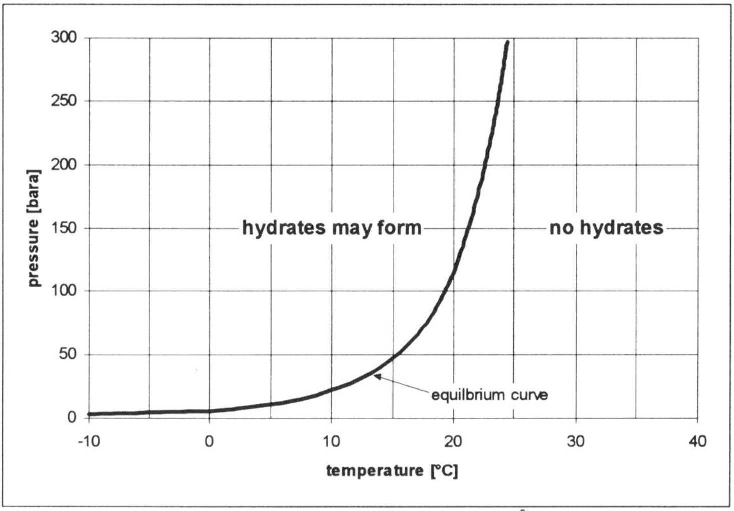

Figure 3 Typical hydrate equilibrium curve for Åsgard well stream

There are, however, a number of reasons why the amounts of chemicals in general, and methanol in particular, should be limited. Today there is great focus on environmental issues, and many common chemicals are poisonous and have negative effects on the wild life. Åsgard is located in a fishery area, and there is close attention to chemical usage and discharges to the sea.

Another reason to keep the methanol usage at a minimum is that it will follow the gas phase through the process plant. Thus a considerable fraction of the used methanol will be exported together with the rich gas to the onshore terminal, where it may contaminate the propane product. Contaminated propane will have a low price. In addition, extensive use of methanol and other chemicals represent a significant operational cost.

The above mentioned factors call for innovative ways of controlling hydrate formation, while keeping the chemical usage at a minimum level.

2.1 Requirements to the hydrate control

On Åsgard there are three principal demands to the hydrate control philosophy:

1. The well stream temperature shall be above the hydrate formation temperature when it arrives at the platform, also during tail production. This is one way to avoid continuous chemical hydrate inhibition, and usually it may be accomplished by thermal insulation of the flowlines.
2. During a production shutdown the flowlines shall maintain a temperature above hydrate formation conditions for at least 10 hours. The main reason for this criterion is the large number of flowlines connected to each platform. When, e.g. a process shutdown occurs it will take some time to restart all the wells, but in many cases it may be possible within the 10 hour limit.

3. The well stream temperature must not exceed 125°C in the flexible risers, even at full capacity production. The Smørbukk reservoirs are relatively hot, and flowing well head temperatures up to 140°C are expected. The risers are constructed of plastic materials that will degrade rapidly at high temperature. The flowline insulation must therefore ensure sufficient cooling from the well head to the riser base.

The last requirement above may be in conflict with the other two, because it puts an upper limit to the quantity of thermal insulation that may be applied on the flowlines. If the flowlines would only be operated with little variation in production rate, and the fluid properties were not changing with time, the three terms above could more easily be satisfied. However, the flow velocity, oil fraction, water cut and other important quantities will change significantly during the production life. This fact makes the selection of equipment and preparation of procedures more complicated.

2.2 Alternative methods

Four different hydrate control concepts have been evaluated for the Åsgard fields. One is continuous inhibitor injection, which has been chosen for the Midgard field. These 20 inch flowlines are about 40 and 50 km long. In practice this is too long for sufficient thermal insulation, and besides it will take valuable time to de-pressurise these pipes. If the Midgard flowlines where not inhibited, pressure relief would be necessary after a certain period of production shutdown, and there could be problems during the subsequent pressurisation and start-up.

An other traditional concept has been chosen for some production lines from Smørbukk and Smørbukk Sør to Åsgard A. Here no hydrate inhibitor is used during normal operation, but before a planned shutdown the lines may be inhibited with methanol. In case of an unprepared production stop, these flowlines must be de-pressurised within about 10 hours, because after that period the temperature may have decreased below hydrate formation temperature. Any movement in a super cooled system, i.e. with temperature below hydrate formations conditions, can cause lumping of the hydrates and possible plugging.

Two other hydrate control methods considered for Åsgard involve the ability to heat up the fluids present in the flowlines, and thus avoid the conditions where hydrates are stable. One of these concepts is applied for four of the production lines to Åsgard A. The flowlines are collected in bundles, and each of the two bundles hold two production lines, one gas injection line and pipes for heating medium. Thus hot fluids may be circulated when necessary, and the bundles act like giant heat exchangers. Details about this concept may be found in [1].

The other flowline heating alternative uses electric current to generate heat in the steel pipe. There exists electric heating methods that is based on electromagnetic induction and others on heating cables. However, for Åsgard a new concept has been developed: direct electrical heating. The idea of the method is to put electric current directly through the steel flowlines. The heat is basically generated because of the electric resistance in the steel. This hydrate prevention concept is applied for all six production flowlines from the Smørbukk field to the Åsgard B gas processing plant, see Figure 2. A brief technical description of the method follows in section 3 below. More details about the electrotechnical issues of the concept are found in [2].

2.3 Selection of method

Selection of the hydrate control concept has been based on technical feasibility of the different systems and life cycle cost. Evaluation of technical feasibility included investigation of the required topside and subsea equipment, operational procedures and ability to avoid hydrates.

As previously mentioned, there will be considerable amounts of produced water from the Smørbukk field. Continuos injection of hydrate inhibitor would require very large storage tanks on the SEMI, which will involve a safety hazard, and in addition the export gas would be contaminated. A combined concept where flowlines where inhibited by methanol prior to a shutdown, and de-pressurisation after an unexpected shutdown was also considered. The de-pressurisation operations are complex and time consuming, may require extra personnel and will affect the regularity of the plant. These operations will also involve an increased risk of human or equipment error, possibly leading to hydrate formation. The heated concepts; bundles or electrical heating, require operation of topside equipment and power generation to function. The power consumption will be relatively high, however, it will only be required when the processing plant is shut down or partly running.

Life cycle cost analyses for the different concepts have included investment cost, operational cost, process regularity, operational safety, environmental pollution, price penalties of contaminated sales products, CO_2 and NO_X taxes and the value of deferred or lost production. A heated concept will have a much higher investment cost than the inhibitor/de-pressurisation concepts, but has lower operational cost, is more environmental favourable and will contribute to a better regularity. For the six Smørbukk production lines routed to Åsgard B the difference in investment cost has been estimated to about 100 MNOK (about $17,000,000). Nevertheless, direct electric heating has been estimated to be the most cost effective for these flowlines during the life cycle of the process plant.

3 DESCRIPTION OF THE CONCEPT

3.1 Requirements and capabilities

The main requirements for the direct electric heating system for the six topical Åsgard flowlines may be summarised as follows:

1. Keep the temperature of the fluids inside the flowline above stable hydrate conditions during planned or unplanned production shutdown situations. This generally makes pressure relief unnecessary, and enables a significant reduction of hydrate inhibitor usage during subsequent start-up.
2. Heat up flowlines that have been shut down for longer periods, and consequently may contain hydrate crystals and lumps. If presence of hydrates can be expected, the flowline should not be restarted, because further lumping of the hydrates may cause plugging.
3. Obtain sufficient arrival temperature during very low production rates.

The heat needed to maintain a steady state fluid temperature above hydrate formation conditions in a flowline depends on several parameters. The most important are pipe diameter, the efficiency of the thermal insulation, sea water temperature and flow and the thermal properties of the well stream fluids. For Åsgard design conditions about 70 - 80 W/m of heat is lost to the surrounding water from the 10 inch flowlines, when the internal well stream temperature is about 30°C. This means that the electric heating system has to supply that amount of heat to maintain a safe temperature, typically 0.5 - 0.6 MW per flowline for relevant flowline lengths.

When the system is in heating mode, ref. item 2 in the list above, the required power supply is governed by the wanted heating time. A typical maximum accepted time to heat up the stagnant well stream from 6 to 30°C may be 24 hours. For Åsgard conditions this corresponds to an effective heat supply of about 105 W/m, or 0.7 - 0.8 MW per flowline. Almost 30% of the total heating time; 6½ hours, is in this case used to increase the temperature from 27 to 30°C. In addition to the required heat to the flowline, the power supply must be dimensioned to make up for losses in the system. This will be discussed later.

If flowlines have been shut down for very long periods, significant formation of hydrates may have occurred. The hydrates have not necessarily formed plugs, but the formation of hydrate structures means that large amounts of energy has to be supplied to melt the hydrates. Typical melting heat for a compact hydrate structure can be around 400 kJ/kg. In a calm system the hydrates will probably not grow to fill more than half the cross section in any part of the pipe. At hydrate equilibrium conditions the heat loss from the flowline to the sea water may be about 70 W/m. At these conditions the amount of hydrates necessary to fill half the cross section of a 10" flowline may be formed in 52 hours.

The process of melting such a volume of hydrates may take longer time, depending on available electrical heat, etc. Assuming a heat supply of 105 W/m from the electrical system, the net power transferred to the hydrates is about 35 W/m (105÷70). In other words, the net heat supply rate during melting is in this case half of what is lost during formation, and melting will therefore take twice as long: more than 4 days. It is therefore important that the field operation procedures are prepared to avoid situations where hydrates are allowed to form in the flowlines. There are also safety aspects related to melting of hydrate plugs, which should be considered carefully. For instance one should avoid applying heat to melt a plug when there may be a high pressure drop across it. Local areas of high pressure due to the large volumes of gas released during the melting process must also be evaluated.

3.2 Main equipment set-up and efficiency

A schematic drawing of the direct electric heating system is shown in Figure 4. At each end of the steel pipe a power cable is connected. The pipe is an active conductor in a single phase electric circuit, and the power is supplied from the platform via two feeder cables. The heat development and the efficiency of the system depend on the current in the pipe, dimensions and magnetic and electrical characteristics.

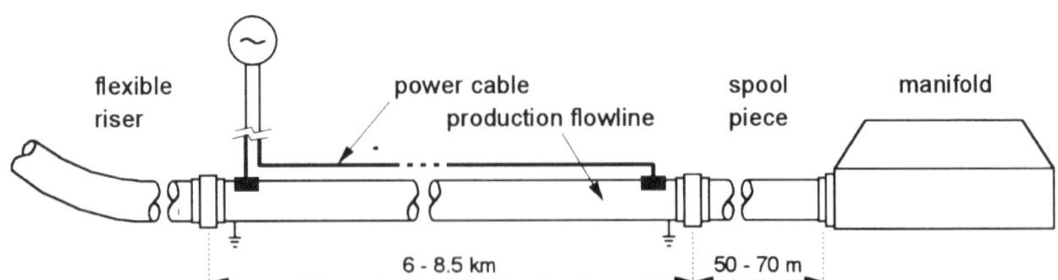

Figure 4 Overview of the production and heating system

The target heat requirement is obtained through the resistive power loss (W/m), due to the pipe current and the resistance. The power loss depends on the resistance and magnetic properties of the steel and the distance between pipe surface and the power cable that is parallel to the pipe. This distance is important, because it influences both the magnitude of the pipe current and the distribution of current in the steel walls. The electric resistance has a maximum value when the power cable is attached with as little distance as possible to the pipe.

Test on an 8" carbon steel pipe have shown that increasing the distance between pipe and cable surface from 5 to 50 cm results in a 30% lower heat development in the pipe. Based on these results it has been decided to strap the power cable to the heated flowlines during installation.

Heat generation is only wanted in the flowline materials, and not in other parts of the system. There will, however, be some resistive loss in the power cables and the feeder cables, which may be reduced to some extend by increasing the cable conductor diameter. In addition less than 10% of the supplied power will be lost in the surrounding sea water. The total loss in cable and sea water has been computed to about 30% of the supplied energy, meaning an overall efficiency of about 70%.

Given the power losses and the required heat in the flowline at different operating modes, calculated above, typical values of required power from the platform may be estimated. To maintain the fluid temperature above hydrate conditions in one 8 km Åsgard flowline about 0.9 MW must be fed to the system. To heat up the content of one such flowline from sea water temperature to 30°C in 24 hours, neglecting hydrate melting, about 1.2 MW has to be delivered from the topside electrical equipment. It should be emphasised that the calculations above are based on a rough data basis, in addition to being simplified for convenience.

3.3 Electric current distribution

If the pipe and connections could be electrically isolated from the sea water, all the current from the supply cables would be traced in the steel pipe. However, it is not regarded a good solution to design the system based on electrical isolation, because a fault then could have severe consequences for the safety and reliability of the system. Electrical faults may arise due to general ageing or accidental damage of the isolation materials. For these reasons the heating system is intentionally connected (grounded) to the sea water at both ends of the flowline.

The connection to the sea water is obtained with sacrificial anodes fitted to both ends of the flowline. These are placed in the transfer zones, less than 100 m away from the power cable connection points. The consequence of this connection is that the sea water acts as an electric conductor in parallel with the flowline. The total system current is thus divided between the pipe and the sea water. As mentioned above, this causes a reduction of the overall system efficiency. The share of the total current that flows into the sea from the grounding electrodes at the connection points is mainly governed by the electromagnetic fields along the pipe and not the conditions locally near the connections.

At the transfer zones electrical current flows in a radial direction between the pipe and the sea water, but apart from these areas the current in the sea water is parallel to the pipe. The current density in the water is highest close to pipe and, according to the electromagnetic penetration law, decreases with the radial distance. When a parallel (non-heated) pipe is located close to the electrically heated pipe, a part of the current will be transferred to this pipe. The current distribution between the pipes is dependent on the distance between them. The effect on the current in the heated pipe is negligible for distances above 10 m.

Electric currents from the heating system may float into the subsea manifold template. To minimise these stray currents it may be necessary to terminate the heating at some distance from the template. It is confirmed by tests that these currents are reduced by iron cores clamped around the pipe, which work as electrical impedance units.

3.4 Magnetic fields

The direct electric heating system will generate a magnetic field that may influence on the operation of electronic equipment. Measurements on a test installation indicate maximum magnetic flux densities of approximately 50 μT, which is of the same magnitude as the international industrial standard exposure for electric equipment. Equipment that will be used close to the heating system should be tested, to ensure that it is not destroyed or malfunctioning because of the magnetic field. An example of such equipment is remotely operated vehicles for valve operations and other purposes.

4 ACKNOWLEDGEMENTS

This work has been supported by Statoil and Saga through the Åsgard Development project. We would like to thank Sintef Energy Research (former Norwegian Electric Power Research Institute) for considerable contributions and support regarding electrotechnical issues.

5 REFERENCES

[1] A. Hærdig, R. Moe and G. Sæther: Dynamic simulations of multiphase flow in flowline bundles on Åsgard, © BHR Group 1997 *Multiphase '97*

[2] H. Kulbotten et. al.: Introduction to Direct Electric Heating of Subsea Pipelines, Sintef Energy Research TR F4630, 1998-01-22, ISBN 82-594-1173-3

Prediction of sand erosion in process and pipe components

A HUSER and **O KVERNVOLD**
Det Norske Veritas, Høvik, Norway

1 ABSTRACT

Prediction of sand erosion in pipe systems is outlined presenting two procedures, one for general complex geometries, and one for standard pipe components.

Firstly, a general procedure to calculate sand erosion in three-dimensional flow geometries is outlined. This procedure is designed for the detailed assessment of sand erosion rates in complex process components such as production choke valves, manifolds, diverter systems, intruding sand probes, multiphase sensors, etc. The procedure is based on the Computational Fluid Dynamics (CFD) programme CFX4. DNV has, however, developed specific models to calculate erosion based on general flow and particle calculations performed within CFX4. Validation examples show good comparison with experiments and an application example demonstrate that the procedure is applicable to visualise the erosion mechanisms, to optimise inspection and maintenance routines as well as provide a valuable design tool.

Based on experimental findings and the above procedure for complex geometries, DNV has also developed a DNV Recommended Practice (RP O501) which can be applied to calculate sand erosion in typical pipe components: Pipe bends, blinded Tee bends, straight pipes, welds and reducers. The new RP provides valuable tools for dimensioning of piping systems and components, optimisation of production, and inspection and maintenance planning. The oil companies Statoil, Norsk Hydro, Saga Petroleum, Conoco, and partly Amoco have supported the development of the new RP, and the RP is now applied in the Norwagian sector in the North Sea,

2 INTRODUCTION

Sand erosion has become an important concern both for design of new field developments and prolongation of existing oil and gas fields. High cost of retrieving process components makes it attractive to design against sand erosion, especially for sub-sea installations. For many existing fields, erosion potential increases toward the end of the well life due to decreasing pressure, increasing GOR, and increasing sand production. For existing fields, inspection and maintenance intervals and methods can be obtained by detailed assessment of the erosion rates of the different components in system. In this assessment, the whole production line has to be considered. The highest velocity in the production line is usually the pressure reduction choke valve, hence the design and material selection of this is crucial /3/. The second most erosion prone component may be a bend, a manifold or a flowmeter. If the choke valve is manufactured in an erosion resistant material, the weakest point may also be the 'second point'.

Design to prevent sand erosion is often done by ad-hoc methods that are independent of the sand production rate. One such method is the API RP 14E /2/, where an "erosional velocity" is calculated. This method may give conservative designs when little sand is produced, whereas non-conservative designs when sand is present. By the present approach, which is based on extensive experiments and testing /3/4/5/6/, design tools are funded on an understanding of the important factors that influence erosion.

3 PREDICTION OF EROSION IN COMPLEX FLOW GEOMETRIES

The method applied to predict erosion rates in complex flow components, such as choke valves, multi-phase flowmeters, manifolds, etc. is here outlined. A general method developed to perform erosion calculations in arbitrary geometries is developed based on flow and particle track calculations performed with a standard Computational Fluid Dynamics (CFD) package. In the present work the CFD programme CFX4 is applied. The prediction method for sand erosion applies the results from flow solution and a particle track calculation to calculate the erosion rate on the internal surfaces of the geometry. Specific models have been developed to calculate erosion rates based on the results from CFD simulations.

3.1 Method of analysis

The sand erosion analysis is performed in four steps; grid generation, flow solution, particle track calculation and erosion rate calculation. The first three steps are performed with CFX4, whereas the models developed by DNV perform the fourth step. The approach applied in the first three steps and the DNV erosion model is outlined below.

The CFD programme CFX4 contains most of the features that are of importance for erosion problems. The grid system is suited to model complex streamlined, or irregular flow domains. With a multi-block grid system, optimised grid may also be created, making computational times friendly. However, care must be taken when creating the grid. The standard k-ε method is applied for the turbulence modelling, and a converged turbulence field must be achieved in order to predict the correct particle movement. Only a grid that has sufficient resolution, orthogonality near the walls, and a reasonable grid expansion ensures a converged turbulence field. If non-orthogonal grids are applied near the walls, a solution may be obtained, but this may give unreasonable values for k and ε which may cause particles to hit the wall with unreasonable speed or particles to be attracted to the walls. This may cause too high erosion rates at these locations.

During the flow calculations a steady state one-phase flowfield is produced. The flow may be either incompressible, or compressible. By assuming mixture quantities for the flow parameters (such as velocity, density, viscosity, etc.), multi-phase flows are approximated.

When the flow- and turbulence- fields are converged, the particle tracks are solved on the steady state flowfield. Up to 10 000 particles are released at arbitrary locations at the domain inlet, where the particle tracking routine within CFX4 is applied. The invoked forces acting on the particles are drag force, gravity force and pressure gradient force. Further, a correction term to the particle velocity is included dependent of the turbulence level. Particles are assumed to be spherical, and the particle diameter is specified together with the sand density to give the particle size and weight. Constant particle diameter or a Gaussian distribution of the size is implemented. The restitution coefficient is *one* for an ideal reflection, and denotes the reduction factor of the wall-parallel velocity after the hit:

$$u_{\|out} = E u_{\|in} \tag{1}$$

Here $u_{\|out}$ is the wall-parallel velocity after the hit, $u_{\|in}$ is the wall parallel velocity before the hit, and E is the restitution coefficient. The particle velocity component normal to the wall keeps its value and only changes sign after the hit. Typically a restitution coefficient of E = 0.8 is applied. Equation (1) is the default formulation applied din CFX4. For applications to erosion it is assessed that the angle of attack, particle shape and size, material type as well as the particle velocity will influence the particle exit angle and velocity. However, this is an area of further research.

Erosion calculations are performed with a general method based on the situation shown below, where u is the hit velocity and α is the hit angle /1/:

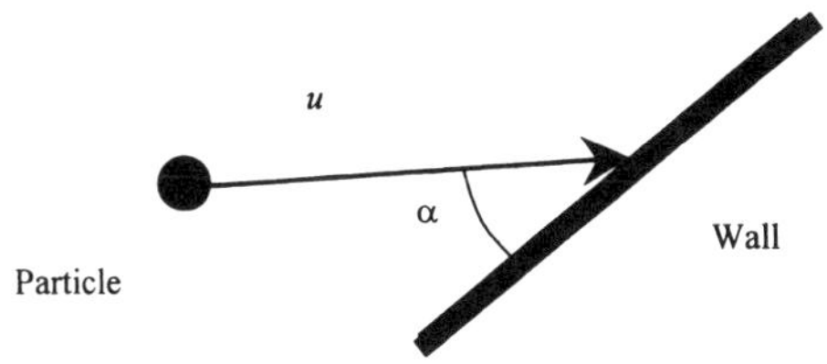

The general equation for the erosion rate is written as follow /7/8/9/:

$$E_L = C_{unit} \frac{K u^n \dot{m}_p F(\alpha)}{\rho_w A} \tag{2}$$

Here E_L is the erosion rate in mm/year, $C_{unit} = 3.15 \cdot 10^{10}$ is a converting factor from m/s to mm/year, K is a material constant, $\dot{m}_p$ (kg/s) is the massflow of sand that hit the area, A (m²) is the size of the area exposed to erosion, ρ_w (kg/m³) is the wall material density, n is the velocity exponent which is dependent of the material, and $F(\alpha)$ is a number between 0 and 1 given by a functional relationship dependant of the material.

The erosion rate in some given small sub-area is found by the summation over all particles that hit within the defined area:

$$R = \frac{10^6 \cdot K}{A \rho_w M} \sum_{i=1}^{nhit} u_i^n F(\alpha_i), \tag{3}$$

where R (μm/kg total sand feed) is the erosion rate in the sub area, M is the total number of particles (total sand feed), and *nhit* is the number of hits in the sub-area. The material dependant parameters have been determined from extensive laboratory tests of a number of different materials given in /3/. Parameters for two frequently used materials are listed in Table 1. The functional relationship, $F(\alpha)$, for two materials is given in Figure 1. It should be noted that the erosion resistance of different materials is varying, and care must be taken when selecting a good erosion resistant material. Parameters applied for tungsten carbid (WC) in Table 1 yield a good WC material with 6% cobalt (Type Sandvik DC-05).

Table 1 Recommended values for material constants for two typical materials to be applied in Equation (3). From ref. /3/.

Material	K	n	ρ_w
Steel grade	$2.0 \cdot 10^{-9}$	2.6	7 800 kg/m³
Tungsten carbide (WC)	$1.1 \cdot 10^{-10}$	2.3	15 250 kg/m³

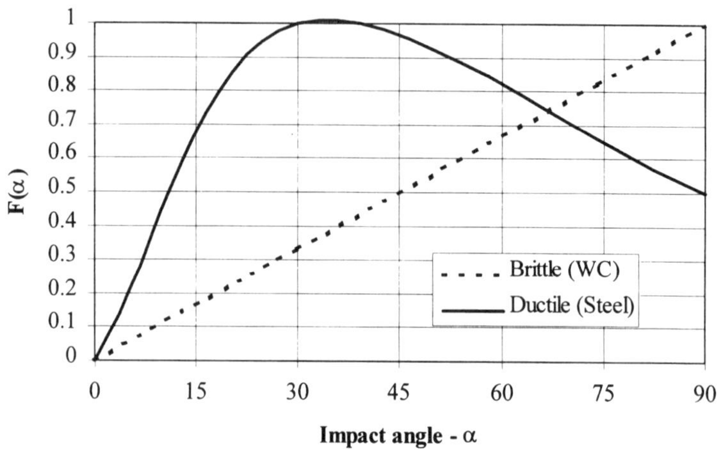

Figure 1 Function *F*(α) for typical 'ductile' and 'brittle' materials /1/.

3.2 Validation cases

Two validation cases are presented indicating that the erosion rate may be predicted for both single-phase and multi-phase fluid flow.

3.2.1 Bean choke

In a bean choke, flow is going through a contraction as sketched to the right. The experimental investigations /4/ were reproduced for carbon-dioxide gas at subsonic speed. The following flow and particle parameters are applied:

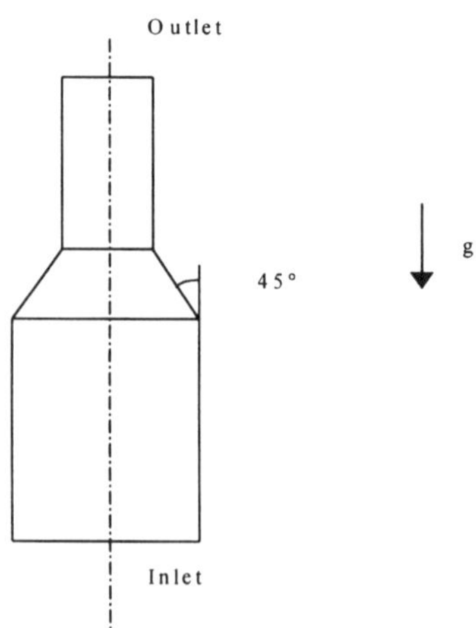

- Inlet velocity: 11.4 m/s
- Inlet pressure: 14.1 bar
- Inlet temperature: 36 °C
- Fluid viscosity: 1.5 10^{-5} kg/(ms)
- Molecular weight: 44
- Inlet ID: 54 mm
- Outlet ID: 20 mm
- Particle diameter: 0.25 mm

Most sand particles hit the 45° contraction and bounces off and hit the second time inside the smaller outlet pipe on the opposite side as the first hit. The maximum erosion rate on the outlet pipe is obtained a small distance from the contraction. In comparison with the experiment, a good agreement on the level of erosion is obtained (Figure 2). The restitution coefficient is $E = 0.8$ is applied this case. The restitution coefficient does influence the results and may give an explanation of why the location of the maximum point is slightly off.

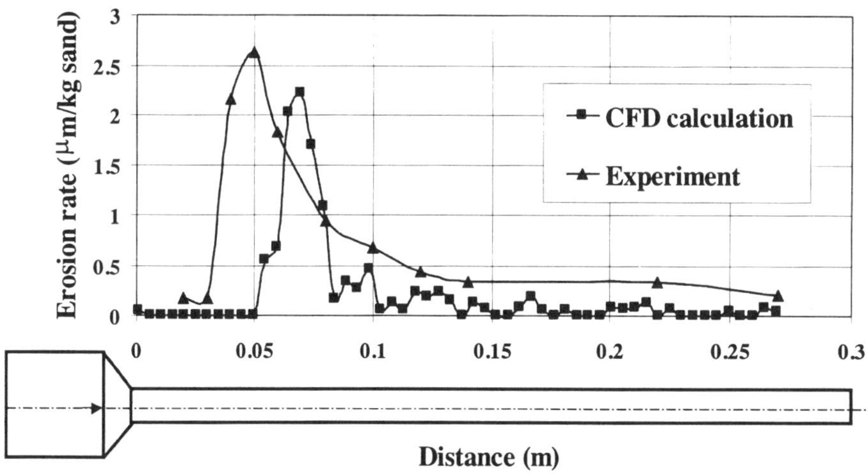

Figure 2 Erosion rate along bean. Comparison with experimental results /4/.

3.2.2 Pipe bend

A 5D-pipe bend with a vertical inlet and horizontal outlet, as shown below in Figure 3, is modelled and compared against experiments. The multi-phase parameters are converted to mixture parameters, which are applied in the model. Assuming incompressible flow, following parameters apply:

- Inlet mixture velocity: 36.3 m/s
- Gas-to-liquid ratio (GLR): 14.5 m^3/m^3
- Mixture viscosity: $1.8 \cdot 10^{-5}$ kg/(ms)
- Mixture density: 72.3 kg/m^3
- Pipe inner diameter, D: 26.5 mm
- Radius of curvature: 5·D
- Particle size: 0.25 mm

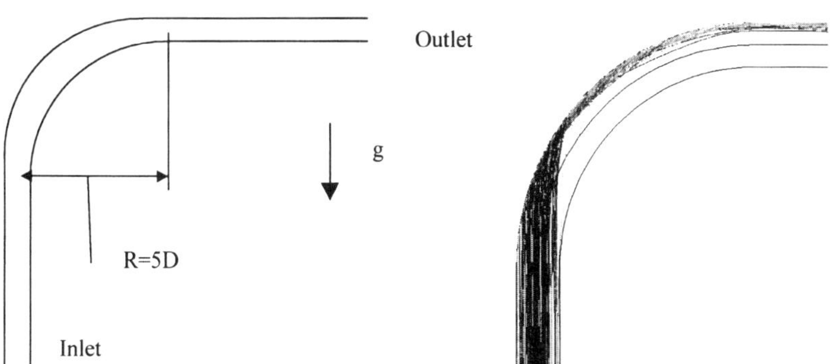

Figure 3 Sketch of flow model (left) and particle tracks from simulations (right).

Excellent agreement is obtained when comparing the CFD results with the experiment (Figure 4), despite the approximation to single phase flow. The DNV RP for calculation of erosion in pipe bends /1/ gives a maximum erosion of 7.3 μm/kg sand in this case. This confirms that the RP is on the conservative side (see paragraph 4 Prediction of erosion in typical pipe components). The DNV developed programme ERBEND is also seen to predict the erosion rate well. This programme employs a 2D model of particle paths to calculate erosion in pipe bends.

The effects of multi-phase flow on particle distribution and erosion patterns are not well understood. The present results indicate that the erosion pattern may be distorted by the multiphase flow, explaining some of the differences between the experiment and the CFD results in Figure 4. However, this should be an area of further research.

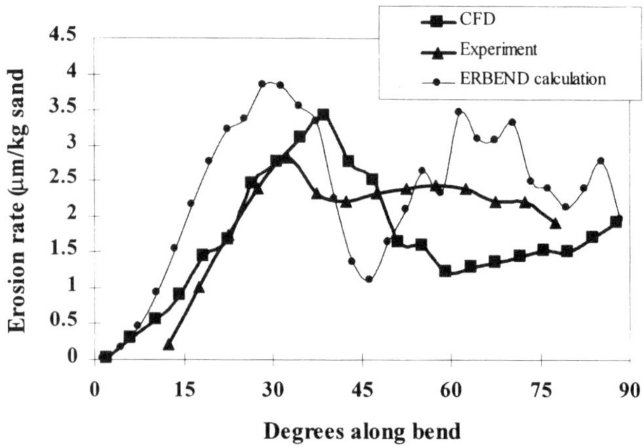

Figure 4 Erosion rate along outer side of bend. Comparison with experiment /5/, and 2D model DNV*ERBEND.

3.3 Application to process components

The production choke valve gives an example of application of the erosion routines to complex geometries. The Multiple Orifice Valve (MOV) in Figure 5 is modelled as shown in Figure 6 containing only the flow domain with a high-pressure part (dark) and a low-pressure part (light). The complex flow domain is shown by the solution in Figure 7 with the inlet pipe, the valve house with the internal turning fork, the two bean holes where the flow is accelerated and the low-pressure outlet spool. Turning the two orifice holes, so that they are non-aligned, regulates the flow. The modelled choke presented here is partly open with the bean holes non-aligned.

Particle tracks are shown in Figure 8 in order to get an understanding of the erosion mechanisms. At small opening angles, the particles are directed toward the sidewalls of the outlet spool with a high velocity. This causes the erosion rate to reach a maximum at two locations caused by the two bean holes as illustrated by the two maximum points which are shifted 180° in Figure 9.

A series of operating ranges have been modelled by changing the opening angle, pressure drop, particle size, and outlet geometry. In Figure 10 are shown results from four cases where the opening angle (or opening percent) is varied. For the largest opening, a reduced erosion rate is obtained. For all cases the erosion rate for both steel walls and WC walls is determined. From each case, the contours of constant erosion on the internal surfaces are found and the point of maximum erosion rate determines the lifetime of the choke. A consolidation of all simulation cases is applied to develop correlation equations that are applied to predict the maximum erosion rates for the relevant operating conditions. Operators can apply this to determine service and inspection intervals as well as optimal operations on fields with a high sand production. The erosion rate is proportional to the sand production rate so that monitoring of the sand production is required to be able to follow the erosion development. On-line sand monitoring is recommended when the sand production is high in order to be able to spot sudden changes in the sand production rate.

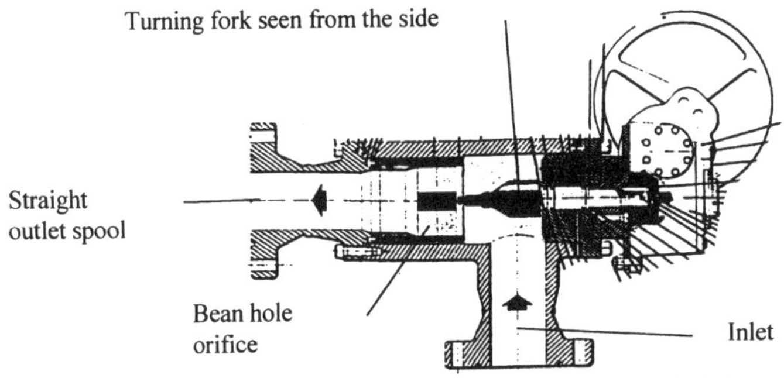

Figure 5 Typical Multiple Orifice Valve (MOV) with two bean holes shown in fully open position.

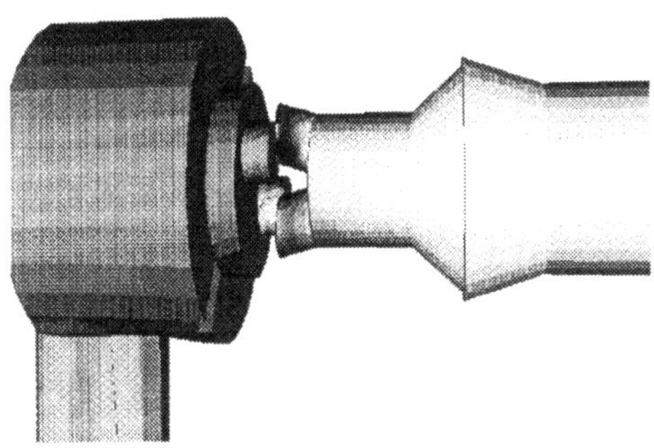

Figure 6 Geometry model showing the flow domain of the Multiple Orifice Valve (MOV) choke applied in the CFD model. This valve is partly open with the bean holes non-aligned.

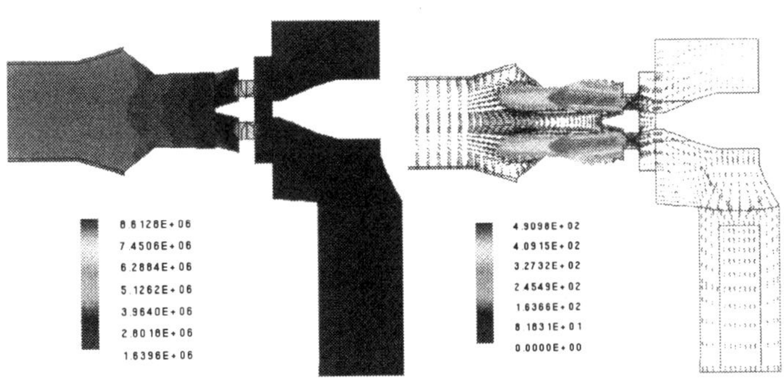

Figure 7 Pressure contours, Pa (left) and velocity vectors, m/s (right) in symmetry plane of MOV choke at partly open valve.

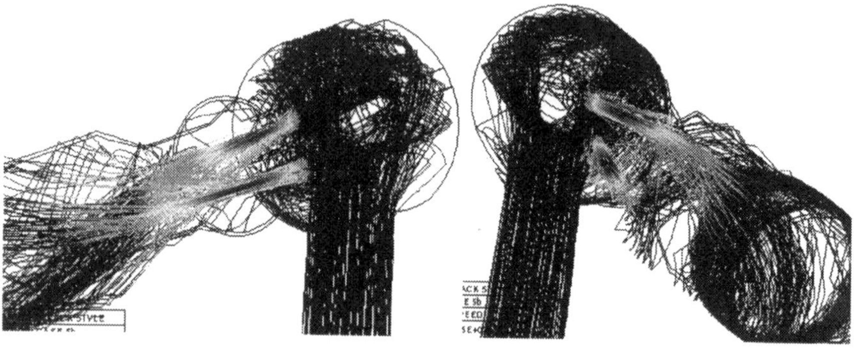

Figure 8 Particle tracks in 3D model of MOV choke seen from two angles. Tracks are shaded by the particle speed showing the slow moving particles in the inlet and the house (dark), and the accelerated light particles in the contraction.

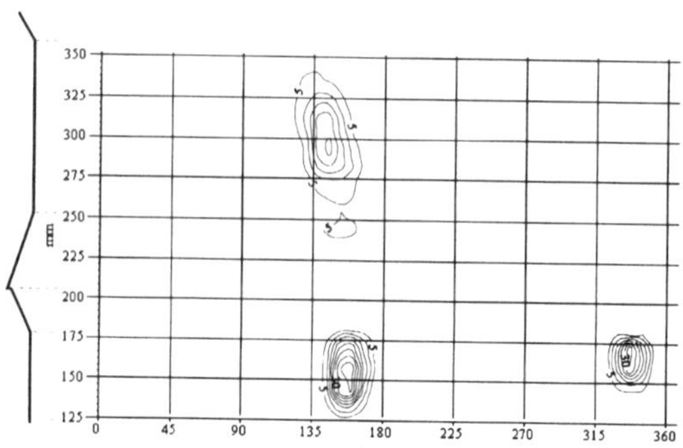

Figure 9 Iso-contours of constant erosion rate on the folded out outlet spool wall. The horizontal axis is the angle around the spool, and the vertical axis is the flow direction. To the left is sketched the profile of the outlet spool.

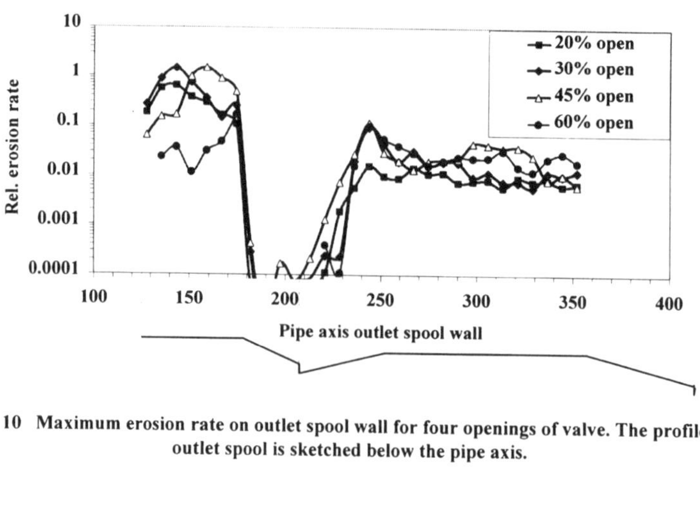

Figure 10 Maximum erosion rate on outlet spool wall for four openings of valve. The profile of the outlet spool is sketched below the pipe axis.

4 PREDICTION OF EROSION IN TYPICAL PIPE COMPONENTS

Procedures for calculation of erosion rates in the pipe components: pipe bends, blinded Tee bends, straight pipes, weld joints and reducers have been developed based on extensive experiments and CFD modelling cases. Results are consolidated in correlation equations for each component type and can be applied to predict the maximum erosion rates for relevant operating conditions. Correlations are of the form given in Eq. (3) where component dependant coefficients and calculation procedures are incorporated to account for the various geometries. Due to spread in the underlying experimental and numerical results, the correlations are made on the conservative side. The complete set of correlation procedures are outlined in a DNV Recommended Practice /1/ which is commercially available from DNV. Also a hand-on PC computer programme is available. This can be applied for e.g. dimensioning of piping systems and to optimise production in an existing system.

Below is given an example demonstrating the use of the RP procedure for estimation of the pipe dimensions, D, for the following production conditions:

- Limiting component in pipe: 90° pipe bend
- Radius of curvature of bend: 2·D
- Production rate: $1.8 \cdot 10^6$ Sm^3/day gas = 20 kg/s
- Well head conditions: 120 bar and 85°C
- Sand production: 0.5 g/s = 32 kg/day
- Particle diameter: 0.35 mm
- Sand particle density: 2600 kg/m^3
- Maximum erosion rate: 0.1 mm/year

The physical properties for the HC mixture at the well head conditions will have to be determined from process simulations by appropriate programs; i.e. PROCESS, MAXISIM, HYSYS etc. Based on the physical properties given in Table 2, the mixture properties are obtained for relevant pipe diameters in Table 3. Based on the parameters in Table 3, the procedure for erosion in a pipe bend can be applied for dimensioning of the pipe system. In the calculations a 90° bend with bend radius equal to 2D is used for all cases.

Based on the RP results in Table 3, a pipe with dimension 6" (D=0.15 m) should be selected for the given case. However, if sand content becomes lower, the production rate can be increased, see Table 4.

Table 2 Physical properties obtained from process simulation programme.

Gas mass flow (kg/s)	Liquid mass flow (kg/s)	Gas density, ρ_g (kg/m^3)	Liquid density, ρ_l (kg/m^3)	Gas viscosity, μ_g (kg/ms)	Liquid viscosity, μ_l (kg/ms)
15	5	100	700	$1.5\ 10^{-5}$	$0.6\ 10^{-3}$

Table 3 Mixture properties and resulting erosion rate for relevant pipe diameters

Diameter (m)	Superficial liquid velocity, V_l^s (m/s)	Superficial gas velocity, V_g^s (m/s)	Mix. velocity, V_m (m/s)	Mix density, ρ_m (kg/m^3)	Mix. viscosity, μ_m (kg/ms)	Erosion rate from RP (mm/year)
0.125	0.6	12.2	12.8	127	$4.5\ 10^{-5}$	0.5
0.15	0.4	8.5	8.9	127	$4.5\ 10^{-5}$	0.1
0.175	0.3	6.3	6.6	127	$4.5\ 10^{-5}$	0.03

Table 4 Recommended maximum production rates for different sand production rates for a 6" pipe with 2D pipe bends in order not to exceed a maximum erosion rate of 0.1 mm/year. The same pressure, temperature, HC composition as in the above analysis is assumed.

Sand feed		Recommended maximum production rate		V_l^s	V_g^s
(g/s)	(kg/d)	(kg/s)	(MSm^3/d)	(m/s)	(m/s)
0.01	0.9	81	7.3	1.6	34
0.05	4.3	44.7	4.0	0.9	18.9
0.1	8.6	34.6	3.1	0.7	14.6
0.5	43.2	20	1.8	0.4	8.1

5 CONCLUSION

Prediction of sand erosion in pipe systems is outlined presenting two procedures, one for general complex geometries, and one for standard pipe components.

A procedure for general three dimensional flow geometries is designed for the detailed assessment of sand erosion in complex process components such as production choke valves, manifolds, blinded Tee bends, intruding sand probes, multi-phase meters etc. The calculation procedure employs the Computational Fluid Dynamics (CFD) programme CFX as a framework. The application of a general CFD programme gives the opportunity to model flow in general geometries. The flowfield and sand particle tracks from the CFD programme are applied to calculate the erosion rates on all internal surfaces. A correlation equation developed by DNV, based on material testing of several materials, is applied when calculating the erosion rate as a function of the particle hit velocity and angle, and material grade. Validation examples show good comparison with experiments and an application example demonstrate that the procedure is applicable to visualise the erosion mechanisms, as well as providing a valuable design tool.

By the aid of experimental findings, and the procedure for complex geometries, is developed a DNV Recommended Practice (RP O501) which can be applied to calculate sand erosion in typical process components: Pipe bends, blinded Tee bends, Straight pipes, welds and reducers.

The present procedures provides a valuable tool for dimensioning of piping systems and components, optimisation of production, inspection and maintenance planning, as well as design of sand monitoring systems and choke valves.

6 REFERENCES

/1/ Det Norske Veritas Recommended Practice "Erosive Wear in Piping Systems" DNV RP O501, 1996.

/2/ American Petroleum Institute "API Recommended Practice for the Design and Installation of Offshore Production Platform Piping Systems" API RP 14E section 2.5b 1991.

/3/ K. Haugen, O. Kvernvold, A. Ronold and R. Sandberg "Sand erosion of wear resistant materials: Erosion in choke valves" Wear 186-187 (1995) 179-188

/4/ P.Jensen, O.Kvernvold & R.Sandberg 1991 "An experimental investigation of the erosion characteristic in bean chokes" VERITEC Report no.: 91-3059

/5/ O.Kvernvold & R. Sandberg 1993 "CRDN 617: Production rate limits in two-phase flow systems: Erosion in piping system for production of oil and gas" DNV Report no.: 93-3252.

/6/ A.K. Cuson & I.M.Hutchings "Influence on erodent particle shape on the erosion on mild steel" Proc. of the 6th Int. Conf. On Erosion by Liq. and Sol. Impact, Cavenish 1983.

/7/ G.P. Tilly "Erosion by impact of solid particles" Treatise on Material Science and Technology, Academic Press, New York, 1979.

/8/ E. Rask "Tube erosion by ash impaction" Wear 13, 301-315, 1969.

/9/ W.F. Adler "Assessment of the state of knowledge pertaining to solid particle erosion" Final Rep. CR79-680, to the U.S. Army Research Office for contract DAAG29-77-C-0039, 1979.

'If you don't mix it, you don't have to separate it'

H McKINZIE, G GRAVES, L STUEBINGER, and **G ELPHINGSTONE**
Texaco Inc, Texas, USA

SYNOPSIS

Traditional water management methods have ranged from "just live with it", to trying "magic bullet" chemical water shut-off treatments, or even trying to develop a market to sell the water. The new view is to manage water in the wellbore and/or in the reservoir. Since nature has spent millions of years separating the water and oil in the reservoir, it seems ironic that only a few seconds in the pump and tubing are required to remix the oil and water – possible into tight emulsions. We then spend millions of dollars on chemicals and equipment, especially on offshore platforms, to re-separate the oil and water. A far better solution is to prevent the mixing by modifying the method of artificial lift. Texaco has now applied for patents on several methods of artificial lift that reduce the amount of water lifted to surface by using the wellbore as a separator, and handling the oil and water separately. These methods also are adaptable to reverse coning applications, giving even further advantages over conventional pumping. The ultimate goal is to dispose the water downhole, rather than lifting it to the surface. The first few methods, such as the Dual Action Pumping System (DAPS), have been field tested, with exciting results. Several of the other newer pumping combinations will also be described is this presentation.

1. INTRODUCTION

The traditional philosophy in the oil industry is to apply any type of artificial lift in oil wells that will maximize oil production, without regard to emulsion formation, (and with very little regard for water coning). Thus, what Mother Nature has spent millions of years separating in the reservoir, we manage to thoroughly mix within seconds in the pump and in the wellbore, forming high quality emulsions, in the extreme cases. We then spend millions of dollars building equipment and purchasing chemicals to separate the oil and water, and even more to clean and dispose of the water. It is time to rethink and revise this philosophy.

Engineers long ago realized that it would be advantageous to leave the water in the reservoir. There is a long history of mechanical and chemical treatments that have been used to try to plug off water producing perforations, ranging from sand, cement or mechanical plugs in the wellbore, to polymers, cements and resins placed outside the perforations. The latter can be near wellbore treatments, or treatments deep within the reservoir. The long-term success rate ranges from dismal to excellent, depending on the reservoir parameters and the water production mechanism. The overall average of these treatments would probably rate only as "fair", but new ideas and technologies continue to advance treatment possibilities.

The most recent view is to manage water either within the wellbore or in the pump, and then dispose of the water into a disposal zone in the well, so that it doesn't have to be lifted to surface. The ultimate goal is to not only leave the water in the ground, but if possible, to use it for pressure maintenance in the reservoir. Many new technologies are being developed to accomplish this goal.

1.1 The Centre for Engineering Research Downhole Oil Water Separator

The Centre for Engineering Research (C-FER), in Edmonton, Alberta, Canada, has developed a method of connecting either one or two motors with two ESP pumps, along with one or more hydrocyclones to form a downhole oil/water separation system (1, 2). A conceptual drawing of one configuration of this system is shown in Figure 1. The water and oil mixture enters the lower pump where it is pumped into the hydrocyclone. The oil and water are separated, and the water is injected directly from the hydrocyclone down into the disposal zone. The separated oil and a small amount water pass through a bypass tube up to the upper pump, where they are pumped together to the surface.

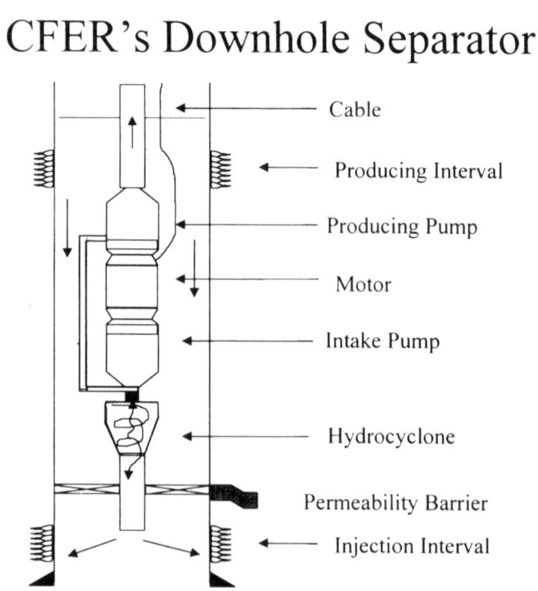

Figure 1. The Centre for Engineering Research Downhole Oil/Water Separation Pump

Other configurations of this technology, using either one or two pumps, have been developed. The ESP, PCP, and beam pump configurations have already been tested. These systems have been installed in more than 35 wells worldwide, with some units in operation more than two years. This technology is now being actively marketed by service companies.

There are currently at least two consortia that are working to push this technology even further. Both are targeting pumps that would be capable of handling up to 3,200 cubic meters (20,000 Bbls) of fluid per day.

1.2 Using the wellbore as a separator

The latest technology to be developed is the concept of using the wellbore as a separator, and reducing the amount of water pumped to the surface (3). Texaco has now submitted more than 18 patent applications concerning artificial lift utilizing these technologies, with the current goal of making these pumps available to the industry as quickly as possible. The technology employed in these devices is the topic for this paper.

As with many discoveries, the current work on using the wellbore as the separator began serendipitously when trying to solve a different problem. In 1994, Texaco's research and operations engineers were using downhole cameras to try to determine where water was entering the wellbore, to allow for more accurate placement of water shutoff treatments. If the source of the water influx could be located, a water shut-off treatment could be placed across this interval. The idea was that if the treatment was placed more accurately, the treatment success rate could be improved, and water production would be greatly reduced (at least until the water again coned into the oil-producing interval).

After looking at the downhole videos from several wells, it was noted that when the camera was below the oil/water interface, the water was virtually clear in almost every well. In every case, the oil was seen to enter the wellbore in fairly large drops, which rapidly rose to the oil/water interface. This was true even for a well that was producing a 95% oil cut, and for wells with fairly high production rates. Additional droplet coalescence was occurring as the oil was rising. In every well, there was virtually no tendency to form an emulsion, and the water and oil remained well separated. Below the perforations, the water remained very clear, and showed no evidence of oil droplets. These observations lead to the ideas which eventually resulted in the development of the Dual Action Pumping System (DAPS) and the subsequent Dual Injection and Lifting Systems (DIALS) technologies.

2. DISCUSSION

2.1 Dual Action Pumping System (DAPS)

The basis for the Dual Action Pumping System, and also for the subsequent Dual Injection and Lifting Systems is that the wellbore is used as the separator. The oil and water are pumped independently using two separate inlets, and handled by two separate pumps connected on the same rod string. Figures 2A and 2B show a schematic of the DAPS system on the upstroke and downstroke. On the upstroke, oil (and some water) is forced upward through the travelling valve in the upper pump, and water is drawn into the lower pump chamber via the lower inlet valve. On the downstroke, oil and some water are drawn into the upper pump through the top inlet, and water is forced downward through the lower valve assembly into an injection zone that is separated from the production perforations by a packer. Oil, gas and water production rates must be known to choose the relative pump sizes.

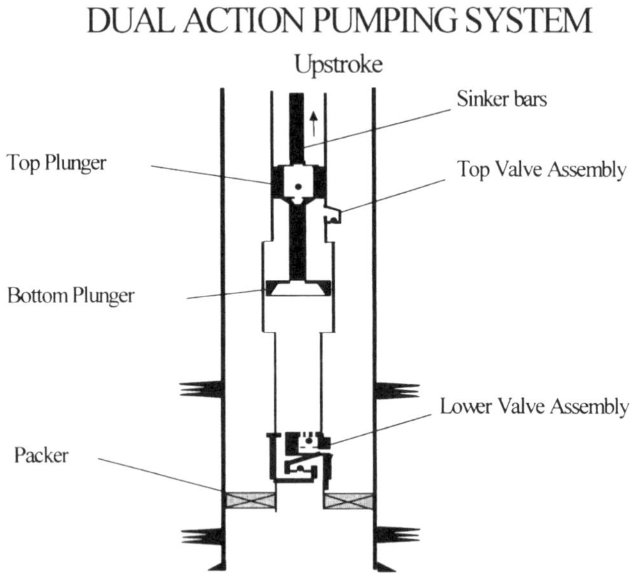

Figure 2A. The DAPS pump on the upstroke, lift oil to surface.

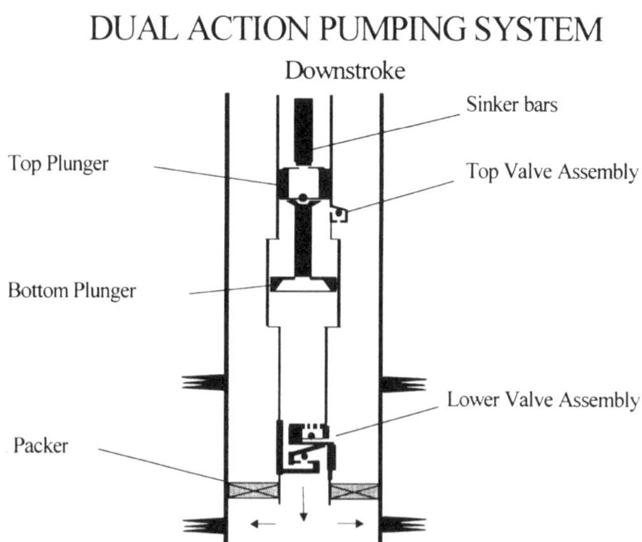

Figure 2B. The DAPS pump on the downstroke, injecting water to the lower disposal zone.

This concept is now thoroughly proven, with 17 pumps using the DAPS configuration installed in various fields throughout the world. Many more are currently planned with interest expressed by more than twenty companies in testing these pumps.

The energy required to run these pumps is often lower than that for standard rod pumps since significantly less water is lifted to the surface. Typically, the energy requirement is about 30% less than that required for the pump which was replaced. The energy requirements to inject the water are normally significantly less than the energy required to lift the water to the surface, unless high injection pressures are required. The above figure does not take into account the energy (and costs) to clean and re-inject the water, so the actual energy and dollar savings are somewhat greater.

Another advantage to this pump is that the pump jack can run at a slower speed, since pumping occurs on the upstroke and downstroke. This will extend rod and tubing lifetimes where wear is a significant problem.

2.2. Dual Injection and Lifting Systems (DIALS)

Virtually all other artificial lift pump combinations were investigated to determine if they would be able to perform the same function as the DAPS system. The concept of keeping the fluids separated in the wellbore is the same, but many problems had to be solved to make the other combinations work. Brief descriptions of several of the DIALS systems are described in the following sections. Most of these have not yet been built and tested. Table 1 provides a tabulation of the technologies which have been conceived by Texaco as a consequence of the DAPS successes.

Table 1. Pump Combinations considered for use as DIALS systems

Fluid Lifted	****************Fluid Injected by****************				
by:	Rod Pump	ESP	PCP/ESPCP	Dumpflood	Other pump methods
Rod Pump	P	P	P	N	N
ESP	M	P	P	N	N
PCP/ESPCP	M	P	P	N	N
Gas Lift	P	P	P	N	N
Natural Flow	N	N	N	M	N
Hydraulic	M	N	N	N	N
Wier Pump	M	N	N	N	N
Plunger Lift	M	N	N	N	N
Velocity String	N	N	N	N	N

Key: P = patent awarded or filed; N = already known or not deemed valuable;
M = mechanically unlikely

This table does not attempt to distinguish between systems which can jointly share the motor or driver system and those that do not (i.e., combinations of ESP's could either use the same motor or have separate motors). Several of the patent applications are concerned with variations or specifics of particular DIALS systems, and the chart does not indicate where multiple applications have been filed. Other potential spin-off technologies are also not covered in this chart.

2.2.1 DIALS with two progressing cavity pumps.

Probably the simplest concept for a DIALS system would be two progressive cavity pumps (usually abbreviated pcpump or PCP) driven by a common rod string in a single completion. There are several ways in which these pumps could be configured, depending on the layout of the well, the location of the disposal zone, and the availability of the ancillary equipment. The simplest configuration is shown in Figure 3.

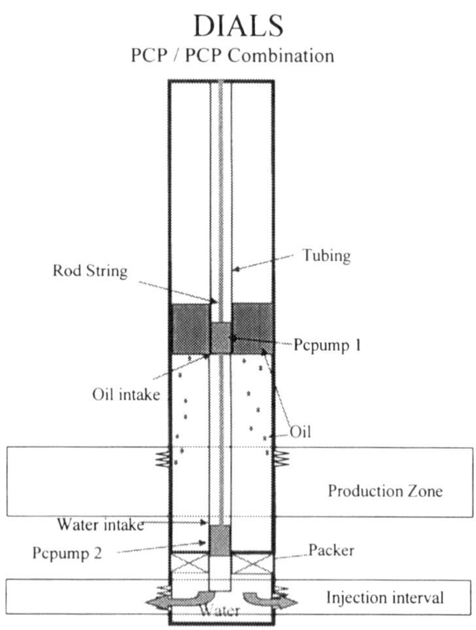

Figure 3. The Dual PCPump configuration of DIALS

The upper pump is a conventional pcpump, except that it has a rod string connected to the bottom of the stator, which extends to connect to the rotor of the water injection pump located at some distance below. While the top pump would normally lift fluids when the rod string is rotated to the right, the lower pump would be constructed to have the opposite (lefthand) rotation, and would pump fluids downward when rotated to the right. The disadvantage to this design is that a thrust bearing (or some other manner of relieving the compression on the connecting rod string) would have to be incorporated into the lower pump design. An alternate design would place the pumps very close together, so that only a short pony rod would be used to connect the rotors. This would alleviate the need for a thrust bearing, but would add the complication of having to "phase" the two pumps, because of the eccentricity of the rotors. This would also require an inlet tube for the lower pump to extend downward to just above the packer.

Alternatively, the lower pcpump could also be configured to pump upward, but this would require a bypass tube running from the pump outlet back down past the pump to the packer below. This configuration might best work best if the water disposal zone is above the production zone, so that the bypass tubing extends upward past the oil pump and above a packer. This configuration might also require a seal between the pumps in the tubing string. PCP applications would be most beneficial when solids problems or gas interference problems are anticipated, although gas is usually not a problem in high water cut wells.

2.2.2 DIALS with two electrical submersible pumps

The simplest version of this DIALS configuration could be achieved with a single motor and motor control system connected to two centrifugal pumps. The major drawback to this configuration is that the pumps would have to be correctly sized because the rates are not independently variable. If the pumps are located close together on either side of the motor, a second tubing string would have to be connected to one of the pump intakes to allow sufficient distance between the two intakes. If the pumps themselves are widely separated, a rod string of at least 10 meters would have to be used to connect the two pumps. This configuration would be significantly less expensive than one with two motors, but it would have the above operational disadvantages. A more versatile configuration is the combination of two complete ESPs that can be independently controlled, as shown in Figure 4 below.

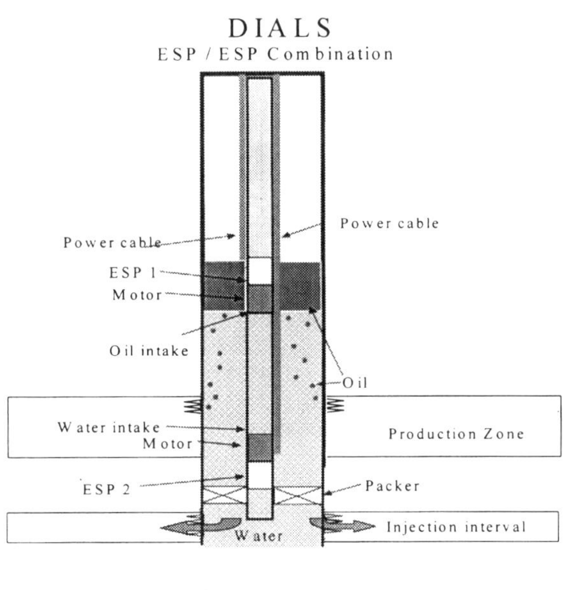

Figure 4. DIALS system using two ESPs with two controllers and power cables.

These combinations of ESPs are called "Splitstream™" systems by the manufacturer. Two are now scheduled for testing in two Texaco fields. The lower ESP is inverted to allow the fluids to be pumped downward. The respective pump intakes are positioned such that the liquid flows would pass by the motors for cooling purposes. Dual power cables are required to allow the pumps to be independently controlled. In this way, the oil/water interface within the connecting tubing can be controlled. There are several other versions of these basic designs that have been conceived to accomplish different objectives within the well. For example, one design incorporates a small diameter sample tubing run to the surface to keep a check on the quality of the water being injected.

A very similar pump combination could employ an electrical submersible progressive cavity pump (ESPCP) and an ESP. This version would have the advantage of being slightly more energy efficient than the dual ESP configuration.

2.2.3. DIALS with a rod-driven PCP and an ESP or ESPCP

Two other pump combinations that would give individual rate control are those of a conventional pcpump and a conventional ESP or an ESPCP. Both configurations would be very similar, using a rod string for the PCP and a power cable for the lower pump. Two other advantages for these systems are that the capital costs would be lower than the previous case above, and that the lifting costs would be reduced because of the higher efficiency of the pcpump. A schematic of this configuration is shown in Figure. 5.

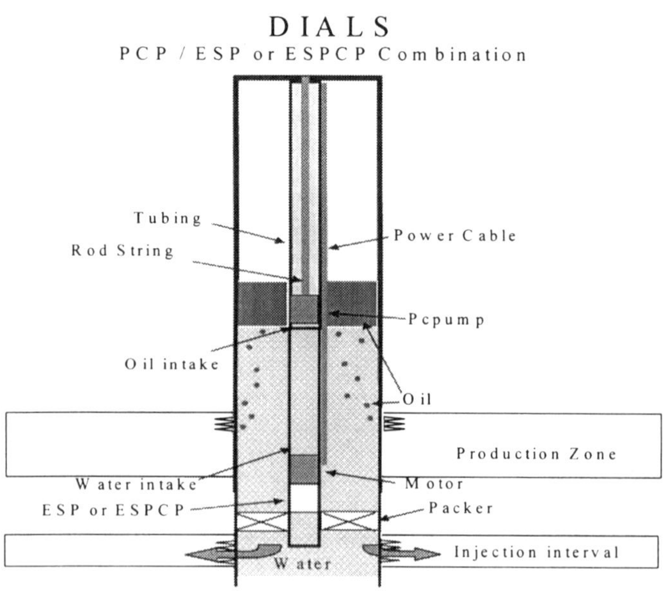

Figure 5. DIALS system using a rod-driven PCP and an ESP or ESPCP

2.2.4. DIALS using a conventional rod pump and an ESP or ESPCP

Another relatively simple DIALS arrangement is the combination of a conventional rod pump with an ESP or ESPCP. This is very similar to the above case shown in Figure 5 in that the upper pump is driven with a rod string, and the lower pump is powered via an electric power cable. The major difference is that in the former case the rod string rotates to power the pcpump, and in this case it reciprocates to power the rod pump. Individual rate control is again possible. The major advantage here is that for many wells it would require only the addition of the ESP to a conventional setup with a rod pump. In many cases the same beam-pump unit could be used by reducing the pumping rate. Again in this case, injection into the lower zone is possible even if the injection pressure is relatively high.

2.2.5. DIALS using Gas Lift (GL) and an ESP

One of the more unusual DIALS combinations is that of using gas lift with another type of pumping method for the injection pump. The combination that appears to make the most sense would be to use either an ESP or an ESPCP with gas lift, since any method requiring a rod string for the injection pump would tend to interfere with the gas lift, unless dual strings were used.

This combination again has the advantage of allowing separate rate control for the lifting and injection. As before, when the lower pump is powered by a downhole motor, the assembly would be installed upside down to ensure that the liquids pass by the motor for cooling purposes, and there would have to be sufficient liquids to achieve the required cooling. This option could be used in locations where gas lift is already being employed, and the lower injection pump could be an "add on".

2.2.6. The Triple Action Pumping System (TAPS)

If the disposal zone is at high pressure, the DAPS system may not be able to inject the water. The injection pressure is limited to the pressure of the hydraulic head in the tubing, plus the weight of the rod string. Sinker bars can be added to increase the weight of the string. However, if the injection pressure in not attainable, the DAPS will cease to operate.

An additional piston and chamber added to the DAPS allows for the amount of pressure generated from the downstroke to be increased significantly. This system is called the Triple Action Pumping System (TAPS). This feat is accomplished by allowing the liquid column in the tubing to act on the upper piston on the downstroke. This piston has a larger surface area than the lower (injection) piston. The pressure on the top piston is transferred to the lower piston via the connecting rod, as shown in Figure 6. The water is injected on the downstroke with the smaller piston, so there is a pressure multiplication that is approximately equal to the ratio of the surface areas of the two pistons. The drawback to this method is that the volume of water that can be injected is smaller than with the DAPS, and the reduction in volume is again approximately equal to the ratio of the areas of the two pistons.

Several of the other DIALS systems could also accomplish injection into a high-pressure zone, and the water volumes could be much larger than those using the TAPS. The drawback to the alternate systems is that the cost would likely be much higher, especially if a surface beam unit and a rod pump were being replaced.

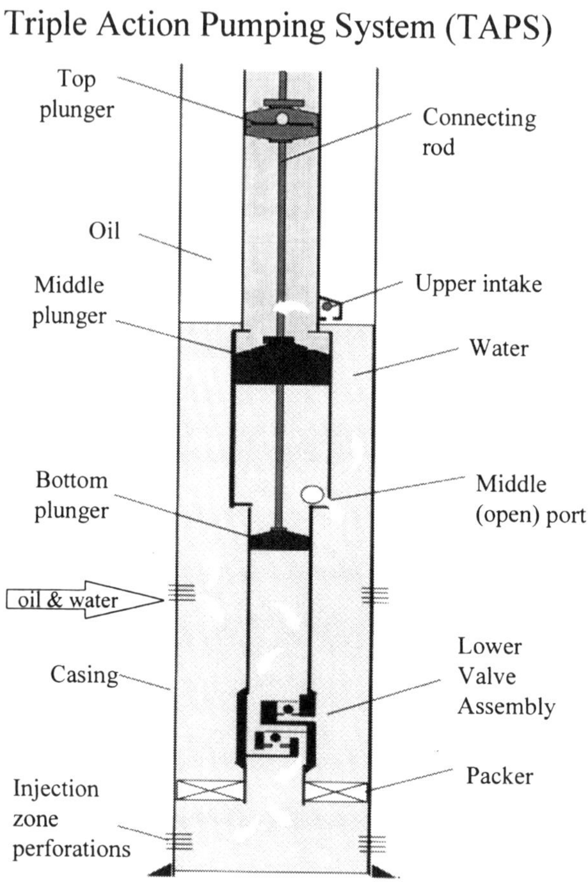

Figure 6. The Triple Action Pumping System, which can inject water into high pressure formations.

3. FIELD RESULTS

3.1 DAPS results

At least seventeen units have now been installed, for eight companies. The results have been quite impressive for a new technology in those wells where casing and cement integrity have been good, or when other mechanical well problems have not been encountered. The average oil and water production before and after the DAPS installation (in M^3/day) are shown in Table 2. More than 20 companies have now indicated an interest in installing pumps.

Table 2. Average Oil and Water Production Before and After the Installation of DAPS in M^3/day.

	Oil		Water			
	Before	After	Before	After To Surface	Injected	Total
Average	2.4	4.8	54.4	18.4	73.9	92.3
Highest Total Rate	2.1	7.0	100.0	29.7	141.1	170.8

One of the most important lessons learned from the field tests is that the well must have good casing integrity and also have good cement behind the casing, to keep the injection zone from communicating with the production zone. A second lesson learned is that the success rate also depends heavily on the quality of the rig operations. With these new and somewhat more complicated pump installations, there are many things that can go wrong, and if given half a chance, Mr. Murphy and his law will take over.

4. CONCLUSIONS

Bringing large amounts of water to surface either mixed with oil or in the form of tight emulsions is a choice to be avoided if at all possible. Many options are now available which may greatly affect the economics of this decision, including water shutoff treatments and downhole separation.

At this time, because of the complicated interaction between the production methodology and the behavior of the reservoir, along with several other factors, such as corrosion treatment costs, water treatment and disposal costs, etc., it is difficult to accurately access the long-term economic impact of downhole separation. However, especially for new wells, it is highly likely that significant economic advantage would be obtained by preplanning for downhole separation and injection before the well is drilled and completed.

Currently, several options are available to separate oil and water in the wellbore, and to allow injection of a large part of the water without having to bring it to the surface. These options will contribute to a long list of items that will tend to make the economics of downhole separation more favorable, including:

- Lower lifting costs.
- Reduced chemical costs (corrosion, scale, demulsifier, de-oiler)
- Reduced water-treating costs.
- Reduced water disposal costs.
- Improved oil production rates.
- Extended economic lifetime of marginal wells.
- Improved pressure maintenance in the reservoir (when the water can be injected into the same zone or a zone with pressure communication).
- Improved recovery factors, by extending the economic lifetime of the wells.

- Increased oil throughput for bottlenecked facilities.
- Smaller topside facilities for new installations, which would reduce Capex.
- Reduced pipeline costs for tiebacks and wet shipping lines.
- Reduced Capex costs for offshore platforms using artificial lift. (This would be especially true for deep-water installations, where platform costs are extremely high; and for platforms designed for heavy crudes and/or high production rates, where the treating facilities are very large.)

It is obvious that there are many opportunities to find new ways to exploit this technology, and many related ideas cannot be discussed here because of their proprietary nature. Other ideas, such as producing and waterflooding from the same wellbore, especially in horizontal wells and in multiple laterals, are quite obvious even to a casual observer. The downhole oil/water separation technology is in its infancy, but we expect its impact on the oil industry will be significant in the near future. Each new application has opened our eyes to new potential for these technologies.

Finally, initial indications from regulatory agencies have been highly favorable so far, again indicating the desirability of leaving the water in the ground rather than bringing it to surface for treatment and disposal.

5. ACKNOWLEDGEMENTS

The authors would like to thank Texaco, Inc. for permission to publish this paper, and Dresser Oil Tools for providing the field data presented herein. We would also like to thank the Centre for Engineering Research for permission to show the drawing for their downhole separator.

6. REFERENCES

1. Peachey, B.R., and Matthews C.M., "Downhole Oil/Water Separator Development", Journal of Canadian Petroleum Technology, Sept. 1994.
2. Peachey, B. R., Solanki, S., Zahacy, T. and Piers, K.D., "Downhole Oil/Water Separation Moves into High Gear", 1997 Petroleum Society of CIM Annual Technical Meeting, Calgary, AB, June 11, 1997.
3. Stuebinger, L., Bowlin, K., Wright, J., Poythress, M., and Watson, B., "Dual Injection and Lifting Systems: Rod Pumps", Paper SPE 38790 presented at the SPE Annual Technical Conference and Exhibition, San Antonio TX, October 5-8, 1997.

Horizontal Wells and Pipes

Wall shear stress and differential pressure in large-diameter horizontal multiphase pipelines

L C MALEY and **W P JEPSON**
Corrosion in Multiphase Systems Center, Ohio University, Athens, USA

Abstract: When a slug is formed a mixing vortex, the mixing zone, is created which contains pulses of bubbles which are shot toward the bottom of the pipe where they may impact and collapse. Bubble collapse creates high localized pressure, temperature, and wall shear stress, which cause a reduction in the corrosion inhibitor efficiency. Wall shear stress instruments cannot withstand the high turbulence found in slug flow, therefore, it would be beneficial to develop a relationship for wall shear stress. This study shows that wall shear stress fluctuations are related to differential pressure fluctuations to the 1.16 power.

1. INTRODUCTION

Wall shear stress is a difficult parameter to measure in the field for slug flow. Wall shear stress instruments cannot withstand the high level of turbulence associated with slug flow conditions. Therefore, another method of measuring the wall shear stress during slug flow is needed. This study will show that the fluctuations in wall shear stress can be related to the fluctuations in differential pressure inside the slug. Therefore, once the average wall shear stress is calculated, a maximum and minimum wall shear stress can be determined once the fluctuations in differential pressure are measured by using a differential pressure transducer.

Wall shear stress and differential pressure have been extensively studied for single-phase flow. An average wall shear stress, τ_w, is commonly calculated from the following equation:

$$\tau_W = \tfrac{1}{2} f \, \rho_s \, v_s^2 \tag{1}$$

where f is the Fanning friction factor, ρ_s is the density of the slug, and v_s is the velocity of the slug. Kaul (5) has shown that the maximum wall shear stress can be approximately 35% higher than the average shear stress. He measured the instantaneous wall shear stress . using a hot film probe in stationary slugs for Froude numbers of 6, 9, and 12. The test fluids were Conoco LVT200, water, and carbon dioxide in a 10.1 cm pipeline. Kaul's results indicated that as the Froude number increased the fluctuations in wall shear stress increased. The Froude number, Fr, is defined as:

$$Fr = \frac{v_t - v_f}{\sqrt{g h_{eff}}} \tag{2}$$

where v_t is the translational velocity, v_f is the liquid film velocity, g is acceleration due to gravity and h_{eff} is the effective height and is calculated by using equations developed by Chow (2).

Kaul also suggested that the peaks in the wall shear stress trace occurred at regular intervals. These intervals represented pulses of bubbles, which impacted and possibly collapsed on the bottom of the pipeline. His results showed that the frequency of the pulses of bubbles ranged from 20 to 30 Hz to 5 Hz, depending upon the Froude number and the distance into the slug body.

Sun (6) used water, oil, and carbon dioxide in a horizontal 10.1 cm plexiglass pipeline to study the average wall shear stress in stationary slugs. Sun measured the wall shear stress at the top, middle and bottom of the pipe with a hot film probe. His results show that as distance into the slug increased, the average wall shear stress decreased, and the maximum wall shear stress occurs at the bottom of the pipeline. The data also suggest that large changes in wall shear stress occurs within the mixing zone of the slug, and the higher the Froude number, the larger the change which occurs.

Zhou (7) also used stationary slugs in a horizontal 10.1 cm plexiglass pipeline to study slug characteristics by using sea water, LVT, and carbon dioxide. A hot film probe was used to measure the wall shear stress. Her results showed that there were large shear stress forces in the mixing zone of the slug, and as the Froude number increased so did the shear stress forces. The results also confirmed that as the distance into the slug increased the wall shear stress decreases, and once past the mixing zone of the slug, the wall shear stress became low and approximately constant.

Beggs and Brill (1) studied the pressure drop in slug flow. Their model assumes that the pressure drop due to kinetic energy is much smaller than the sum of the pressure drop due to hydrostatic head and the pressure drop due to friction. The model that was developed is:

$$\Delta P = \frac{g D \rho_M \, \Delta L \sin\vartheta + 2 f_{tp} v_S^2 \rho_M \, \Delta L}{g_c D} = \frac{g D \rho_M \, \Delta L \sin\vartheta + 4\tau_W \, \Delta L}{g_c D} \tag{3}$$

where g_c is a conversion factor, ϑ is the degree of inclination, ΔL is the length of pipe over which the pressure drop is calculated, D is diameter, g is acceleration due to gravity, and v_s is the slug velocity, the sum of the superficial liquid velocity and the superficial gas velocity.

Dukler and Hubbard (3) developed a similar correlation by using a 3.8 cm i.d. horizontal pipeline to conduct experiments with slug flow. To calculate the shear stress in the film region pseudo parallel flow was assumed and an equation that has the same format as Equation 1 was used for the calculation. To calculate the pressure drop due to friction, the following equation was used:

$$\Delta P = \frac{2f_s\left[\rho_L H_s + \rho_G(1-H_s)\right]v_s^2(l_s - l_m)}{2g_c D} = \frac{\left[2\tau_W H_s + f_s\rho_G(1-H_s)v_s^2\right](l_s - l_m)}{g_c D} \tag{4}$$

where H_s is the liquid holdup in the slug, l_s is the length of the slug, l_m is the length of the mixing eddy, and f_s is the friction factor for the slug.

This study showed that the fluctuations in wall shear stress could be related to the fluctuations in differential pressure. A stationary slug was used so that the wall shear stress and pressure can be studied at different locations within the slug. Jepson (4) has shown that within the mixing zone of a stationary slug the slug characteristics are equivalent to a moving slug at the same film Froude number. No literature has been published to date that relates the fluctuations in pressure to fluctuations in wall shear stress for slug flow conditions.

2. EXPERIMENTAL SETUP

2.1 Description of the Flow Loop

Figure 1 shows the experimental layout sections. The system consists of an 18-m long pipeline constructed from PVC and acrylic, and a 1.3 m^3 stainless steel tank used to store liquid. A 7.4 kW centrifugal pump was used to pump the liquid into a 7.6 cm i.d. PVC pipeline. An orifice plate was used to measure the flow rate, where the pressure drop was measured with a 0 to 5 psi Omega differential pressure transducer. The flow then passes into a 10.1 cm i.d. acrylic pipeline where the liquid is forced under a gate, and a fast moving liquid film is produced. The ratio of the height of the liquid film to the diameter of the pipe was 0.33.

Nitrogen was then injected immediately after the gate from a 142 m^3 storage tank to a pressure regulator set at 1 MPa. A slug is formed and then moved to the test section and held stationary by adjusting the nitrogen flowrate. Measurements were then taken, and experiments were repeated until the end of the slug mixing zone was reached. The water/nitrogen mixture then flowed to the tank where the water was recycled and the nitrogen was released into the atmosphere. To measure the temperature, a type-K thermocouple was connected to an Omega DP3200-TC thermometer. To maintain the temperature at 40°C, a Wiegand heater was used with a Wellman Thermal System thermostat. The pressure was controlled and maintained at 0.136 MPa by setting the backpressure in the storage tank.

Figure 2 shows the diagram of the test section. The ports labeled A are used to measure corrosion. Ports B and C are a pressure probe adapter and a liquid holdup sample port, respectively. The ports A through C were not used in this study. To measure pressure, the pressure transducer could be connected to any port labeled D. The ports labeled E could be used to measure the wall shear stress. However, in this study the wall shear stress port on the bottom of the pipe was used in the wall shear stress measurements.

2.2 Wall Shear Stress Measurements

Wall shear stress measurements were taken at several axial positions within the mixing zone of the slug for water at film Froude numbers ranging from 8.6 to 14.0. A slug was held stationary over a TSI hot film probe, which sent a voltage reading to the data acquisition board connected to a 486 DX2 computer. The voltage was calibrated to the wall shear stress by using different velocities of water in full pipe flow. For each velocity, the wall shear stress was calculated from Equation 1. The calculated wall shear stress was raised to the one-third power and plotted against the square of the voltage, and a relationship was observed. This relationship was determined to be:

$$\tau^{1/3} = 2.59V^2 - 3.65 \tag{5}$$

The voltage data was logged at a sampling frequency of 1000 Hz for a duration of one second. The wall shear stress was then calculated from Equation 5 and a wall shear stress versus time plot was created. The average, minimum, and maximum wall shear stress were determined from the traces. The fluctuations in wall shear stress were then plotted against distance into the slug.

2.3 Pressure Drop Measurements

The pressure drop measurements were taken at the same axial positions within the mixing zone of the slug for a given film Froude number as the wall shear stress measurements. The measurements were made by a high speed Omega type PX820-002DV transducer at a sampling rate of 512 Hz. The differential pressure transducer was more sensitive to noise than the wall shear stress probe. Therefore, the sampling frequency was reduced to help reduce the amount of noise in the pressure trace. The slug was held stationary over the positive leg of the differential pressure transducer, and the negative leg was connected to a tank, which was held at system pressure. The data from the transducer was sent through an Omega filter model DMD-465WB; then it was sent to the same 486 DX2 computer used in the wall shear stress measurements. A fast Fourier transform was performed while the data was being logged to the computer. The computer software used to collect the differential pressure data was Notebook Pro for Windows version 9. The slug was then moved to another axial position within the mixing zone of the slug and the experiments were repeated until the end of the mixing zone was reached.

3. RESULTS AND DISCUSSION

For each set of experimental conditions, measurements of wall shear stress and pressure drop with time were taken and then analyzed and compared.

The variation of wall shear stress in the slug with time at 60 cm from the slug front for a Froude number of 14 is shown in Figure 3. It is clearly seen that there are large, periodic fluctuations in the values. At this sampling rate of 1000 Hz, the maximum and minimum wall shear stresses were determined to be approximately 75 Pa and 155 Pa, respectively. The average value being 117 Pa. It should be noted that, even at this sampling frequency of 1000 Hz, each point is an average of several data. If the individual data are examined, much larger instantaneous values of wall shear stress are found. Local values as high as 1000 Pa have been seen.

The frequency of the fluctuations were examined and compared to the frequency of the pulses of bubbles generated at the slug front. The latter was obtained from video pictures using a stroboscope as the light source. As shown later, the frequencies are very similar. Figure 3 shows that there is a frequency of approximately 6 Hz in the wall shear stress trace. When generated, these pulses of bubbles are forced towards the bottom of the pipe. Above a Froude number of 4, approximately, the bubble can impact the pipe wall and sometimes collapse. This bubble impact has been shown to significantly increase the corrosion rate and decrease the efficiency of corrosion inhibitors.

Figure 4 shows the equivalent pressure trace at 60 cm into the slug for a film Froude number of 14.0. The average pressure for this film Froude number was 1,794 Pa with a minimum and maximum pressure of –2,000 Pa and 6,200 Pa, respectively. This Figure also indicates the presence of periodic fluctuations which again shows evidence of bubbles impacting and collapsing on the bottom of the pipeline. Figure 4 also shows that there is a frequency of approximately 6 Hz, which is the same frequency shown in the wall shear stress plot.

A film Froude number of 8.6 was also studied. The fluctuations in pressure and wall shear stress were plotted against distance into the slug, as shown in Figure 5. This shows that, as the distance into the slug increases the fluctuations in pressure and wall shear stress decrease. The maximum pressure fluctuation is 2,550 Pa at 15 cm into the slug and decreases to 1,050 Pa at 60 cm into the slug. The maximum wall shear stress fluctuation is 14 Pa at 15 cm into the slug and decreases to 5 Pa at 60 cm into the slug.

Figure 6 shows that for a film Froude number of 12.0, the fluctuations in pressure and wall shear stress also decrease with increasing distance into the slug. This figure also shows that for a given distance into the slug, the fluctuations in pressure and wall shear stress increase as the film Froude number increases. The range of the maximum pressure fluctuation is 5,080 Pa at 30 cm into the slug and 3,710 Pa at 60 cm into the slug. The wall shear stress also decreases from a value of 35 Pa at 30 cm into the slug to a value of 32 Pa at 60 cm into the slug. The maximum fluctuations in pressure and wall shear stress at the beginning of the slug is approximately doubled when compared to the lower film Froude number of 8.6. At the end of the slug, the maximum fluctuations in pressure are more than three times the value of film Froude number 8.6. The fluctuations in wall shear stress are more than six times the value for film Froude number 8.6.

The fluctuations from the wall shear stress were then plotted against the fluctuations in differential pressure for all film Froude numbers studied, as shown in Figure 7. The figure shows that when the fluctuations in wall shear stress and the fluctuations in pressure are plotted on a log-log plot, a relationship occurs between these two variables. The relation that was developed from this Figure 7 is:

$$F_\tau = 0.0019(F_p)^{1.16} \tag{6}$$

This relation, Equation 6, will allow the range of the wall shear stress to be calculated when it is added to the average wall shear stress, Equation 1. The addition of Equations 1 and 6 is summarized in the following equation:

$$\tau_W = \frac{1}{2} f \rho_s v_s^2 \pm 0.0019 (F_P)^{1.16} \tag{7}$$

The fluctuations in pressure and wall shear stress decrease as the distance into the slug increases because the frequency of the pulses of bubbles which impact on the bottom of the pipe decreases as the distance increases. The frequency of the pressure fluctuations, or pulses of bubbles, was determined by using a fast Fourier transform. The fast Fourier transform calculates a power spectrum for range of frequencies, which is the sampling frequency divided by two. Two frequencies always showed up on the power spectrum, the frequency of the pump, 30 Hz, and the natural frequency of the apparatus, 0.5 Hz.

The power spectrum for a film Froude number of 8.6 at 15 cm into the slug is shown in Figure 8. This figure shows that there is a dominant frequency of 5.3, which has a power spectrum magnitude of 0.0025 and a secondary frequency at 4.5 with a magnitude for the power spectrum of 0.0015. Figure 9 shows that as the distance into the slug is increased to 30 cm, the dominant frequency decreased to 4.8. The magnitude of the power spectrum also decreased for the dominant frequency to 0.0021. When the distance was further increased to 45 cm, as shown in Figure 10, the power spectrum showed several small peaks. The largest peak was at 5.6 with a power spectrum magnitude of 0.0009. Three more small peaks were also shown at 1.4, 2.4, and 4.1, which had power spectrum magnitudes ranging from 0.0004 to 0.0006. Figure 11 shows that when the end of the mixing zone was reached, the dominant frequency disappeared. This plot shows that there are several small peaks with a power spectrum value of 0.0007 or less.

4. CONCLUSIONS

This study has shown that if a differential pressure transducer is used to measure the fluctuations in differential pressure, the fluctuations in wall shear stress can be calculated. Once the average wall shear stress is calculated by using a normal fluid mechanics equation, the fluctuations in wall shear stress can be added to the average wall shear stress to obtain a true maximum and minimum wall shear stress. This study has shown that as the film Froude number is increased the fluctuations in wall shear stress are increased. At high film Froude numbers the maximum wall shear stress can be approximately 34% higher than the calculated average obtained from a fluid mechanics equation.

The results also show that a dominant frequency appears in the pressure traces. As the distance into the slug is increased, the frequency in the pressure trace decreases until the end of the mixing zone is reached. The dominant frequency, which appears in the pressure trace is believed to be the frequency of the pulses of bubbles impacting on the bottom of the pipe. The pulses of the bubbles are strong in the beginning of the slug and become less frequent as the bubbles become distributed throughout the body of the slug. Once past the mixing zone of the slug, the high turbulence of the slug front is no longer present and the dominant frequency disappears.

5. NOMENCLATURE

D = Pipe Diameter
F = Fluctuations
H = Liquid Holdup
L = Length of pipe
P = Pressure
V = Bridge Voltage
f = Fanning friction factor
g = Acceleration due to gravity
g_c = Conversion factor
h = Height of the liquid film
l = Length within the slug
v = Velocity

Greek Letters
Δ = Change in
ρ = Density
τ = Shear Stress

Subscripts
L = Liquid
P = Pressure
W = Wall
eff = Effective height
f = Film
m = Mixing Eddy
s = Slug
t = Translational

6. REFERENCES

(1). Beggs and Brill. Transactions of Petroleum Society of A.I.M.E., 256, 607 (1973).

(2). Chow, V. Te, "Open-Channel Hydraulics", McGraw-Hill, (1959).

(3). Dukler, A. E. and Hubbard, M. G. "A Model for Gas-Liquid Slug Flow in Horizontal and Near Horizontal Tubes" Ind. Eng. Chem. Fundam., Vol. 14, No. 4, 1975.

(4). Jepson, W. P., "The Flow Characteristics in Horizontal Slug Flow", 3rd Int. Conf. Multiphase Flow, Paper F2, 187, 1987.

(5). Kaul, Ashwini, "Study of Slug Flow Characteristics and Performance of Corrosion Inhibitors, in Multiphase Flow, in Horizontal Oil and Gas Pipelines", Thesis, Ohio University, March 1996.

(6). Sun, Jyi-Yu, "Multiphase Slug Flow Characteristics and Their Effects on Corrosion in Pipelines", Dissertation, University of Illinois at Urbana-Champaign, 1991.

(7). Zhou, Xianling, "Experimental Study of Corrosion Rate and Slug Flow Characteristics in Horizontal, Multiphase Pipeline", Thesis, Ohio University, June 1993.

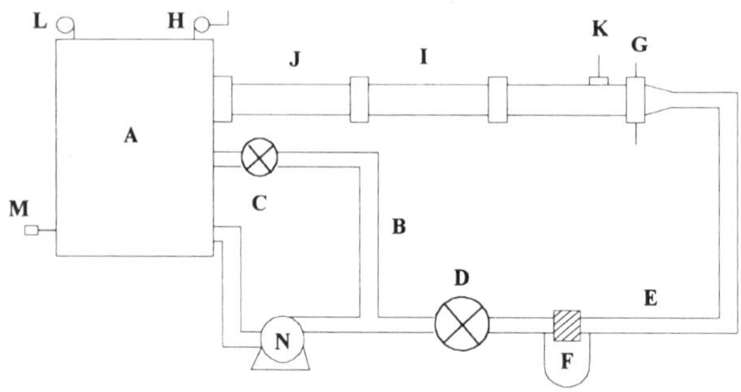

A. Storage Tank for Liquid
B. Liquid Recyle Line
C. Liquid Recycle Valve
D. Liquid Feed Valve
E. Liquid Feed Line - 7.5 cm PVC Pipe
F. Orifice Plate to Pressure Transducer
G. Liquid Height Control Gate
H. Pressure Gauge and Back Pressure Regulator
I. Test Section - 10 cm Plexiglass Pipe
J. 10 cm Plexiglass Section
K. Nitrogen Feed Line
L. Safety Valve
M. Heater
N. Pump

Figure 1: Experimental Layout

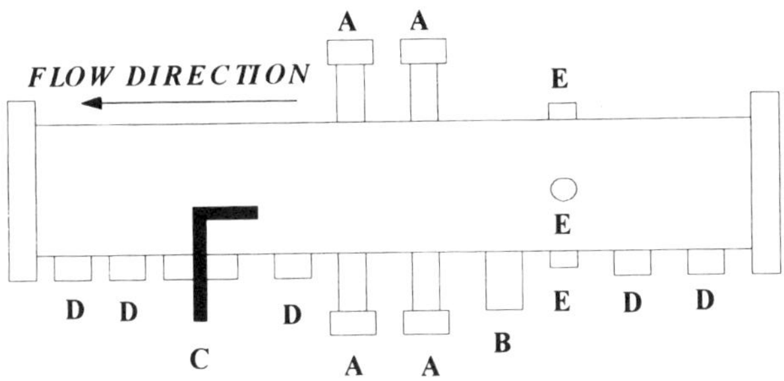

A. LPR/ER Probe Adapter

B. Pressure Probe Adapter

C. Liquid Holdup Sample Port

D. Pressure Tappings

E. Shear Stress Port

Figure 2: Test Section

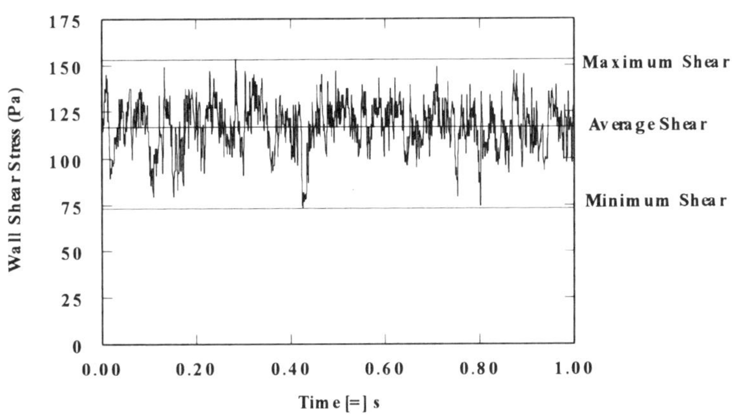

Figure 3: Wall Shear Stress Vs. Time for 60 cm into the slug and a film Froude number 14.0

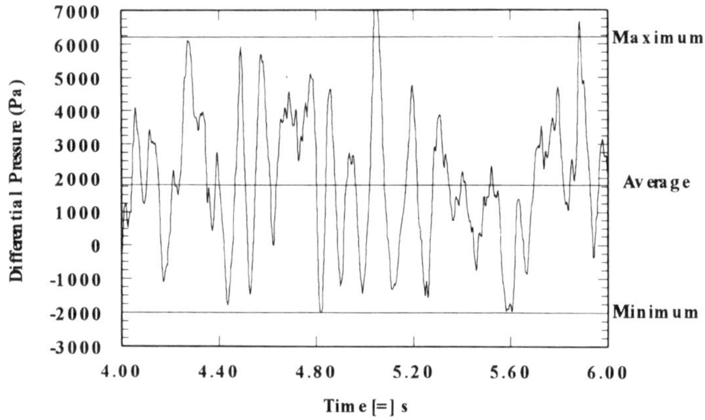

Figure 4: Differential pressure Vs. Time 60 cm into slug for film Froude number 14.0

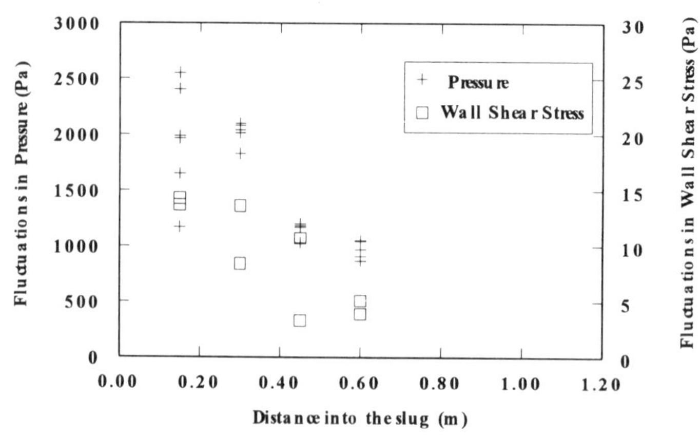

Figure 5: Fluctuations in pressure and wall shear stress Vs. Distance into the slug for film Froude number 8.6

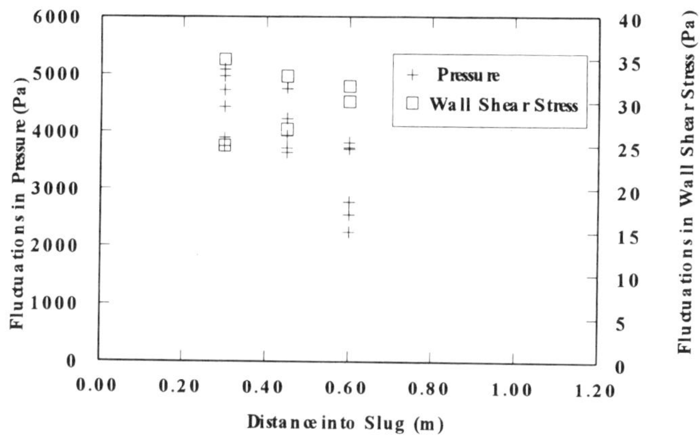

Figure 6: Fluctuations in pressure and wall shear stress Vs. Distance into the slug for film Froude number 12.0

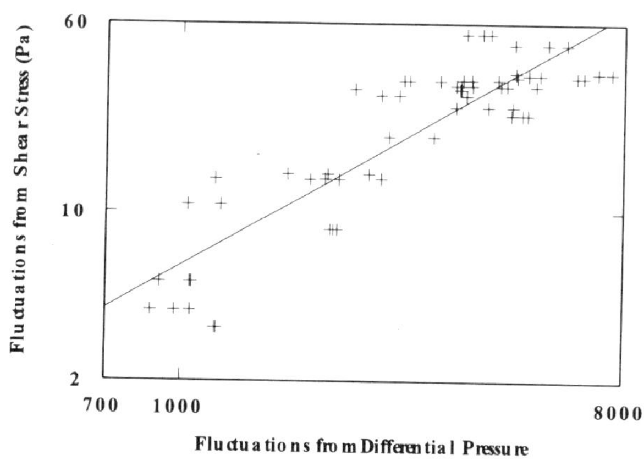

Figure 7: Fluctuations in wall shear stress Vs Fluctuations in differential pressure

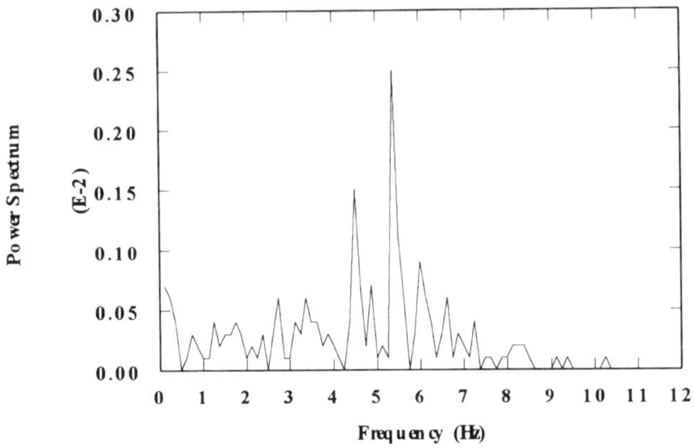

Figure 8: Power spectrum Vs. Frequency 15 cm into slug for film Froude number 8.6

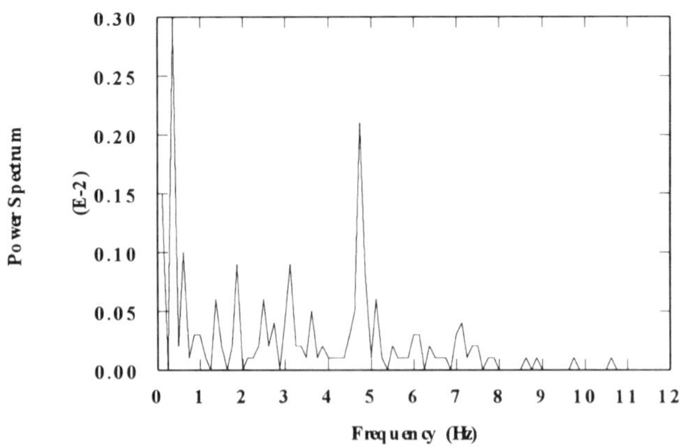

Figure 9: Power spectrum Vs. Frequency 30 cm into slug for film Froude number 8.6

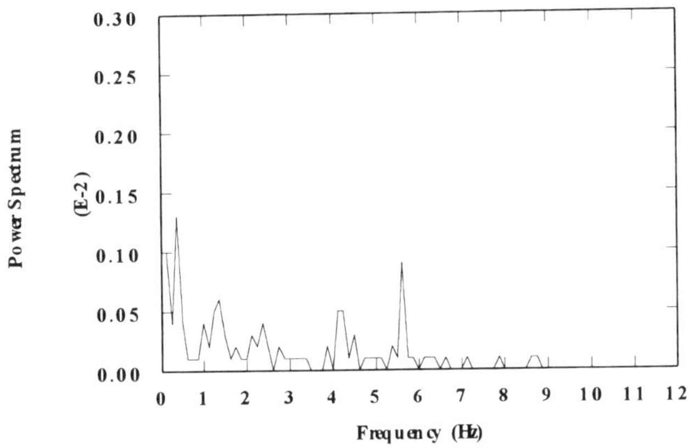

Figure 10: Power spectrum Vs. Frequency 45 cm into slug for film Froude numer 8.6

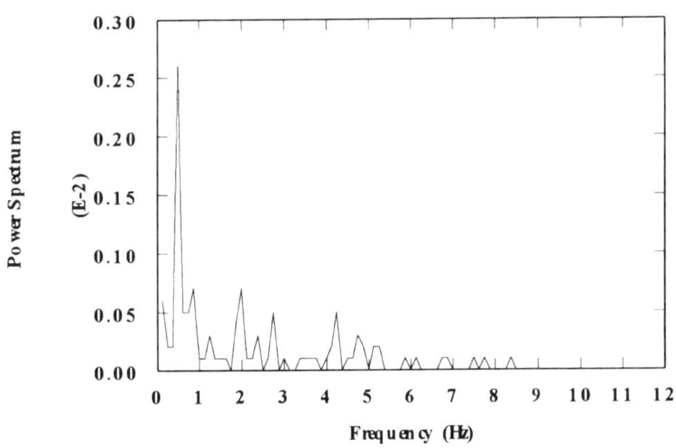

Figure 11: Power spectrum Vs Frequency 60 cm into slug for film Froude number 8.6

Application of multiphase flow methods to horizontal underbalanced drilling

S P SMITH and **G A GREGORY**
Neotechnology Consultants Limited, Calgary, Canada
N MUNRO
Northland Energy Services Inc., Canada
M MUQEEM
UBD Engineering Inc., Canada

SYNOPSIS

Multiphase flow can be present in all aspects of underbalanced drilling. This paper outlines the ways in which multiphase flow pressure loss calculations can be used by drilling engineers in the design and optimization of underbalanced drilling operations. Then, detailed field measurements for several horizontal wells drilled underbalanced with coiled tubing are used to evaluate the application of existing pressure loss calculation methods to this unique application.

1.0 INTRODUCTION

Underbalanced drilling (UBD) is rapidly gaining popularity in the oil and gas industry. The Alberta Energy and Utilities Board ID94-3 (1) defines underbalanced drilling as follows:

> "When the hydrostatic head of a drilling fluid is intentionally designed to be lower than the pressure of the formation being drilled, the operation will be considered underbalanced drilling. The hydrostatic head of the drilling fluid may be naturally less than the formation pressure or it can be induced. The induced state may be created by adding natural gas, nitrogen, or air to the liquid phase of the drilling fluid. Whether induced or natural, this may result in an influx of formation fluid which must be circulated from the well and controlled at surface."

1.1 Benefits of underbalanced drilling

Maintaining the pressure in the wellbore below the reservoir pressure allows reservoir fluids to enter the wellbore while UBD operations proceed, thus preventing flow of drilling fluids (and associated solids) into the formation, thereby minimizing or even eliminating formation damage. This is of particular importance in the drilling of horizontal wells as the formation is exposed to the drilling fluids for an extended period of time.

Although formation damage reduction is the most widely recognized benefit of underbalanced drilling, several additional benefits are outlined below:

- Increased Penetration Rates
  Underbalanced drilling can achieve higher rates of penetration due to reduced "chip holdown" and decreased hydrostatic pressure at the bit face. Reduction in "chip holdown" refers to easier removal of drilled solids from the vicinity of the drill bit due to the flow of drilling and reservoir fluids, thus allowing the bit to drill into fresh rock continuously. The decreased hydrostatic pressure at the bit face reduces stress in the rock being drilled, allowing it to fail more easily. The experience of drilling engineers (2) familiar with UBD is that the rate of penetration can be increased by between three and ten times that of conventional drilling

- Minimal Lost Circulation
  Underbalanced drilling gives better control in situations where fractured, low pressure, or high permeability formations may lead to the loss of drilling fluids and the associated problems that can cause.

- Evaluation While Drilling
  Data acquired in real time during underbalanced drilling operations allow for both the short term on site optimization of the UBD operation, and the longer term assessment of the well's potential. On site, the data acquired can be used to optimize drilling parameters such as the well's horizontal length, vertical depth, and orientation. Other data that can be obtained, useful both on site and long term, include fluid properties, productivity, and geological interpretations of the formation.

- Potential Payback
  It is possible that as a result of collecting and selling reservoir production during drilling that the cost of the drilling operation may be partially or fully recovered.

1.2 Multiphase flow in horizontal underbalanced drilling

To achieve an underbalanced condition, a gas-liquid mixture may be used in place of traditional drilling mud. In horizontal underbalanced drilling this gas-liquid mixture is injected through the drill string, and ultimately passes through the bottom-hole assembly (BHA) (see Figure 1) where it provides energy to drive the downhole motor, which in turn drives the drill bit. The injected fluids then flow from the bit into the annulus formed by the drill string and the open hole. As they flow in the near horizontal rough-walled asymmetric annulus, the injected fluids are commingled with any production from the reservoir. The resulting multiphase stream then flows from the heel of the well up the vertical or near-vertical casing-drill string annulus to the surface. The resulting presence of multiphase flow in every aspect of the underbalanced drilling operation presents unique challenges to the drilling engineer.

Field experience has shown that multiphase hydraulics calculations can be used in both the design and near real-time optimization of underbalanced drilling operations. Previous authors (3) have outlined multiphase pressure loss calculations that may be considered in horizontal underbalanced drilling applications. These include, but are not limited to:

- vertical downward flow,
- flow through a bottom-hole assembly,
- three/four phase flow in an asymmetric, rough-walled, horizontal or near-horizontal annulus (possibly with flow increasing with length),
- three/four phase flow in a vertical annulus.

The following sections of this paper will outline how multiphase flow hydraulics calculations can be used to support underbalanced drilling operations. The work will focus on the production of injected and produced fluids through the horizontal and vertical annulus, as these are the calculations most used by drilling engineers. Actual field data are used to validate multiphase flow calculation methods used in these applications.

2.0 APPLICATION OF MULTIPHASE FLOW HYDRAULICS CALCULATIONS TO HORIZONTAL UNDERBALANCED DRILLING

Underbalanced drilling of horizontal wells is a technically complex operation, and its successful completion depends upon detailed engineering analysis both before and during the drilling operations. Simulation of the multiphase flow streams is useful in both the preliminary design and the on site optimization of underbalanced drilling operations. These simulations must account for such factors as temperature effects, oil and gas volume effects, viscosity effects, wall roughness, and commingled production. The lack of consideration for any of these factors can lead to deceptive results.

2.1 Preliminary design calculations

Appropriate multiphase simulations can help to determine the technical and economic feasibility of underbalanced drilling, optimize the gas and liquid injection rates, optimize surface back pressure, optimize the wellbore geometry (i.e., drill string size, hole size, casing size), establish an operating envelope and ultimately, determine the horizontal drilling limit. Obviously, each of these aspects are interrelated. For example, gas and liquid injection rates optimized for a particular wellbore geometry, will no longer be optimal if the size of the drill string is changed.

Each design parameter is evaluated with respect to the operating envelope for underbalanced drilling (see Figure 2). The operating envelope is bounded by two curves. The first is the reservoir pressure. The second is the pressure performance curve for the annulus. The performance curve is a plot of flowing bottom-hole pressure versus injection gas rate for a given design case (i.e., fixed wellhead pressure, wellbore geometry and liquid injection rate, and an assumed reservoir influx). The intersection points of these two curves mark the two gas injection rates that will give balanced drilling; the actual operating limits are typically defined by a required drawdown that will be determined based on the reservoir. If the minimum pressure of the performance curve (i.e., the point where the slope changes from negative to positive) exceeds the reservoir pressure, then underbalanced drilling is not possible for the specific case being investigated.

Further, the operating envelope allows the drilling engineer to determine, for a particular gas injection rate, whether the flow is dominated by hydrostatic or frictional pressure loss. Any point on the performance curve with a negative slope is dominated by hydrostatic pressure losses. These points are inherently unstable, show large pressure changes with small changes in gas flow rate, and exhibit increasing bottom-hole pressure with decreasing flow rate. Points on the performance curve with a positive slope are dominated by frictional pressure loss. These points are inherently stable and exhibit increasing bottom-hole pressure with increasing gas flow rate. Note that "dominated by frictional pressure loss," does not necessarily imply that the frictional pressure loss is greater than the hydrostatic pressure loss; instead, this means that the reduction in hydrostatic pressure loss associated with an increase in the gas injection rate is less than the increase in frictional pressure loss due to the increased gas rate.

This information can be used in several ways. If a reduction in bottom-hole pressure is required, a decrease in gas injection, the obvious answer to someone only familiar with single phase flow, will lead to an increase in bottom-hole flowing pressure if the flow is hydrostatically dominated. Further, the cost of nitrogen (as the injection gas), if bulk liquid nitrogen is used, can be one of the most significant costs associated with UBD operations. If on-site nitrogen generation is used, increasing nitrogen demand can lead to increased contamination by oxygen, further increasing the risk of corrosion.

One of the most common misconceptions in underbalanced drilling is that more nitrogen (i.e., gas) injection is better. This stems from observations of drilling operations that are hydrostatically dominated, where an increase in the gas injection rate can lead to significant decreases in the bottom-hole pressure. However, if the drilling operation is frictionally dominated, increasing the gas injection rate will not only increase the bottom-hole pressure, but may dramatically increase the cost, or the risk of corrosion, associated with nitrogen used while drilling.

Thus, for a specific design case, the operating envelope can not only confirm the feasibility of underbalanced drilling, but also offer valuable insights into both the acceptable and optimal gas injection rates, and the influence of those rates on the bottom-hole flowing pressure. Operating envelopes should be developed for a range of design parameters.

However, the operating envelope cannot tell the entire story. Each point on the operating envelope corresponds to a single wellbore calculation for a specific gas injection rate. For each such calculation valuable additional information can be gathered by analysing profiles of the in-situ liquid holdup, actual gas and liquid velocities, pressure, and temperature.

In underbalanced drilling, it is the liquid phase that provides the medium to transport the drilled cuttings. As such, the profiles of both the in-situ liquid holdup and the actual liquid velocity in the annulus are important in determining the hole cleaning efficiency. For any specific design case, the in-situ liquid holdup fraction decreases as the gas injection rate increases. For the same case, the actual liquid velocity increases as the gas injection rate increases, up to the point where the gas and liquid velocities are identical. Thus, the optimal liquid injection rate is a compromise between the requirements for liquid holdup and liquid velocity. As well, limits on the acceptable range of gas injection rates may be imposed by these requirements.

In addition to decreasing the overall density of the drilling fluids, the injection gas is valuable because of the gas lift effect. In the annulus, the injected gas provides some of the energy required to lift liquids to the surface. Review of the liquid holdup profile, and comparison of the gas and liquid velocity profiles can aid in assessing the effectiveness of the gas lift.

Analysis of the pressure profile will ensure that the entire open hole section is maintained in an underbalanced condition. Review of the temperature profile is also particularly useful in cases of high reservoir temperature. For example, some Indonesian wells that are programmed to be drilled underbalanced have had reservoir temperatures that exceeded 300 °F (2).

In addition to the cases investigating variation of design parameters (i.e., wellhead pressure, wellbore geometry, and liquid rate), a range of reservoir inflow rates should be investigated, as this parameter is the most difficult to predict prior to the actual UBD operation. It has been recommended (3. 4) that design cases based on 50%, 100%, and 150% of the expected reservoir inflow be modelled and analysed to ensure that on site personnel are prepared should the productivity of the reservoir be different from anticipated. Note, however, that in cases of flush production, the reservoir inflow can greatly exceed even 150% of the expected stabilized production rates.

Finally, an accurate simulation will allow the drilling engineer to assess the horizontal drilling limit. The horizontal drilling limit is the horizontal length, for any design case, beyond which increased pressure losses due to friction in the wellbore would be such that the "toe" of the well would no longer be underbalanced.

A substantial volume of information can result from the application of multiphase flow hydraulics calculations to underbalanced drilling operations. With analysis and interpretation provided by drilling engineers experienced in UBD operations, this information can prove valuable in the development of successful underbalanced drilling programs.

2.2 On site optimization

When drilling has commenced, the success of an underbalanced drilling operation is dependent on interpretation of data, such as the design parameters outlined above, that are typically monitored in real time while drilling proceeds.

Variation from expectations during drilling is possible for any of these parameters. It is important that hydraulic simulations be used to analyse these variations while the well is being drilled. Such simulations may uncover problems such as overbalanced conditions in the wellbore or poor cuttings transport. To quote one drilling engineer, "If it [measured pressures] can't be explained by the hydraulics, then something else is likely wrong," (i.e., "bridging" of the annulus with cuttings that are not being effectively transported to surface). Discrepancies between measured and predicted pressures can act as warning signs to the drilling engineer and allow corrective action to be taken before a significant problem develops. In the extreme case, if a BHA becomes stuck in the hole, it could result in the loss of hundreds of thousands of dollars.

A common misunderstanding in underbalanced drilling operations is the assumption that the production of hydrocarbons to the surface confirms that the well is being drilled in an underbalanced state. Hydrocarbon production at surface only indicates that some portion of the well is underbalanced. Pressure losses along the horizontal portion of the well may be such that the pressure near the vertical portion of the well is underbalanced, and hydrocarbons are produced, while the pressure at the "toe" of the well, near the drill bit, exceeds the reservoir pressure. In this situation the actual drilling occurs in an overbalanced state, and significant formation damage may occur. If bottom-hole pressure measurement is not available, only multiphase hydraulics calculations can confirm that the underbalanced state is maintained throughout the wellbore.

As drilling progresses, and the length of the well and the contribution to flow from the reservoir increase, the optimum gas and liquid injection rates must be continuously analysed and modified. Thus, the on site personnel can perform hydraulic simulations with the latest real time data to support changes in the injection rates. The gas injection rate must be monitored to optimize the gas lift effect. The liquid injection rate must be controlled to provide sufficient liquid velocity and liquid holdup in the annulus to ensure effective cuttings transport and thereby prevent bridging. Given the high cost of the downhole tools used in underbalanced drilling, this phenomenon must be recognized and effective cuttings transport reestablished in a timely fashion, as those tools can be lost if bridging becomes severe.

In addition, since cuttings are transported with the liquid phase, fluid flow analysis can give the drilling engineer the transit time of the liquid phase between the drill bit and the surface, and thus, an estimate of the transit time for the cuttings. This information will allow the geologist on site to correlate cuttings samples with a location downhole, valuable information in the geological interpretation of the reservoir.

3.0 VALIDATION OF MULTIPHASE FLOW METHODS FOR APPLICATION TO HORIZONTAL UNDERBALANCED DRILLING

3.1 Well data

Detailed measurements gathered during the underbalanced drilling of four wells in Western Canada were provided to the authors by the producing company that drilled the wells. At the request of the producing company, the wells will be referred to in generic terms. The first three wells (Wells A, B, and C) are oil wells, the fourth well (Well D) is a natural gas well.

Prior to the start of underbalanced drilling operations, Well A was drilled to a measured depth of 1058 m (971.6 m TVD) by conventional means, and completed with 139.7 mm production casing at a bottom-hole inclination of 89.8 degrees. A coiled tubing (73.0 mm O.D.) drilling unit was then brought in to drill the horizontal leg underbalanced.

Underbalanced drilling of the 127.7 mm open hole section was initiated with nitrogen (20 scm/min) and native crude oil (250 L/min). UBD continued to a final measured depth of 1540 m, where drilling operations were discontinued due to failing reservoir quality. The reservoir pressure was 8000 kPa. Throughout the UBD operations, bottom-hole pressures were maintained between 3000 and 4500 kPa, and continuous production of reservoir oil averaged

120 m^3/day. Corresponding annulus surface pressures were between 200 and 1000 kPa.

Prior to the start of underbalanced drilling operations, Well B was drilled to a measured depth of 1044 m (939 m TVD) by conventional means, and completed with 139.7 mm production casing at a bottom-hole inclination of 90 degrees. A coiled tubing (73.0 mm O.D.) drilling unit was then brought in to drill the horizontal extension underbalanced.

Underbalanced drilling of the 120.7 mm open hole section was initiated with nitrogen (15 scm/min) and native crude oil (250 L/min). At a measured depth of 1067 m the crude oil circulation rate was increased to 300 L/min to improve hole cleaning. At a measured depth of 1182 m the nitrogen rate was increased to 20 scm/min, and the oil rate decreased to 250 L/min to decrease the bottom-hole pressure. UBD continued to a final measured depth of 1255 m, where drilling operations were terminated due to difficulties with wellbore stability. The reservoir pressure was 7200 kPa. Throughout the UBD operations, bottom-hole pressures were maintained between 3200 and 6000 kPa, and continuous production of reservoir oil averaged approximately 29 m^3/day. Corresponding annulus surface pressures were between 150 and 1000 kPa.

Prior to the start of underbalanced drilling operations, Well C was completed with 139.7 mm production casing to a measured depth of 1089 m (946 m TVD) and a bottom-hole inclination of 85.7 degrees. A coiled tubing (73.0 mm O.D.) drilling unit was then brought in to drill the horizontal leg underbalanced.

Underbalanced drilling of the 120.7 mm open hole section was initiated with nitrogen (15 scm/min) and native crude oil (250 L/min). At a measured depth of 1161 m the nitrogen circulation rate was increased to 30 scm/min to reduce the bottom-hole pressure and increase production. At a measured depth of 1240 m the nitrogen rate was decreased to 25 scm/min to reduce the production rate due to concerns about on-site tank volume. UBD continued to a final measured depth of 1255 m, where drilling operations were terminated as the horizontal drilling objective had been reached. The reservoir pressure was 5700 kPa. Throughout the UBD operations, bottom-hole pressures were maintained between 4000 and 6000 kPa (i.e., near the end of the drilling operation, bottom-hole pressures were actually overbalanced), and continuous production of reservoir oil averaged approximately 360 m^3/day. Corresponding annulus surface pressures were between 150 and 1000 kPa.

Prior to the start of underbalanced drilling operations, Well D was drilled to a measured depth of 2550 m (2440 m TVD) by conventional means, and completed with 139.7 mm production casing at a bottom-hole inclination of 91.5 degrees. A coiled tubing (73.0 mm O.D.) drilling unit was then brought in to drill the horizontal leg underbalanced.

Underbalanced drilling of the 120.7 mm open hole section was initiated with nitrogen (25 scm/min) and diesel oil (500 L/min). UBD continued to a final measured depth of 3085 m, where drilling operations were terminated as the horizontal drilling objective had been reached. Throughout the UBD operations, bottom-hole pressures were maintained between 7000 and 12000 kPa, and continuous production of reservoir gas averaged approximately 265 10^3m^3/day. Corresponding annulus surface pressures were between 500 and 1700 kPa.

3.2 Multiphase flow calculation methods

Pressure drop calculations were performed for each of the wells using the following procedures:

Aziz, Govier, and Fogarasi (5, 6)
Modified Aziz, Govier, and Fogarasi (5, 6) (interpret annular-mist flow prediction to be dispersed bubble flow; use homogeneous flow model instead of Duns and Ros (7) correlation)
Gregory (8), with Duns and Ros (7) in the annular-mist flow regime
Gregory (8), with Gray (9) in the annular-mist flow regime
Beggs and Brill (10)

Two sets of calculations were performed using each method. In the first case, the calculations were begun at known bottom-hole conditions, and the wellhead pressure was computed. These calculations will be referred to as based on the measured bottom-hole pressure. In the second case, the calculations were begun with an assumed bottom-hole pressure, and the wellhead pressure was computed. Iterative calculations were then performed until the bottom-hole pressure corresponding to the measured wellhead pressure was found. These calculations will be referred to as based on the measured wellhead pressure.

4.0 RESULTS

4.1 Results for oil wells

A total of fifteen sets of measurements were evaluated for the three oil wells; 6 points measured during the drilling of Well A, 5 measured during the drilling of Well B, and 4 measured during the drilling of Well C. The average and standard deviation of the percent error for each multiphase flow method, and each calculation procedure, are summarized in Table 1. Figures 3 to 7 show a comparison of the calculated and measured pressure losses for each multiphase method where the calculations were based on the measured bottom-hole pressure. Figures 8 to 12 show similar comparisons where the calculations were based on the measured wellhead pressure.

The method of Aziz et al (5, 6), modified by using a homogeneous flow model when annular-mist flow is predicted to occur, is, for the data used in this study, the most accurate method for use in the simulation of multiphase flow pressure losses in underbalanced drilling of oil wells (this is a modification that is also recommended for high rate crude oil wells under gas lift). The second most accurate method is the combination of the Gregory (8) model, with the Gray (9) correlation for annular-mist flow.

The original method of Aziz et al (5, 6), and the combination of the Gregory (8) model with the Duns and Ros (7) correlation for annular-mist flow, both give acceptably accurate results when the calculations are based on the measured bottom-hole pressure. However, these methods significantly under-predict the measured pressure losses when the calculations are based on the measured wellhead pressure.

The Beggs and Brill (10) correlation consistently over-predicts the measured pressure drops. Note that in Table 1, no results are presented for this correlation when calculations are based on the measured bottom-hole pressure. This is because in thirteen of the fifteen calculations, the computed pressure loss exceeded the measured bottom-hole pressure (see Figure 7).

Table 1

Comparison of the Accuracy of Various Multiphase Flow Calculation Methods for Oil Wells

Calculation Procedure	Measured Bottom-hole Pressure		Measured Wellhead Pressure	
Multiphase Method	Mean % Error	Standard Deviation	Mean % Error	Standard Deviation
Aziz, Govier, and Fogarasi	-9.0	6.7	-33.4	21.1
Aziz, Govier, and Fogarasi (Modified)	-2.0	8.8	-4.4	19.9
Gregory / Duns and Ros	-12.5	7.7	-39.2	16.1
Gregory / Gray	-5.7	8.6	-13.2	18.2
Beggs and Brill	-	-	32.0	22.0

For all of the multiphase flow calculation methods investigated, the calculated pressure drops are more accurate when based on the measured bottom-hole pressures. This is most pronounced for the original Aziz et al (1972) method applied to Well B. In this case, the ratio of the errors when calculations are based on the measured wellhead pressure to those when calculations are based on the measured bottom-hole pressure is approximately 5 : 1.

The explanation of this phenomenon is best understood by reviewing the pressure profiles for the well for the two calculation procedures as shown in Figure 13. For an oil well, the pressure gradient is typically highest at the bottom of the vertical portion of the well. As the fluids are produced up the annulus, and the pressure decreases, the volume of the gas expands and the liquids are transported more efficiently. In an oil well, the volume of the gas is typically such that the decrease in the pressure loss due to hydrostatic head is greater than any increase in frictional pressure loss associated with the expansion of the gas, and the combined pressure gradient decreases. Thus, the calculation error is minimized when calculations are based on the measured pressure at the point where the gradient is highest (i.e., for oil wells, the bottom-hole pressure).

Figure 14 shows the measured annulus pressures for Well B throughout the underbalanced drilling operation. Also shown are the bottom-hole pressures calculated based on the measured wellhead pressures using the modified method of Aziz et al (5, 6). The periodic *wiper trips* indicated on the plot refer to the use of the drill string as a tool to reduce bridging. During a wiper trip, drilling is stopped and the drill string is pulled back through the wellbore in an effort to physically remove any accumulated cuttings. Note the significant increase in the bottom-hole pressure prior to the second wiper trip, and the corresponding reduction in bottom-hole pressure after the wiper trip.

Figure 15 is a similar plot for Well C. Because of its short horizontal length, Well C was drilled without wiper trips. At 1166 m, communications with the bottom-hole pressure sensor became unreliable. Prior to this point, it can be seen that the bottom-hole pressure steadily increased throughout the drilling operation. Of the four points analysed for this well, the first two match significantly better than the remaining two. For the modified method of Aziz et al (5, 6), the average error for the first two points is approximately -4.5%, while for the final two points it is approximately -15.5%. As this well was drilled without wiper trips, it is possible that bridging had restricted the annular space and led to the higher measured pressure losses for the final two points. It is exactly this type of marked change in the ability of multiphase hydraulics models to predict the measured pressure loss that may warn the on-site drilling engineers, and allow them to diagnose a problem before it becomes severe.

The statistical analysis presented in Table 1 was revised to eliminate the two suspect points. The revised results are presented in Table 2.

Table 2

Comparison of the Accuracy of Various Multiphase Flow Calculation Methods for Oil Wells (Revised)

Calculation Procedure	Measured Bottom-hole Pressure		Measured Wellhead Pressure	
Multiphase Method	Mean % Error	Standard Deviation	Mean % Error	Standard Deviation
Aziz, Govier, and Fogarasi	-7.3	5.5	-29.3	19.6
Aziz, Govier, and Fogarasi (Modified)	0.1	7.4	-0.2	17.7
Gregory / Duns and Ros	-10.6	6.0	-36.0	14.9
Gregory / Gray	-3.5	6.8	-9.2	15.9
Beggs and Brill	-	-	36.9	19.2

4.2 Results for gas wells

A total of nine sets of measurements were evaluated for the gas well, Well D. The average and standard deviation of the percent error for each multiphase flow method, and each calculation procedure, are summarized in Table 3. Figures 3 to 7 show a comparison of the calculated and measured pressure losses for each multiphase method where the calculations were based on the measured bottom-hole pressure. Figures 8 to 12 show similar comparisons where the calculations were based on the measured wellhead pressure.

The original method of Aziz et al (5, 6), and the combination of the Gregory (8) model with the Duns and Ros (7) correlation for annular-mist flow are, for the data used in this study, the most accurate methods for use in the simulation of multiphase flow pressure losses in underbalanced drilling of gas wells. The only difference between these two methods is that the Gregory (8) model uses a mechanistic model for droplet flow to determine the transition to annular-mist flow. As both models predict annular-mist flow for the data points analyzed, the identical predictions of the two models are to be expected.

The modified method of Aziz et al (5, 6), the combination of the Gregory (8) model with the Gray (9) correlation for annular-mist flow, and the Beggs and Brill (10) correlation consistently over-predict the measured pressure drops. Note that in Table 3, no results are presented for these methods when calculations are based on the measured bottom-hole pressure. This is because in all nine of the calculations, the computed pressure loss exceeded the measured bottom-hole pressure (see Figures 4, 6, and 7).

Table 3

Comparison of the Accuracy of Various Multiphase Flow Calculation Methods for Gas Wells

Calculation Procedure	Measured Bottom-hole Pressure		Measured Wellhead Pressure	
Multiphase Method	Mean % Error	Standard Deviation	Mean % Error	Standard Deviation
Aziz, Govier, and Fogarasi	-6.7	6.6	-6.6	8.0
Aziz, Govier, and Fogarasi (Modified)	-	-	38.9	8.0
Gregory / Duns and Ros	-6.7	6.6	-6.6	8.0
Gregory / Gray	-	-	30.5	6.5
Beggs and Brill	-	-	59.8	5.3

For the original method of Aziz et al (5, 6), and the combination of the Gregory (8) model with the Duns and Ros (7) correlation for annular-mist flow, the calculated pressure drops are slightly more accurate when based on the measured wellhead pressures. Unlike those of the oil wells, the calculated pressure gradients for Well D (Figure 16) are slightly higher at the wellhead. This is because the produced gas volume is such that the increase in the frictional pressure loss due to gas expansion as fluids flow up the wellbore is greater than any corresponding decrease in the elevation pressure loss. Thus, although the difference is small for Well D, the calculation error is minimized when calculations are based on the measured pressure at the point where the gradient is highest (i.e., for gas wells, the wellhead pressure).

4.3 Results for BHA Pressure Loss

Butler and Gregory (3) presented a flow rate dependent equation for pressure loss across the bottom-hole assembly, but noted that the pressure loss across the motor would also be influenced by the load applied to the motor. Measured field data confirm this, and show that the pressure loss due to motor loading can comprise a significant portion of the overall BHA pressure loss.

Figure 17 shows the pressure loss across the BHA and the instantaneous rate of penetration during the drilling of a four meter segment of Well B. It can be clearly seen that at two points during the drilling of this segment the rate of penetration falls to zero (i.e., the load on the motor is removed). At the first point, the pressure loss across the BHA falls from approximately 4100 kPa to approximately 2000 kPa. At the second point, the pressure loss across the BHA falls from approximately 3750 kPa to approximately 2000 kPa. Throughout the drilling of this well segment, the injection rates were constant. Thus, if the variations in the pressure loss are attributed to changes in the load on the motor, the load dependent pressure loss across the motor makes up approximately half of the overall BHA pressure loss.

5.0 CONCLUSIONS

The application of multiphase flow pressure loss calculations to horizontal underbalanced drilling have been evaluated using measured data for three oil wells and one gas well in Western Canada drilled underbalanced with coiled tubing. The conclusions based on this data can be summarized as follows:

- The modified method of Aziz et al (5, 6) gave the most accurate results for the three oil wells, which would suggest that this model be used in the simulation of multiphase flow pressure losses in underbalanced drilling of oil wells. This is consistent with what has been, for many years, the procedure recommended by of one of the authors (G.A.G.) in dealing with producing oil-gas wells.

- The original method of Aziz et al (5, 6), and the combination of the Gregory (8) model with the Duns and Ros (7) correlation for annular-mist flow gave the most accurate results for the gas well, which would suggest that this model be used in the simulation of multiphase flow pressure losses in underbalanced drilling of gas wells. This is consistent with what has been, for many years, the procedures recommended by of one of the authors (G.A.G.) in dealing with producing gas-condensate wells (11).

- The accuracy of the calculations is sensitive to the pressure basis of the calculations (i.e., are the calculations begun with the measured bottom-hole pressure, or are they done iteratively until the bottom-hole pressure corresponding to the measured surface pressure is found). This is particularly true of calculations for oil wells, for which the ratio of the error in calculations based on the measured surface pressure and the error in calculations based on the measured bottom-hole pressure is seen to be as high as 5 : 1. The drilling engineer must be aware of this fact as many preliminary design calculations are based on the assumption of a fixed wellhead pressure.

- The calculation of the pressure loss across the bottom-hole assembly shows a sensitivity to the load placed on the motor, in addition to a sensitivity to flow rate. The pressure loss due to the load on the motor can make up a significant portion of the overall BHA pressure loss.

- Bottom-hole pressure measurement is extremely valuable where it is technically and economically feasible. Its uses include confirmation of underbalanced conditions throughout the wellbore and calibration of multiphase modelling efforts on site which may allow detection of phenomena like bridging that might not be detected based on measured surface pressures alone.

- The optimization and validation of multiphase flow pressure loss calculations for application in horizontal underbalanced drilling is an ongoing process and further investigations are required.

6.0 REFERENCES

(1) Alberta Recommended Practices For Underbalanced Drilling, Interim Directive ID 94-3, Calgary, Canada, July (1994)

(2) Personal communication, Northland Group of Companies.

(3) Butler, S.D., and Gregory, G.A., "Multiphase Flow Considerations in Underbalanced Drilling of Horizontal Wells", presented at the BHR Group 7th International Conference - Multiphase 95, Cannes, France, June (1995)

(4) Butler, S.D., Rashid, A.U., and Teichrob, R.R., "Underbalanced Drilling - Considerations and Methodology", presented at the First American Horizontal Underbalanced Drilling Conference, Denver, U.S.A., August (1996)

(5) Aziz, K., Govier, G.W., and Fogarasi, M., "Pressure Drops in Wells Producing Oil and Gas", J. Can. Petrol. Technol., Vol. 11, p. 38, July (1972)

(6) Govier, G.W., and Aziz, K., The Flow of Complex Mixtures in Pipes, Van Nostrand-Reinhold Pub. Co., New York (1972)

(7) Duns, H. Jr., and Ros, N.C.J., "Vertical Flow of Gas and Liquid Mixtures in Wells", Sixth World Petroleum Congress, Section 2, Paper No. 22, Frankfurt (1963)

(8) Gregory, G.A., "Comments on the Prediction of Minimum Unloading Velocities for Wet Gas Wells", Technical Note No. 14, Neotechnology Consultants Ltd., Calgary, Canada, December (1989)

(9) American Petroleum Institute, "Subsurface Controlled Subsurface Safety Valve Sizing Computer Program", API Manual 14BM, Second Ed., p. 38, API, January (1978)

(10) Beggs, H.D., and Brill, J.P., "A Study of Two-Phase Flow in Inclined Pipes", J. Petrol. Technol., p. 607, May (1973)

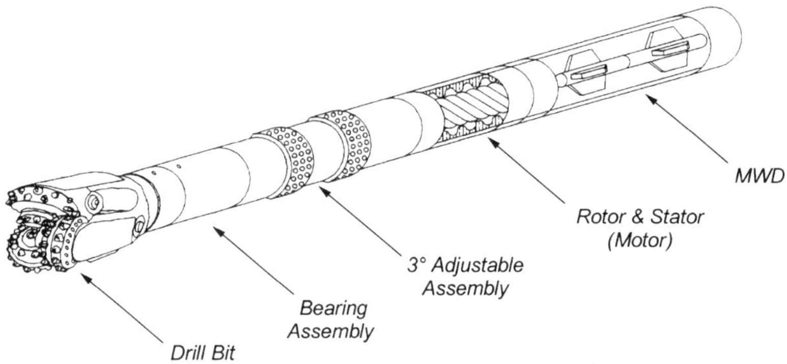

Figure 1

Typical bottom-hole assembly (BHA) for Horizontal Drilling
(courtesy of Computalog Drilling Services Limited)

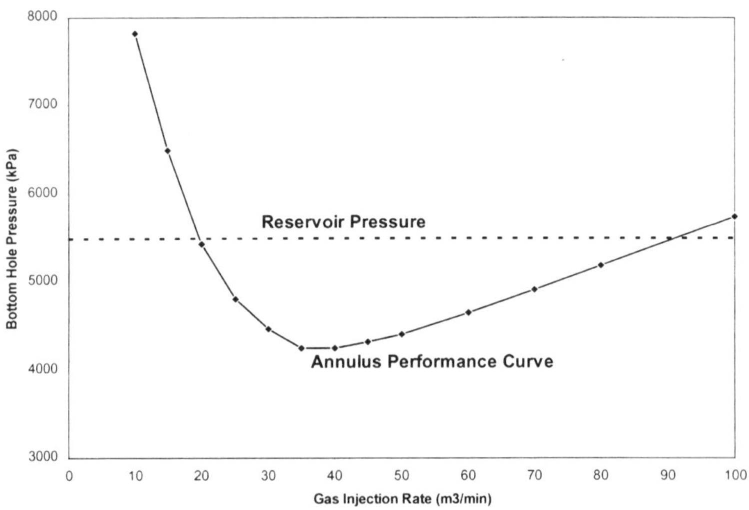

Figure 2
Operating Envelope for Underbalanced Drilling

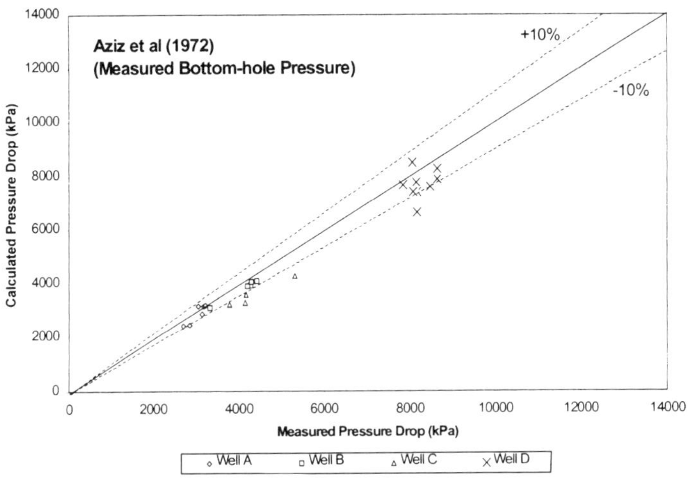

Figure 3

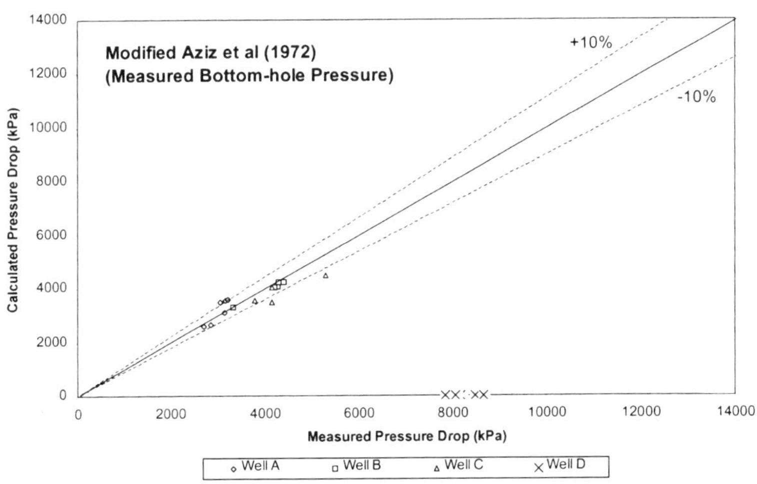

Figure 4

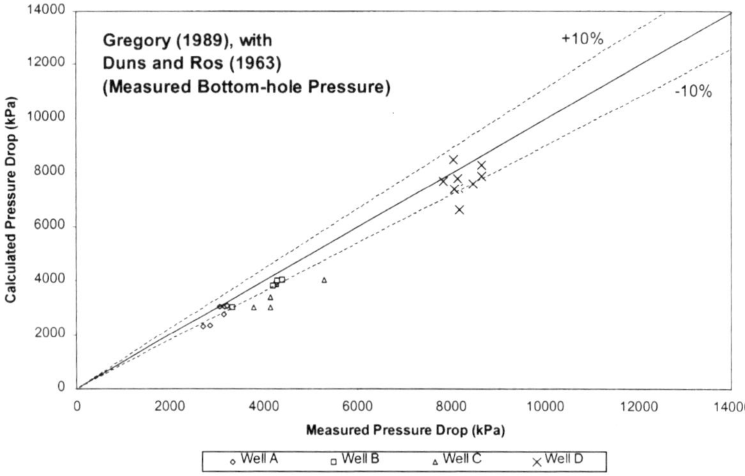

Figure 5

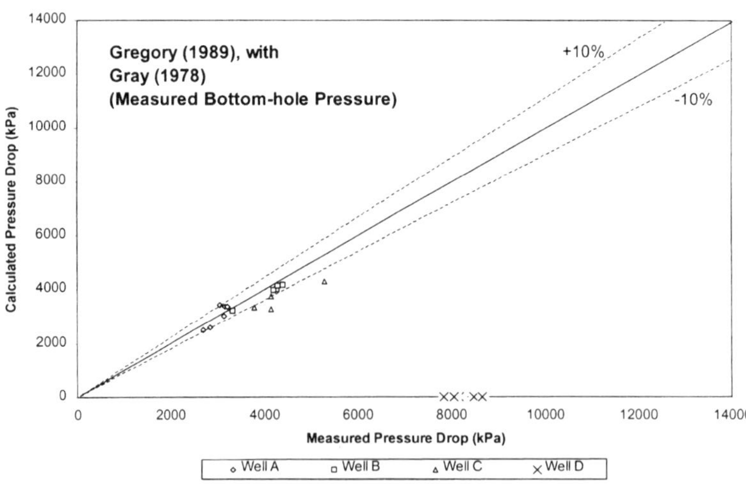

Figure 6

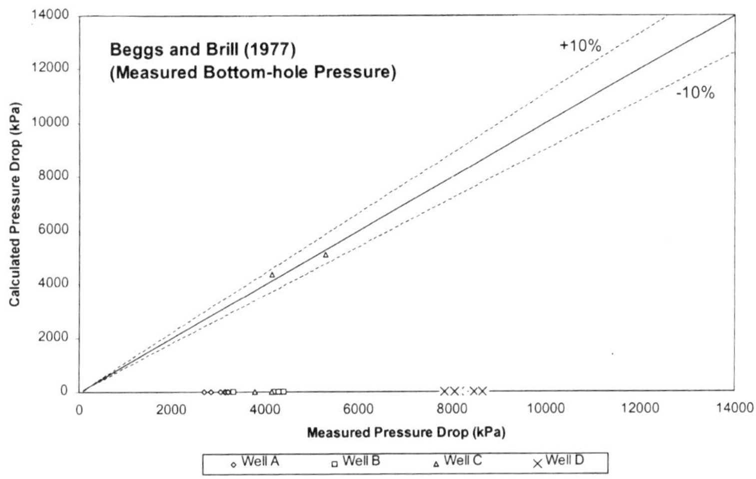

Figure 7

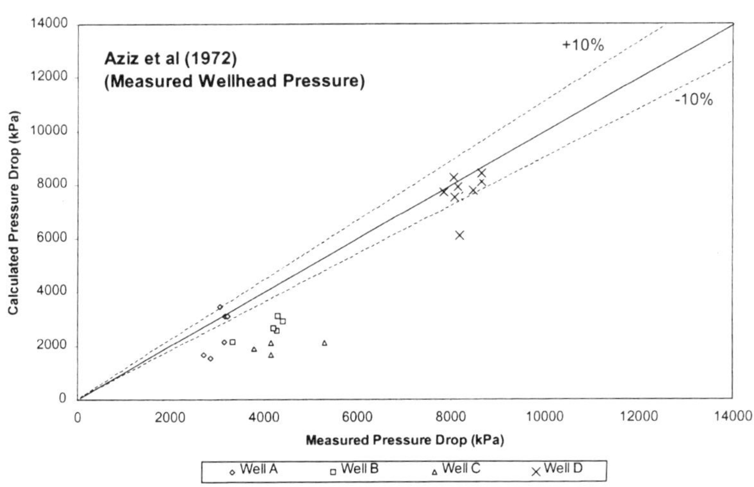

Figure 8

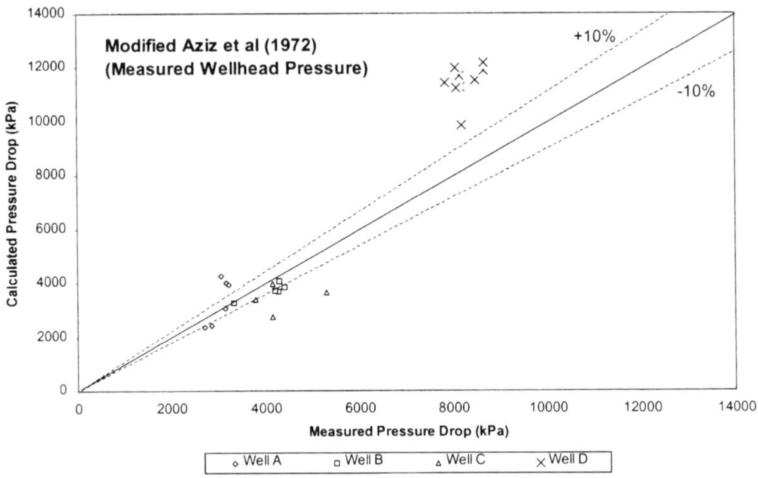

Figure 9

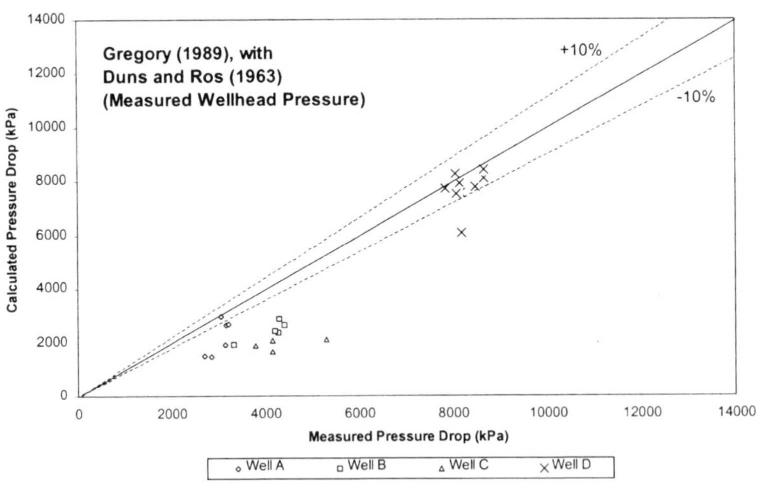

Figure 10

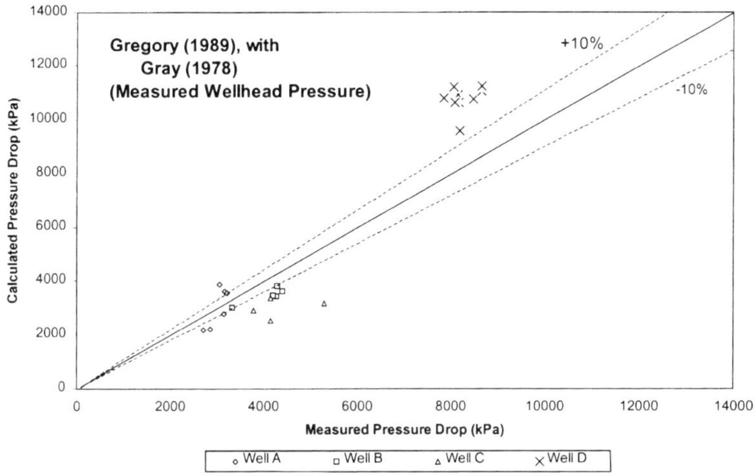

Figure 11

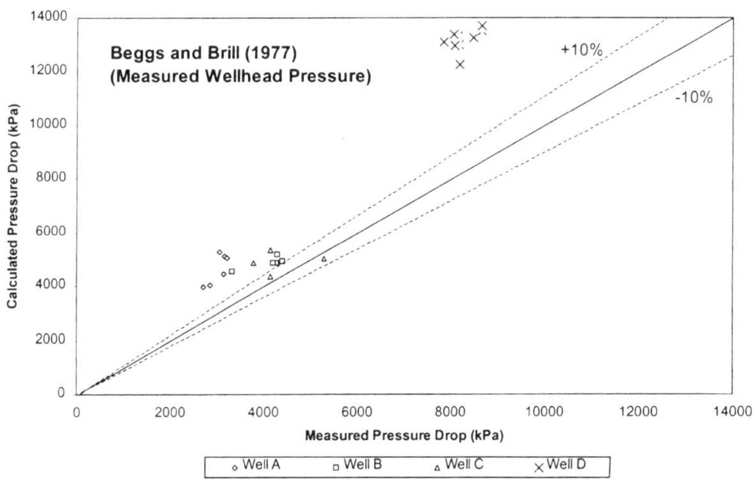

Figure 12

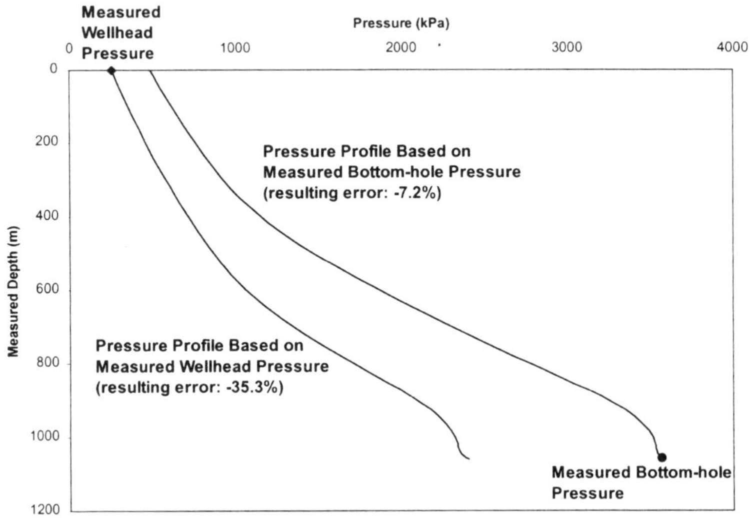

Figure 13
Predicted oil well pressure profiles by both calculation methods

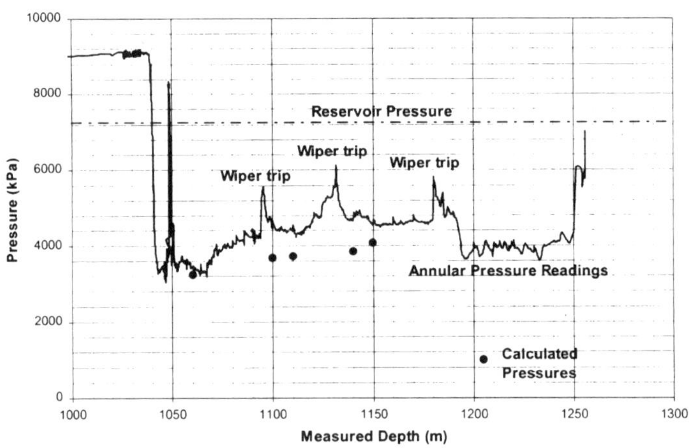

Figure 14
Annular pressure vs. depth for Well B

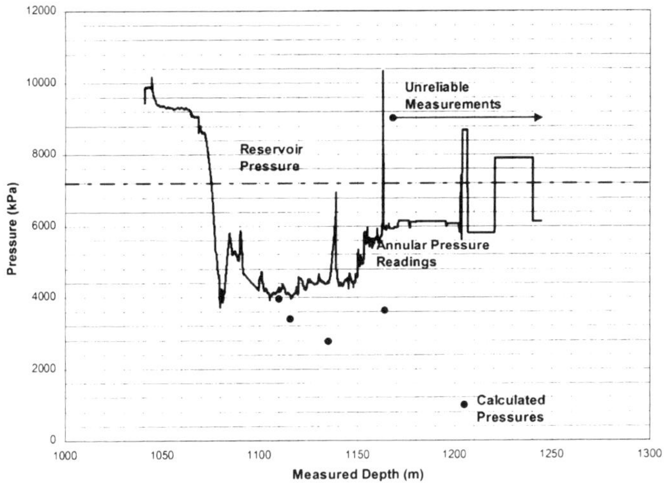

Figure 15
Annular pressure vs. depth for Well C

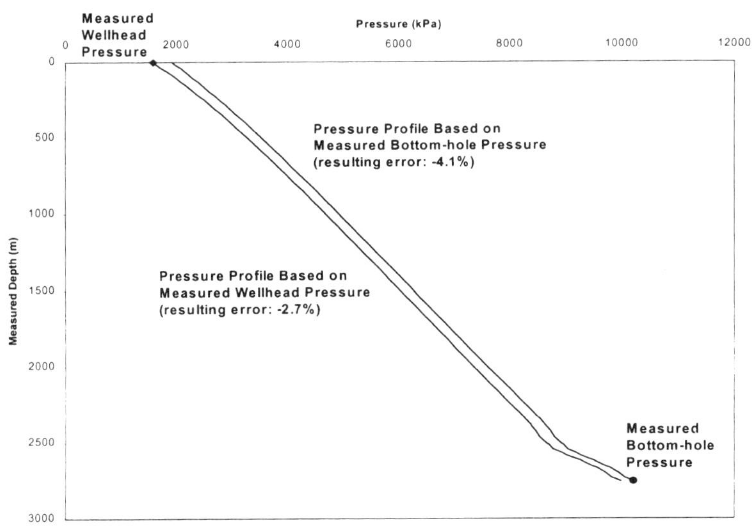

Figure 16
Predicted gas well pressure profiles by both calculation methods

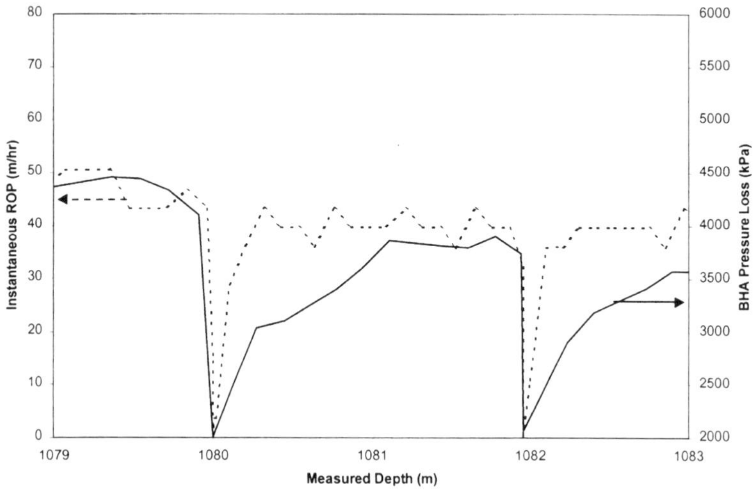

Figure 17
Motor-dependent BHA pressure loss

Using the perforation shot density to secure uniform inflow to horizontal wells with multiphase production

B STRAND
Norsk Hydro ASA, Oslo, Norway
T YTREHUS
Department of Applied Mechanics, Norwegian University of Science and Technology, Trondheim, Norway

1 ABSTRACT

This work presents two-phase gas and liquid flow in horizontal wells with perforation completion and radial inflow. The aim is to establish rational criteria for a perforation shot density which will allow the reservoir to be drained uniformly, in order to avoid gas or water coning. We consider only stratified wellbore flow, and the pressure loss between the reservoir and the well is assumed to take place in annular concentric zones surrounding the perforation cavities. This concept is based on the so-called horizontal-microwell model.

The wellbore flow is modeled as one-dimensional, and the resulting equations are integrated numerically. Mean gas and liquid wellbore velocities, wellbore flowing pressure, liquid holdup fraction and perforation shot density as a function of position along the wellbore are calculated, and effects of production rates and formation permeability are included in the analysis.

The results demonstrate that a uniform perforation shot density can cause the reservoir to be drained at a higher rate at the downstream end of the wellbore than at the upstream end, and that the perforation shot density must decrease in a unique manner downstream in order to maintain a uniform production.

2 INTRODUCTION

The production from horizontal wells and reservoirs with thin oil columns is now commonplace, and the resulting production zones are typically of a considerable length, of the order of one kilometer. Inside the wellbore, there is a pressure loss in the direction of flow, and the wellbore pressure gradient causes the pressure difference between the well and the surrounding reservoir, the so-called drawdown, to increase downstream. As a result of this, the reservoir may be drained at a considerably higher rate at the low-pressure downstream end of the wellbore (hereafter called the toe of the well) than at the high-pressure upstream end (hereafter called the heel of the well).

Effects of pressure drop in horizontal wells on production have been studied by several authors. Folefac, Archer, Issa and Arshad (1) have analyzed the effects of pressure drop along horizontal wellbores, concluding that the pressure drop can be significant. This can lead to coning, meaning that gas above or water underneath the oil column can break through to the wellbore in the low-pressure region towards the heel of the well. This may happen even if the reservoir is otherwise homogeneous, and is not always a desired effect. Dikken (2, pp. 1426 - 1433), has simulated the effect on production performance of pressure drop in horizontal wells with single-phase production, claiming that a proper description of horizontal well flow needs to be included in the design of such wells, especially in situations where coning of water and gas may result. It should be noted however, that according to Novy (3), long horizontal well can offer a large contact area with the productive layers. This can in some instances allow production with a low drawdown, which in turn can reduce the extent of gas or water coning. A solution to the problem of undesired coning may be to choke the inflow from the reservoir to the well. If the degree of choking is a function of the pressure difference between wellbore and reservoir, an axially uniform reservoir drainage can be achieved despite the wellbore pressure gradient. Landman and Goldthorpe (4) and Landman (5, pp. 179 - 188) have modeled selectively perforated wells, using single- or two-phase pressure drop correlations. Marett and Landman (6) suggest a method where the well is divided into a number of discrete regions (four), each with a constant perforation shot density and consider multiphase flow with frictional pressure drop based on standard multiphase pressure drop correlations.

The aim of the present work is to outline a concept for uniform production from horizontal wells with perforation completion and two-phase flow. The idea is to control the inflow by introducing an axially non-uniform and continuous distribution of the wellbore perforations through which the reservoir is drained. Basically, this optimization problem reduces to three separate tasks, all of which are treated in the following: First, the two-phase flow of a well must be described, second, the flow from the reservoir to the well via the perforations must be expressed mathematically, and, third, the desired perforation distribution must be determined. The study undertaken is subject to the following simplifying assumptions:

- Stratified two-phase flow (liquid and gas) only
- Horizontal well
- No mass transfer between the liquid and gas phases
- Quasi-steady flow
- Homogeneous reservoir

The basic mathematical treatment of this work can be extended to include other wellbore flow regimes such as dispersed or annular flow, mass transfer between the phases inside the wellbore and around the perforation cavities, and the problem of three-phase flow. The concept framework and well design philosophy may still remain unchanged. Based on fluid and geometry properties for a typical North Sea wells, the stratified flow assumption can be justified for lower gas and liquid velocities, e.g. based on flow regime prediction by a multiphase simulator such as OLGA.

3 THEORY

3.1 Two-phase well flow

The basic equations of motion for a two-phase flow situation as indicated in Figure A1, may be obtained from the control volume form of the general mass- and momentum conservation laws, i.e. Thompson (7)

$$\frac{d}{dt}\int_V \rho\, dV + \int_S \rho \cdot \left(\vec{u} - \vec{b}\right) \cdot \vec{n}\, dS = 0 \qquad [1]$$

$$\frac{d}{dV}\int_V \rho \cdot \vec{u}\, dV + \int_S \rho \cdot \vec{u} \cdot \left(\vec{u} - \vec{b}\right) \cdot \vec{n}\, dS = -\int_S p \cdot \vec{n}\, dS + \int_S \vec{\tau}\, dS + \int_V \rho \cdot \vec{f}\, dV \qquad [2]$$

where $\vec{u}$ is the local flow velocity, $\vec{b}$ is the velocity of the control surface, $\vec{n}$ is the local normal to that surface, $\vec{\tau}$ is the viscous stress vector, and $\vec{f}$ is an external force per unit mass, i.e. gravity (ignored in the following). Application to the control volumes corresponding to the gas and liquid phases as indicated in Figure A1, yields respectively

$$\frac{d}{dx}(\rho_j \cdot U_j \cdot \alpha_j) = \frac{\dot{m}_j(x)}{\pi \cdot r_w^2} \qquad [3]$$

$$\frac{d}{dx}\left(\rho_j \cdot U_j^2 \cdot \alpha_j\right) = -\frac{dp_{wf}}{dx} - \frac{\tau_j}{\pi \cdot r_w^2} \cdot S_j \pm \frac{\tau_i}{\pi \cdot r_w^2} \cdot S_i \qquad [4]$$

in which

$$U_j = \frac{1}{A_j} \cdot \int_{A_j} u_j\, dA \qquad [5]$$

is the intrinsic area-averaged velocity of phase j, and in which we have put to unity the momentum coefficient γ_j in the expression

$$\gamma_j \cdot U_j^2 = \frac{1}{A_j} \cdot \int_{A_j} u_j^2\, dA \qquad [6]$$

This latter assumption is natural since we consider turbulent flow and practically flat velocity profiles over the cross sections A_j of either of the two phases. Also, the density ρ_j in each phase is considered to be constant across the respective phase areas. In the above expressions,

m_j is the influx of mass per unit axial length of phase j, and α_j is the phase fraction defined as

$$\alpha_j = \frac{A_j}{A_{C.S.}} \qquad [7]$$

where A_j is the cross-sectional area of the wellbore occupied by phase j and $A_{C.S.}$ is the cross-sectional area of the wellbore, see Figure A1. From this definition, we obviously have

$$\alpha_l + \alpha_g = 1 \qquad [8]$$

where the subscripts l and g denote liquid and gas phases, respectively.

Furthermore

- p_{wf} is the wellbore flowing pressure
- x is the axial coordinate, Figure A2
- τ_j is the wall shear stress acting on phase j, Figure A1
- τ_i is the interfacial shear stress, Figure A1
- r_w is the internal radius of the well, Figure A1
- S_j is the wetted perimeter of the wellbore in contact with phase j, Figure A1
- S_i is the width of the interface, Figure A1

The sign of the last term in Equation 4 reflects the fact that the gas "pulls" the oil, and that $(\tau_i)_g = -(\tau_i)_l$.

In wells with perforation completion, the radial inflow from the reservoir will disturb the velocity profile in the pipe. One may expect the inflow through the individual perforations to change the local flow structures in the wall boundary layers, and thereby change the actual friction factor and affect the pressure drop in the wellbore. This phenomenon has been studied by Asheim, Kolnes and Oudeman (8, pp. 97 - 104) (modeling and experiments with single-phase flow in 50 - 150 mm pipes), Su and Gudmundsson (9) (single phase experiments with radial inflow in small diameter pipes), Schulkes, Utvik and Rinde (10, pp. 45 - 55) (experimental and modeling work with 160 mm pipes aimed at quantifying the effect of radial inflow on pressure drop in both single-phase and two-phase pipe flow) and by Kristoffersen, Sveen and Ytrehus (11) (numerical simulations). The results of modeling and experiments found in the open literature render no complete model for the prediction of the effect of radial inflow on two-phase flow, and in the present work, the frictional pressure drop will therefore be calculated expressing shear stresses in terms of friction factors in the standard manner:

$$\tau_j = f_j \cdot \frac{\rho_j \cdot U_j^2}{2} \qquad [9a]$$

$$\tau_i = f_i \cdot \frac{\rho_g \cdot \left(U_g - U_i\right) \cdot \left|U_g - U_i\right|}{2} \qquad [9b]$$

where U_i is the axial velocity of the interface and ρ_j is the density of phase j. For simplicity, we consider a smooth surface and adopt the formulas

$$f_j = 0.046 \cdot Re_j^{-0.2}$$ [10a]

$$f_i = f_g \cdot \left(2 + 2.5 \cdot 10^{-5} \cdot \frac{Re_l}{2 \cdot r_w}\right)$$ [10b]

where

$$Re_j = \rho_j \cdot \frac{D_j \cdot U_j}{\mu_j}$$ [11]

is a Reynolds number and D_j is the hydraulic diameter for phase j and μ_j is the dynamic viscosity of the same phase. D_j is defined as

$$D_g = \frac{4 \cdot A_g}{S_g + S_i}$$ [12a]

for the gas phase, and as

$$D_l = \frac{4 \cdot A_l}{S_i}$$ [12b]

for the liquid phase.

The friction factor f_i in Equation 10b is presented by Laurinat, Hanratty and Dallmann (12), and r_w must be provided in meters. f_j represents turbulent flow in a smooth pipe, as the flow is assumed to be turbulent along the full length of the wellbore.

3.2 Pressure drop across perforations

The well's perforation cavities extend into the formation, and do generally have a rather irregular circumference. A possible approach is to model the perforation cavities as cylindrical cavities, a method which allows for a fairly simple mathematical treatment. Figure A3a displays how the flow from the reservoir is assumed to converge towards the perforations. The distribution of flow along the length of a perforation has been studied both numerically and experimentally by Deo, Tariq and Halleck (13). Their objective was to find out whether a "radial" or "linear" laboratory target, Figures A3b and A3c, could best describe the flow along a perforation cavity. Their results are not conclusive as to which description is more correct, as the flow along the perforations was found to be represented by either the "radial" or the "linear" targets depending, among other factors, on whether the perforation shot density was high or low.

The simplest way to model a cylindrical perforation cavity, however, is probably to include only a strictly radial inflow mode. This concept has been pursued by Perez and Kelkar (14) for two-phase flow across perforations, and has been denoted the horizontal-microwell model, see Figure A3d. Since the aim of this work is to outline a concept for uniform production from horizontal wells, this approach is adopted due to its relative simplicity. Perez and Kelkar make use of Forchheimer's equation (15)

$$\frac{dp}{dr} = \frac{\mu}{k} \cdot v + \beta \cdot \rho \cdot v^2 \qquad [13]$$

in which

- r is the radial coordinate parallel to the direction of flow
- μ is the dynamic viscosity
- v is the superficial fluid velocity through the porous formation
- k is the absolute permeability of the porous medium (a measure of the ease by which a fluid flows through a porous medium)
- β is the velocity coefficient
- p is the pressure

With $\beta = 0$, Equation 13 becomes the well known Darcy's equation applicable to low-velocity cases, and according to Golan and Whitson (16, p. 276), "Most authors consider permeability as the main correlating parameter for β." Based on a literature review, Perez and Kelkar conclude that it appears that the Forchheimer equation "... provides the most appropriate relation to represent the behavior of fluids at high velocities." Equation 13 will be used to model the flow in the proceeding sections of this work.

To develop expressions for the mass fluxes $\dot{m}_j(x)$ of Equation 3, we need to consider the flow through the individual perforations that connect the reservoir to the wellbore. Introducing a continuity requirement for the radial velocity component in polar coordinates

$$\frac{d}{dr}(\rho \cdot v \cdot r) = 0 \qquad [14]$$

conservation of mass for a cylindrical geometry surrounding the perforation cavities is expressed as

$$\hat{\dot{m}}_j = 2 \cdot \pi \cdot r \cdot L_p \cdot \rho_j \cdot v_j \qquad [15]$$

where L_p is the length of the perforation cavities, Figure A2, and $\hat{\dot{m}}_j$ is the mass flow rate across one perforation.

Whenever more phases flow through a porous medium simultaneously, phase j experiences an effective permeability which is smaller than the absolute permeability k, because the permeability of a given phase depends on the fractions of total pore space occupied by that phase and the remaining phases. Relative permeability k_{rj} is in most cases a function of saturation, which is the volume fraction of phase j occupying the pore volume, and effective and relative permeabilities are correlated as

$$k_j = k_{rj} \cdot k \qquad [16]$$

where k_{rj} is the relative permeability of phase j. Generally, $\sum_{j=1}^{n} k_{rj} < 1$, where n is the number of phases occupying the pore volume (16). Equations 13, 15 and 16 combine to

$$2 \cdot \pi \cdot \rho_j \cdot r \cdot L_p \cdot \frac{dp}{dr} = \frac{\mu_j}{k_{rj} \cdot k} \cdot \hat{\dot{m}}_j + \beta_j \cdot \frac{\hat{\dot{m}}_j^2}{2 \cdot \pi \cdot r \cdot L_p} \qquad [17]$$

as the expression for the pressure gradient $\frac{dp}{dr}$ across the porous medium surrounding one perforation cavity.

3.3 Liquid flow

For the flow of liquid, with assumed constant density across perforations, Equation 17 reduces to

$$2 \cdot \pi \cdot \rho_l \cdot L_p \cdot \frac{dp}{dr} = \frac{\mu_l \cdot \hat{\dot{m}}_j}{k_{rl} \cdot k} \cdot \frac{1}{r} + \beta_l \cdot \frac{\hat{\dot{m}}_l^2}{2 \cdot \pi \cdot L_p} \cdot \frac{1}{r^2} \qquad [18]$$

In the introduction, we stated that no phase mass transfer between the liquid and gas phases would be considered. Since this effect is neglected for the wellbore flow, it will not be taken into account for the flow across the perforations either, since the latter is subject to a pressure drop of the same magnitude as the first. Thus, effective permeability $k_{rl} \cdot k$ and velocity coefficient β_l are considered to be constants across the perforations, since k is constant (homogeneous formation assumed), saturation is considered to be constant, and β_l is assumed to be a function of permeability. For the sake of reference, Perez et al. (14) offer a discussion of flow across perforations including pressure dependence of properties such as k_{rl}.

A zone of reduced permeability surrounds the perforation cavities, and this reduced permeability is caused by a change in rock properties which takes place when the perforations are shot. With reference to Figure A2, the inside radius of the perforation cavities is denoted by r_p and the corresponding pressure by p_p. Further, the radius outside the damaged zone, considered as a concentric annulus, is denoted by r_d, and the pressure outside the damaged zone as p_d. Equation 18 may be integrated across the damaged zone to yield

$$(p_d - p_p) = \frac{\mu_l}{2 \cdot \pi \cdot \rho_l \cdot L_p \cdot k_{rl} \cdot k} \cdot ln\left(\frac{r_d}{r_p}\right) \cdot \hat{\dot{m}}_l + \frac{1}{\rho_l \cdot \left(2 \cdot \pi \cdot L_p\right)^2} \cdot \beta_l \cdot \left(\frac{1}{r_p} - \frac{1}{r_d}\right) \cdot \hat{\dot{m}}_l^2 \qquad [19]$$

An inspection of Equation 19 reveals that the pressure drop across the reduced permeability zone is a function of the liquid mass flow rate across one perforation, $\hat{\dot{m}}_l$, and that the pressure drop may be written on the form

$$p_d - p_p = \Delta p\left(\hat{\dot{m}}_l\right) = \xi_1\left(\hat{\dot{m}}_l\right) + \xi\left(\hat{\dot{m}}_l^2\right) \qquad [20]$$

The function ξ_1 in Equation 20 is identified as the Darcy contribution to the pressure loss, whereas ξ_2 is the finite-velocity contribution. The non-negative solution to Equation 19 with respect to $\hat{\dot{m}}_l$ is

$$\hat{\dot{m}}_l = \frac{ln\left(\frac{r_d}{r_p}\right) - \sqrt{\left[ln\left(\frac{r_d}{r_p}\right)\right]^2 - \frac{4\cdot\rho_l\cdot\left(k_{rl}\cdot k\right)^2}{\mu_l^2}\cdot\beta_l\cdot\left(\frac{1}{r_d}-\frac{1}{r_p}\right)\cdot\left(p_d-p_p\right)}}{\frac{\rho_l\cdot k_{rl}\cdot k}{\pi\cdot\mu_l\cdot L_p}\cdot\beta_l\cdot\left(\frac{1}{r_d}-\frac{1}{r_p}\right)} \qquad [21a]$$

or

$$\hat{\dot{m}}_l = \Psi_1 + \sqrt{\Psi_2 + \Psi_3\cdot(p_d - p_p)} \qquad [21b]$$

where Ψ_1, Ψ_2 and Ψ_3 are functions of well completion and reservoir properties. Note how the mass inflow of liquid per perforation varies with the drawdown $p_d - p_p$.

3.4 Gas flow

Whereas the liquid was assumed to have constant density, the following equation of state is introduced to represent the relation between the density of the gas phase and remaining thermodynamic variables.

$$p_{wf} = \rho_g\cdot Z\cdot R\cdot T \Rightarrow \rho_g = \frac{p_{wf}}{Z\cdot R\cdot T} \qquad [22]$$

where Z is a compressibility factor, T is the wellbore temperature and R is the gas constant of the gas phase. T and Z are taken as constants. Equations 17 and 22 now combine to

$$2\cdot\pi\cdot L_p\cdot\frac{p}{Z\cdot R\cdot T}\cdot\frac{dp}{dr} = \frac{\mu_g}{k_{rg}\cdot k}\cdot\hat{\dot{m}}_g\cdot\frac{1}{r} + \beta_g\cdot\frac{\hat{\dot{m}}_g^2}{2\cdot\pi\cdot L_p}\cdot\frac{1}{r^2} \qquad [23]$$

for the flow of gas across the perforations. Again, k_{rg} and k are considered to be constant across the perforation cavities (as for the liquid), and Equation 23 integrates to

$$\frac{k_{rg}\cdot k}{2\cdot\pi\cdot\mu_g\cdot L_p}\cdot\beta_g\cdot\left(\frac{1}{r_d}-\frac{1}{r_p}\right)\cdot\hat{\dot{m}}_g^2 - ln\left(\frac{r_d}{r_p}\right)\cdot\hat{\dot{m}}_g \ldots$$

$$\ldots + \pi\cdot L_p\cdot k_{rg}\cdot k\cdot\frac{1}{\mu_g\cdot Z\cdot R\cdot T}\cdot\left(p_d^2 - p_p^2\right) = 0 \qquad [24]$$

The non-negative solution to Equation 24 with respect to is

$$\hat{\dot{m}}_g = \frac{ln\left(\frac{r_d}{r_p}\right) - \sqrt{\left[ln\left(\frac{r_d}{r_p}\right)\right]^2 - \frac{2\cdot\left(k_{rg}\cdot k\right)^2}{\mu_g^2\cdot Z\cdot R\cdot T}\cdot\beta_g\cdot\left(\frac{1}{r_d}-\frac{1}{r_p}\right)\cdot\left(p_d^2-p_p^2\right)}}{\frac{\rho_g\cdot k_{rg}\cdot k}{\pi\cdot\mu_g\cdot L_p}\cdot\beta_g\cdot\left(\frac{1}{r_d}-\frac{1}{r_p}\right)} \qquad [25a]$$

or

$$\hat{m}_g = \Psi_4 + \sqrt{\Psi_5 + \Psi_6\cdot\left(p_d^2 - p_p^2\right)} \qquad [25b]$$

where Ψ_4, Ψ_5 and Ψ_6 are functions of well completion and reservoir properties. We note the resemblance between Equations 21b and 25b: Whereas the pressure parameter for the liquid phase in Equation 21 is the drawdown $p_d - p_p$, the corresponding parameter for the gas phase is $p_d^2 - p_p^2$. Notice also how $\hat{\dot{m}}_g$ may be written as a function of the drawdown, because the pressure parameter $p_d^2 - p_p^2$ in Equation 25a can be rewritten as

$$p_d^2 - p_p^2 = \Delta p\left(\hat{\dot{m}}_l\right) \cdot \left[2 \cdot p_d - \Delta p\left(\hat{\dot{m}}_l\right)\right] \quad [26]$$

as a consequence of Equation 20.

3.5 Perforation shot density

To determine the perforation distribution, two parameters must be taken into account: The perforation shot density (perforations per unit length of wellbore) and the phasing, (the angular pattern of shots around the wellbore). Comprehensive studies on perforation phasing have been undertaken (17), emphasizing the influence of the phasing on a well's productivity. According to Golan et al. (16), "For a given shot density, the phasing that provides the greatest distance between the perforations, and thus the least interference between them, is the most flow-effective." By introducing the fractional perforation $\varphi(x)$ such that

$$2 \cdot \pi \cdot r_w \cdot \int_0^x \varphi(z)\, dz = A_p(x) \quad [27]$$

where $A_p(x)$ is the total perforated area of the well's tubing between axial positions 0 and x, the perforation shot density $n(x)$ may be expressed as

$$n(x) = \frac{\varphi(x) \cdot 2 \cdot \pi \cdot r_w}{\pi \cdot r_p^2} = 2 \cdot \frac{r_w}{r_p^2} \cdot \varphi(x) \propto \varphi(x) \quad [28]$$

where r_p is the inner radius of an individual perforation considered as a cylindrical cavity, Figure A2. Thus, via the fractional perforation $\varphi(x)$, the perforation shot density $n(x)$ is determined without regard to the phasing. By making use of the fractional perforation concept, we shall leave the phasing optimization out of the remaining discussion.

If M_j is the total mass flow rate of phase j produced from the reservoir, a requirement for uniform production along the length of the wellbore is

$$\dot{m}_j = \frac{\dot{M}_j}{L_w} \quad [29]$$

where L_w is the total producing length of the well. The perforation shot density $n(x)$ defined in Equation 28 relates to mass flow rates via the equation

$$\hat{\dot{m}}_j \cdot n(x) = \dot{m}_j \quad [30]$$

and Equations 29 and 30 may be written

$$n(x) = \frac{\dot{M}_j}{L_w \cdot \hat{\dot{m}}_j(x)} \propto \frac{1}{\hat{\dot{m}}_j(x)} \qquad [31]$$

Thus, the perforation shot density $n(x)$ is inversely proportional to the mass flow rate of phase j across one perforation, $\hat{\dot{m}}_j(x)$. We also note that $n(x)$ is both a linear function of total production $\hat{\dot{M}}_j$ and inversely proportional to wellbore length L_w.

Equations 29, 30 and 31 yield

$$\dot{m}_l = \frac{\dot{M}_l}{L_w} \qquad [32a]$$

$$\dot{m}_g(x) = \frac{\hat{\dot{m}}_g(x)}{\hat{\dot{m}}_l(x)} \cdot \frac{\dot{M}_l}{L_w} = \frac{\hat{\dot{m}}_g(x)}{\hat{\dot{m}}_l(x)} \cdot \dot{m}_l \qquad [32b]$$

for uniform production of liquid per unit wellbore length. In the remaining, emphasis will be lent to the case of uniform production of oil, results of which are based on the Equations 32. From Equation 32b, it is noted how the mass flow rate of gas $\dot{m}_g(x)$ from the reservoir per unit length of wellbore equals the liquid mass flow rate multiplied by a scaling factor $\frac{\hat{\dot{m}}_g(x)}{\hat{\dot{m}}_l(x)}$, which in turn generally is a function of x.

4 SOLUTION

4.1 Method of solution

The conservation Equations 3 and 4 with the Equation 8 for the phase fractions and Equation 22 for the density of the gas phase incorporated, constitute a system of coupled, non-linear ordinary differential equations in x. Algebraic manipulation renders a system on the form

$$\mathbf{A} \cdot \frac{d\mathbf{X}}{dx} = \mathbf{B} \qquad [33]$$

where

$$\mathbf{A} = \begin{bmatrix} -\rho_l \cdot U_l & \rho_l \cdot \alpha_l & 0 & 0 \\ p_{wf} \cdot U_g & 0 & U_g \cdot (1-\alpha_l) & p_{wf} \cdot (1-\alpha_l) \\ -\rho_l \cdot U_l^2 & 2 \cdot \rho_l \cdot U_l \cdot \alpha_l & 1 & 0 \\ p_{wf} \cdot U_g^2 & 0 & U_g^2 \cdot (1-\alpha_l) + Z \cdot R \cdot T & 2 \cdot p_{wf} \cdot U_g \cdot (1-\alpha_l) \end{bmatrix}$$

$$\mathbf{B} = \begin{bmatrix} \dfrac{\dot{m}_l(x)}{\pi \cdot r_w^2} \\ \dfrac{\dot{m}_g(x)}{\pi \cdot r_w^2} \cdot Z \cdot R \cdot T \\ \dfrac{\tau_i \cdot S_i - \tau_l \cdot S_l}{\pi \cdot r_w^2} \\ -\dfrac{\tau_i \cdot S_i + \tau_g \cdot S_g}{\pi \cdot r_w^2} \cdot Z \cdot R \cdot T \end{bmatrix}$$

$$\mathbf{X} = \begin{bmatrix} 1 - \alpha_l \\ U_l \\ p_{wf} \\ U_g \end{bmatrix}$$

By inserting Equation 9 for the interfacial and wall shear stresses, Equation 32a for the mass flow rate of liquid per unit length, and Equation 32b for the mass flow rate of gas per unit length, the system of Equations 33 can be integrated using a fourth order Runge-Kutta solver.

4.2 Boundary conditions

It is assumed that the pressure p_d outside the zone of reduced permeability (the reservoir pressure) is independent of axial position,

$$p_d \neq p_d(x) \qquad [34]$$

It is recognized that the reservoir pressure may change with axial position along the well. If the variation $p_d(x)$ is known, this may equally well be used as a boundary condition.

The boundary condition for the inside of the perforations is the assumption

$$p_p(x) = p_{wf}(x) \qquad [35]$$

The boundary condition represented by Equation 35 must be provided simultaneously with the calculation of the wellbore pressure, since $p_{wf}(x)$ is an unknown prior to calculations.

Initial wellbore conditions (at x = 0) must be provided for wellbore flowing pressure p_{wf}, wellbore gas and liquid mean velocities U_g and U_l, wellbore holdup fraction α_l, and perforation shot density n. Since the well is sealed in the far upstream end, the obvious boundary condition is zero velocity (and thus zero mass flow rate) at x = 0. However, the system of Equations 33 is singular at x = 0 for $U_l = U_g = 0$. To cope with this, initial velocities are chosen to be small but non-zero, such that $\frac{U_j(0)}{\bar{U}_j} \ll 1$, $\bar{U}_j$ being the velocity of phase j averaged along the wellbore. The far end pressure $p_{wf}(0)$ may, for instance, be chosen as the pressure for which a given production rate generates a specified pressure at the downstream end of the producing section (at $x = L$). It is, however, difficult to find a physical condition to represent the far end value of the holdup fraction $\alpha_l(0)$, and consequently this parameter will be chosen arbitrarily.

5 EXAMPLE CALCULATIONS

The proposed method for establishing an optimized perforation distribution is demonstrated through an example, based on a 1000 m long horizontal well with 159.4 mm inner diameter, and an initial number of 20 perforations per unit length in the toe of the well. The complete set of input data is given in Appendix B. The numerical integration of the above system of equations was carried out for two-phase flow in the wellbore and across the perforations.

5.1 General observations

Observing equations 21 and 31, it is seen that the perforation shot density $n(x)$ for uniform inflow of liquid is

- proportional to total liquid production $\dot{M}_l$
- inversely proportional to wellbore length L_w and liquid mass inflow per perforation $\hat{\dot{m}}_l$
- decreasing with increased effective liquid phase permeability $k_{rl} \cdot k$
- decreasing with increased drawdown $p_d - p_p$

5.2 Effect of initial holdup

As previously mentioned, it is difficult to relate the liquid holdup $\alpha_l(0)$ at the toe end of the well to a physical condition. Figure 1 shows liquid holdup profiles along the wellbore for initial values of 0.04 and 0.64, respectively. Position zero is at the toe of the well, whereas position 1000 m is the downstream end of the perforated section. As can be seen from Figure 1, the two holdup curves converge towards the downstream end of the perforated section.

Shown in Figure 2 is wellbore flowing pressures for the same two initial holdup values. An increase in initial holdup from 0.04 to 0.64 gives an increase in the pressure at the end of the perforated section of approximately 0.2 bar. Thus, the sensitivity of the pressure profile, and in turn the perforation shot density distribution, to the choice of upstream holdup appears to be relatively limited for the chosen case. The remaining results are obtained using an upstream holdup of 0.04.

An increase in upstream holdup from 0.04 to 0.64 also causes a shift in the superficial liquid and gas velocities of approximately plus and minus 0.03 m/s, respectively. Profiles of superficial gas and liquid velocities are provided in Figure 3.

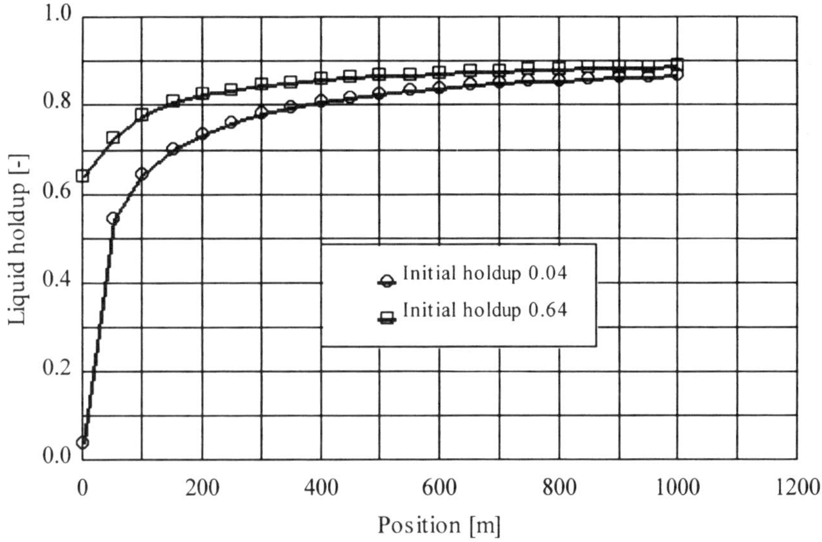

Figure 1 Liquid holdup for different initial holdup values

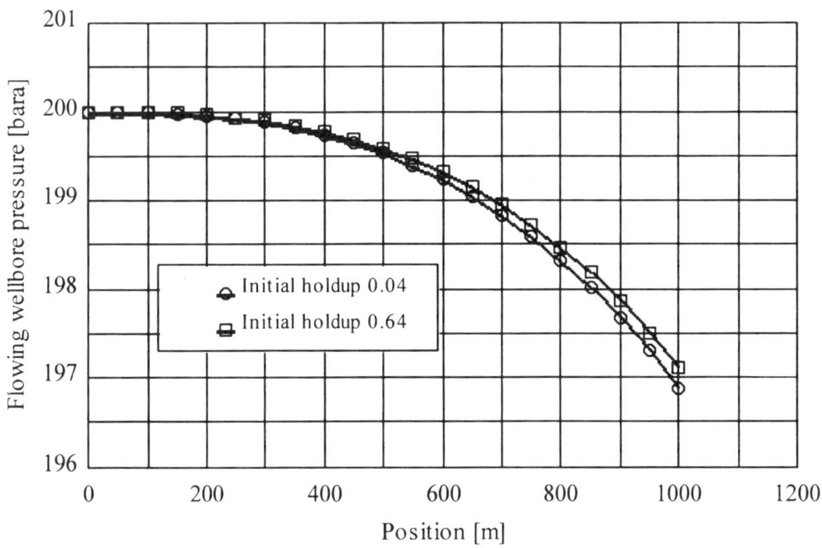

Figure 2 Wellbore flowing pressure for different initial holdup values

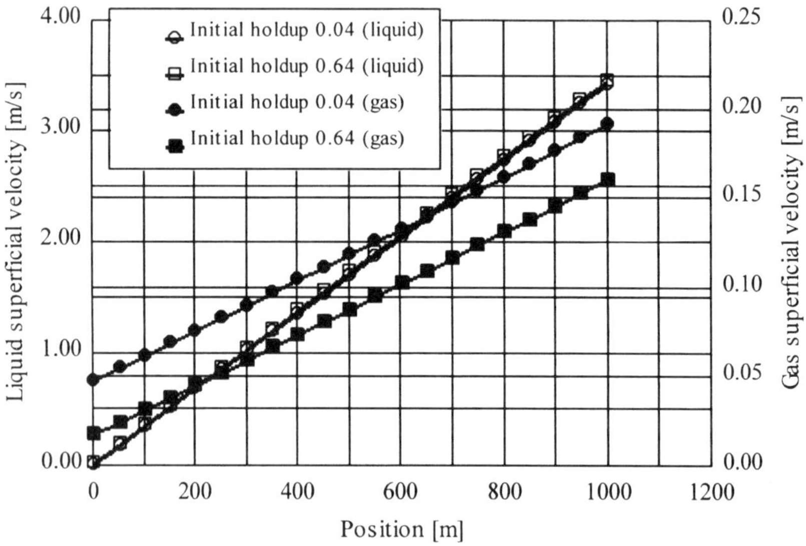

Figure 3 Superficial velocities for different initial holdup values

5.3 Optimized vs. uniform perforation shot density

Figure 4 shows the calculated optimized perforation shot density for the chosen example case (see Appendix B) together with a uniform perforation distribution. For this case, the reservoir pressure p_d is 211 bara, and the perforation shot density must decrease from 20 to approximately 15.5 perforations per unit length from the toe to the heel of the well in order to maintain a uniform liquid production.

Figure 5 shows the calculated liquid and gas inflow rates per unit wellbore length for the example case, based on both optimized and uniform perforation distribution. It is seen how the optimized perforation shot density yields a uniform liquid inflow per unit wellbore length, whereas the uniform perforation distribution in this case leads to an almost 30% increase in inflow along the producing zone of the well.

5.4 Effect of liquid production rate

In Figure 6, the pressure profile along the perforated section of the well is shown based on the example case (70 kg/s liquid production) and on a production of 110 kg/s. The increase in liquid production causes the flowing wellbore pressure to decrease from approximately 197 to 192.5 bara in the heel of the well. As displayed in Figure 7, this increase in production and drawdown calls for a reduction in the perforation shot density along the well's axis. At the downstream end of the producing section, the perforation shot density must decrease from approximately 15.5 to 14 perforations per unit length.

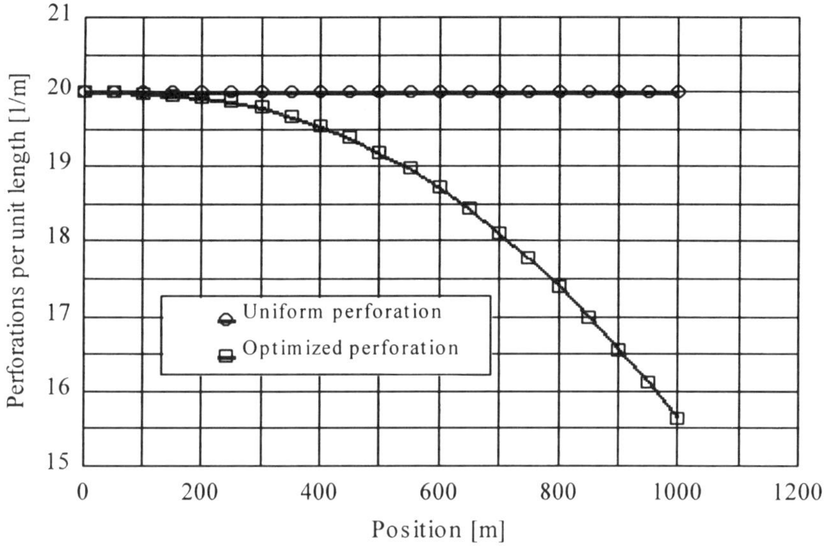

Figure 4 Uniform and optimized perforation shot density

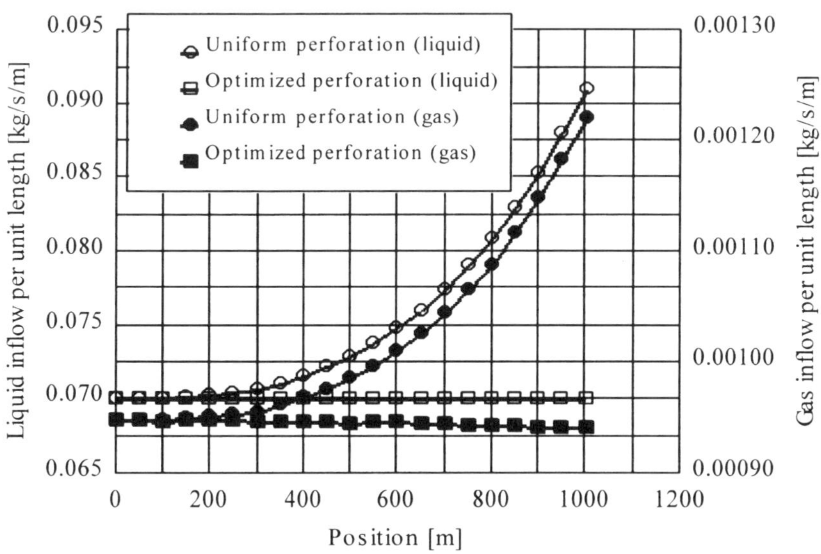

Figure 5 Inflow per unit length for uniform and optimized perforation shot density

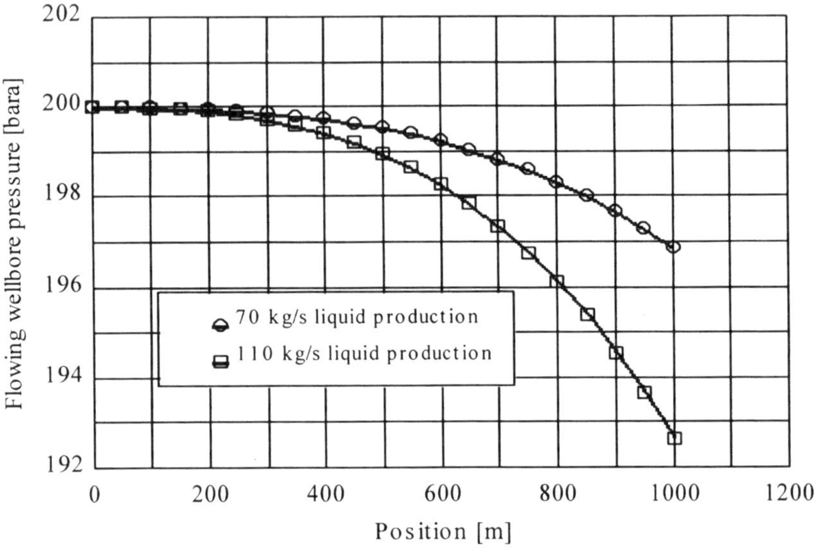

Figure 6 Flowing wellbore pressure for two liquid production rates

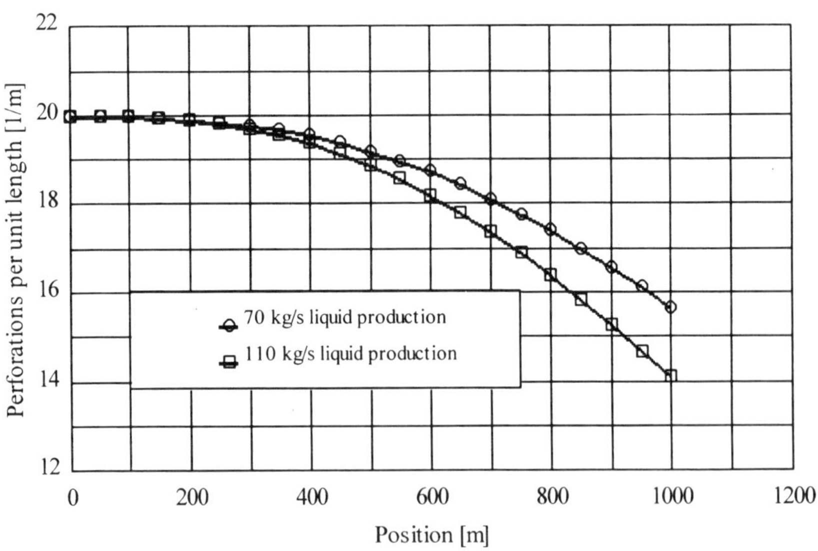

Figure 7 Perforations per unit length for two liquid production rates

5.5 Effect of permeability

The effect in the inflow rate per perforation of increasing the permeability of the zone surrounding the perforations is shown in Figure 8.

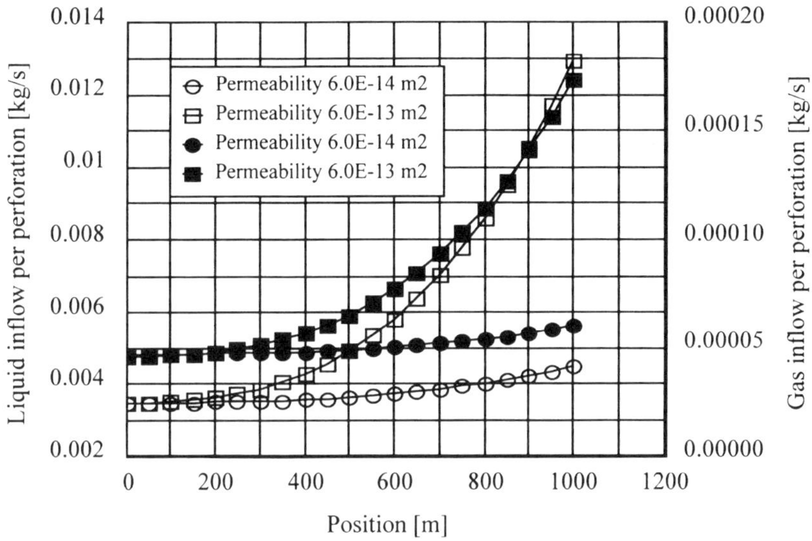

Figure 8 Inflow per unit length for two permeability values

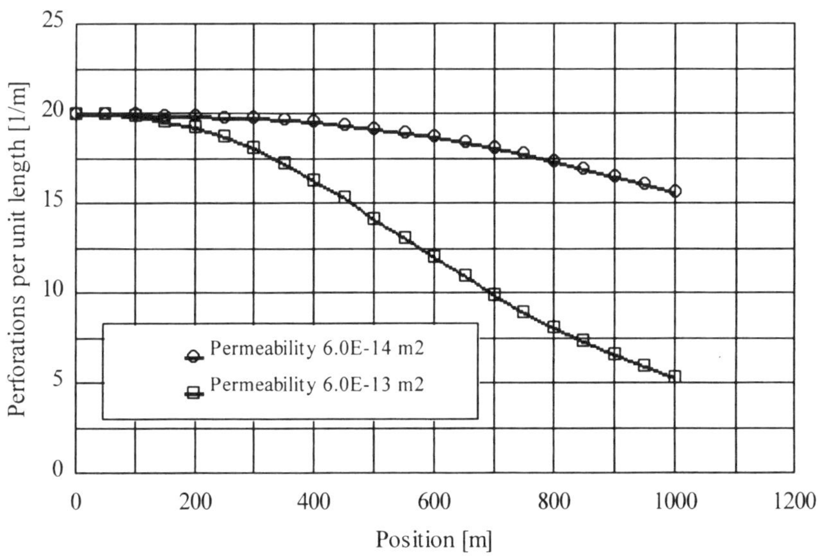

Figure 9 Perforation shot density for two permeability values

To compensate for the increase in inflow brought on by the increase in permeability, the perforation shot density must decrease, see Figure 9.

For the chosen case, the perforation shot density at the downstream end of the perforated section decreases from approximately 15.5 to 5 perforations per unit length. Note that the liquid production rate is the same for these two cases.

6 CONCLUSIONS

The fluid dynamics of a two-phase horizontal well has been modeled, and the well model has been coupled to the reservoir via acknowledged models for inflow across perforations. The model presented allows the optimal perforation shot density for a horizontal well producing from a homogeneous reservoir to be quantified. Based on simulations of an example case, the following is concluded:

- Uniform production of liquid requires a perforation shot density which decreases from the upstream towards the downstream end of the well. Depending on the well's completion, production rates of liquid and gas, wellstream properties and the formation's characteristics, the decrease in perforation shot density along the producing section of the well can be significant in order to maintain uniform production
- The perforation shot density is found to be proportional to total oil production, inversely proportional to wellbore length, and to be decreasing with increased permeability and increased pressure difference (drawdown) between the reservoir and the wellbore
- More sophisticated flow models could be incorporated in order to allow for mass transfer between the liquid and the gas phases, three-phase flow, other wellbore flow regimes, and slant wells
- The quantitative results presented here must be regarded as tentative, and more evaluation of formation, fluid and flow properties is needed to establish relations suitable for engineering use

7 REFERENCES

1 Folefac, A.N., Archer, J.S., Issa, R.I., and Arshad, A.M., *Effect of Pressure Drop Along Horizontal Wellbores on Well Performance*, SPE paper 23094 presented at the Offshore Europe Conference, Aberdeen, September 1991

2 Dikken, B.J., Pressure Drop in Horizontal Wells and Its Effect on Production Performance, *Journal of Petroleum Technology*, November 1994

3 Novy, R.A., *Pressure Drops in Horizontal Wells: When Can They Be Ignored?*, SPE paper 24941 presented at the 67th Annual Technical Conference and Exhibition of the Society of Petroleum Engineers, Washington, D.C., October 1992

4 Landman, M.J., and Goldthorpe, W.H., *Optimization of Perforation Distribution for Horizontal Wells*, SPE paper 23005, prepared for presentation at the SPE Asia-Pacific Conference, Perth, Western Australia, November 1991

5 Landman, M.J., Analytic modelling of selectively perforated horizontal wells, *Journal of Petroleum Science and Engineering*, No. *10*, 1994

6 Marett, B.P., and Landman, M.J., *Optimal Perforation Design for Horizontal Wells in Reservoirs With Boundaries*, SPE paper 25366, prepared for presentation at the SPE Asia Pacific Oil & Gas Conference & Exhibition, Singapore, February 1993

7 Thompson, P.A., *Compressible-fluid dynamics*. New York: Mc-Graw Hill Book Company, 1972

8 Asheim, H., Kolnes, J., and Oudeman, P., A flow resistance correlation for completed wellbore, *Journal of Petroleum Science and Engineering*, No. *8*, 1992

9 Su, Z. and Gudmundsson, J.-S., *Friction Factor of Perforation Roughness in Pipes*, SPE paper 26521, 68th Annual Technical Conference and Exhibition, Houston, October 1993

10 Schulkes, R., Utvik, O.H., and Rinde, T., Pressure drop in horizontal wells: multiphase flow with radial inflow. Proceedings Multiphase '97, BHR Group Limited, Cannes, France, 1997

11 Kristoffersen, R., Ytrehus, T., and Sveen, F., *Flow simulations in perforated sections* (Applied Mechanics/IPT Report 7/92, The Norwegian Institute of Technology, Trondheim, Norway)

12 Laurinat, J.E., Hanratty, T.J., and Dallmann, J.C., Pressure Drop and Film Height Measurements for Annular Gas-Liquid Flow, *International Journal of Multiphase Flow*, No. *3*, 1984

13 Deo, M., Tariq, S., and Halleck, P.M., *Linear and Radial Flow Targets for Characterizing Downhole Flow in Perforations*, paper SPE 16896 presented at the 62nd Annual Technical Conference and Exhibition of the SPE, Dallas, September 1987

14 Perez, G., and Kelkar, B.G., *A New Method To Predict Two-Phase Pressure Drop Across Perforations*, paper SPE 18248 presented at the 63rd Annual Technical Conference and Exhibition of the SPE, Houston, October 1988

15 Dake, L.P., *Fundamentals of Reservoir Engineering*, Amsterdam: Elsevier Science Publishers B.V., 1978

16 Golan, M., and Whitson, C.H., *Well Performance, 2nd edition*, New Jersey: Prentice Hall, Inc., 1991

17 Harris, M.H., *The Effect of Perforating on Well Productivity*, paper presented at the 1965 SPE Annual Fall Meeting, Denver, October 1965

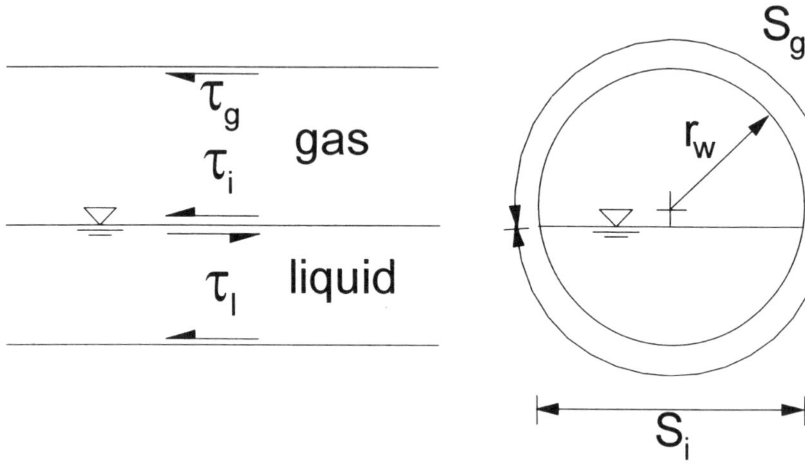

Figure A1 Geometry for two-phase flow

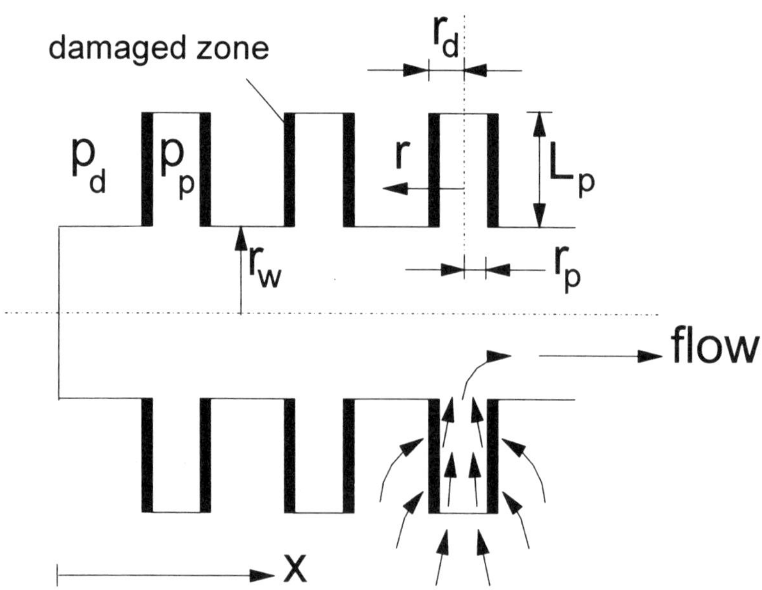

Figure A2 Perforation completion geometry

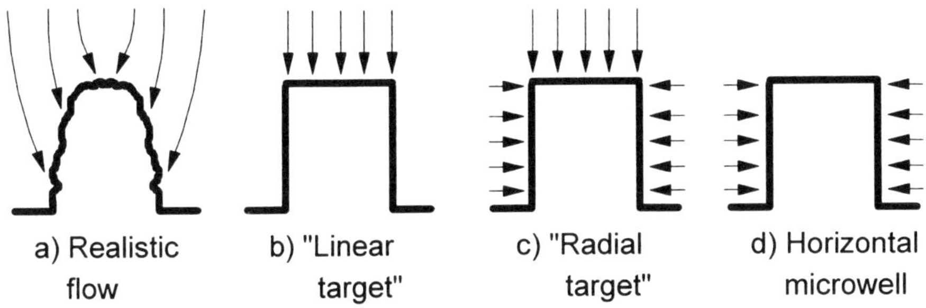

Figure A3 Perforation inflow models

APPENDIX B DATA FOR SIMULATION OF EXAMPLE WELL

Producing length of wellbore, L_w	1000 m
Inner diameter of wellbore	0.1594 m
Inner radius of perforations, r_p	$6.0 \cdot 10^{-3}$ m
Damaged-zone radius, r_d	$2.0 \cdot 10^{-2}$ m
Perforation penetration depth, L_p	0.10 m
Upstream perforation shot density, $n(0)$	20 m^{-1}
Damaged-zone permeability, k	$6.0 \cdot 10^{-14}$ m^2
Wellbore temperature, T	380 K
Viscosity of gas phase, μ_g	$2.0 \cdot 10^{-5}$ Pa s
Gas phase compressibility factor, Z	1
Gas constant, R	200 J/kg/K
Relative gas permeability, k_{rg}	$1.0 \cdot 10^{-4}$
Gas phase velocity coefficient, β_g	$6.0 \cdot 10^{-3}$ m^{-1}
Viscosity of liquid phase, μ_l	$4.0 \cdot 10^{-3}$ Pa s
Liquid phase density, ρ_l	800 kg/m^3
Relative liquid permeability, k_{rl}	$5.0 \cdot 10^{-1}$
Liquid phase velocity coefficient, β_l	$2.0 \cdot 10^{-2}$ m^{-1}
Upstream holdup fraction, $\alpha_l(0)$	0.04
Upstream wellbore pressure, $p_{wf}(0)$	200 bara
Total liquid production rate, $\dot{M}_l$	70 kg/s

Transient operations and modelling of multiphase pigging in pipes

P C R LIMA
Petrobras SA, Rio De Janeiro, Brazil
H YEUNG
Cranfield University, Bedford, UK

Synopsis

Transient gas-liquid processes and operations involving pigging have been successfully applied in hard environments offshore Brazil.

These pigging operations involve: wax control in deep water multiphase flowlines, water and condensate removal from complex offshore pipeline systems and the use of pigging inside the well to lift the oil. Despite the tremendous success of these operations in the last years, they were performed just based on experience. A new subsea boosting system using gas and pig to transport the accumulated oil in the flowline is also proposed and evaluated. This paper describes some of these operations and a new model to simulate them.

A two-fluid model was developed to determine the transient behaviour of fluids during these operations. The derived one-dimensional set of equations for each flow pattern describe the flow of fluids in all regions. Semi-implicit finite difference schemes were used to solve the initial and boundary value problem for each phase of the pigging process: gas/pig injection, gas shut-in, slug production and gas flow out of the system. A significant improvement upon previous work makes unnecessary to track the position of the slug front. Lab and field data are compared to the results from the theoretical model.

NOMENCLATURE

Variables	Description
A	pipe cross sectional area
C_o	distribution coefficient
c	shear or friction coefficient
g	acceleration caused by gravity
p	pressure
t	time
v	velocity
z	position from pipeline inlet
α	volume fraction
Θ	pipe inclination angle
ρ	density

Subscripts	Description
b	bubble
d	drift
f	film
g	gas
i	gas-liquid interface
l	liquid
m	mixture
s	slug
w	pipe wall

Superscripts	Description
n	time index

1. INTRODUCTION

Transportation of multiphase well products through pipeline-riser systems and from horizontal wells pose a number of design constraints, mainly for deep water installations which cannot be accounted for in existing design methods.

The pigging operations for wax control (Lima and Alves[10] - Fig. 1) , for liquid removal (Lima[7]) or oil production as an artificial lift system (Lima[8]) involve a complex transient phenomena. In deep water and even inside the well, such transient operations cause tremendous concerns from the operational and safety viewpoint.

There is a need to improve current understanding of the pigging hydraulics within the pipeline-riser system to ensure an efficient and safe operation.

Some commercial computer flow simulators are said to predict the flow behaviour of pipeline-riser or well pigging. However, they have shown themselves to be inappropriate for estimating the two-phase flow pigging hydraulics, specially in the new systems described below.

In spite of some advantages of the conventional gas lift methods, it has some inconveniences, such as, the high the back pressure exerted by the gas and oil accumulated in vertical or inclined pipes, which is a limiting factor in oil production. Lima[8,9] proposes two methods of intermittent gas lift systems that eliminate this inconvenience, in which accumulated oil does not cause back pressure to the production zone and a pig can be utilised to increase the efficiency of the pneumatic pumping system.

The first method, named pig lift, is shown in Fig. 2. Two production strings, one shorter than the other, stretch from the well head to one joint piece which is responsible for their interconnection downhole. The connection tool is run together with the longest string, which also carries a standing valve in its lower end. As a consequence of the utilisation of two production strings, several surface components are utilised pair wise. Certain components that appear in the portions interconnected with one of the tubing strings have equivalent ones in the portion interconnected with the other tubing string, with both components performing one single function. The pig lift operation begins with the opening of the pig valve of the launching device, allowing the introduction of at least one pig. Production valves are open while gas valves are closed. This procedure assures the accumulation of oil in both production strings. After a certain period of time, the pig is launched. For this, either one of the production valves is closed and either one of the gas valves is opened; they must be the ones installed in the same portion where the pig was introduced before. Inasmuch as the pressure of the supply gas is higher than the pressure of the existing production strings, the pig is pushed by the gas. The pig descends through the whole production string until it reaches the special coupling; it then initiates the rise through the other production string, causing the displacement to the surface of the liquid accumulated in both production strings. The upward movement of the pig is naturally interrupted right after its passage through first derivation , because the pressures upstream and downstream of the pig become equal; the pig then stops between derivations. A pressure sensor installed in the production line detects the moment at which oil starts flowing out of the well. At this moment, this sensor commands the closure of the gas feeding valve. After certain adjusted time, that assures the pig arrival, the production valve that was closed is opened, thus enabling the beginning of a new filling cycle of the two production strings. After the amount of oil has reached an adequate accumulated value, the pig is launched again, now in the opposite direction and the same operations has to be done in the opposite side.

In the second method, named gas pump, the oil is accumulated in the horizontal region of a subsea production flowline. Fig. 3 shows the subsea gas pump system. This gas pump can be similar to pig lift or can use a gas lift line to introduce the pig in the production flowline. Both, pig lift and gas pump, can be operated without pig. It is very important to predict the overall efficiency of these processes.

It is the aim of this paper to present a model which can predict the transient two-phase flow behaviour in pipeline-riser system and horizontal oil wells and show some comparison with experimental data. A new multiphase pigging model is also shown.

2. LITERATURE REVIEW

During concurrent gas-liquid flow in pipes, a variety of flow patterns can exist. Each pattern results from the particular manner by which the liquid and gas distribute in the pipe. Authors differ somewhat in the name they assign to each of the flow patterns. However, the differences are small and most agree on four flow regimes: stratified, slug, annular and dispersed bubble. Theoretical developments for transient two-phase gas-liquid flow in pipes can be divided into three categories: no-slip flow or homogeneous models, slip mixture flow or drift-flux models, and separated flow or two-fluid models. A comprehensive description of gas-liquid flow categories was given by Scoggins[15] , who was one of the first investigators on this subject in the oil industry.

A number of two-phase transient computational codes are commercially available: OLGA, PLAC and TACITE. OLGA, a dynamic, one-dimensional, extended Two-Fluid Model has been jointly developed by IFE and SINTEF (Bendiksen et. al. [3,4]). PLAC, (Pipeline Analysis Code[1]), has been developed by AEA Petroleum Services. It is based on the structure of TRAC. TACITE (Pauchon et al.[14]), a drift flux model simulator, has been developed under a joint research program between IFP, TOTAL and ELF AQUITANE.

Minami[12] developed a pigging model and coupled it with the Taitel et al.[16] simplified transient model assuming quasi-steady state gas flow. An Eulerean-Lagrangean approach using a fixed and moving co-ordinate system is used. He used mechanistic models for predicting flow pattern, the slippage between phase and the pressure drop. Minami performed an extensive experimental program showing this simplified approach is physically sound.

The work of Liao[11] deserves special attention, since it represents a complete modelling effort of the whole intermittent cycle. The result of this model were compared with the experimental data of Brown and Jessen[6] , Brill et al.[5] and Neely et al .[13], and good matches were obtained.

3. PROPOSED TRANSIENT MODEL EQUATIONS

Different set of equations can be derived for different flow patterns in transient gas-liquid flow in pipes. The proposed model assumes local equilibrium between the gas and liquid phases based on Taitel et al.[16]. The horizontal flow pattern is determined by Beggs and Brill[2] criteria. In the vertical flow it is assumed annular flow for liquid hold-up below 0.2 and bubble flow for liquid hold-up above 0.8. Otherwise it is assumed slug flow. The mass and momentum equations for each phase are solved using a one-dimensional semi-implicit finite difference scheme.

3.1 Annular Flow

Gas momentum equation

$$\frac{1}{\rho_g}\frac{dp}{dz}+\frac{c_i}{\alpha_g}(v_g-v_l)^2+gsin\Theta=0 \tag{1}$$

Liquid momentum equation

$$\frac{1}{\rho_l}\frac{dp}{dz}+\frac{c_{wl}}{(1-\alpha_g)}v_l^2-c_i\frac{\rho_g}{(1-\alpha_g)\rho_l}(v_g-v_l)^2+gsin\Theta=0 \tag{2}$$

Gas continuity equation

$$A\frac{\partial(\alpha_g\rho_g)}{\partial t}+\frac{\partial[(\alpha_g\rho_g A)\ v_g]}{\partial z}=0 \tag{3}$$

Liquid continuity equation

$$A\frac{\partial[(1-\alpha_g)\rho_l]}{\partial t}+\frac{\partial[[(1-\alpha_g)\rho_l A]v_l]}{\partial z}=0 \tag{4}$$

3.2 Stratified Flow

In the case of stratified flow the term $\frac{c_{wg}}{\alpha_g}v_g^2$ should be added to equation (1).

3.3 Bubble or Mist Flow

For bubble flow it is assumed that there is no slippage between gas and liquid phases, so the homogeneous model equations can be used.

- Mixture momentum equation

$$\frac{1}{\rho_m}\frac{dp}{dz}+c_m v_m^2+gsin\Theta=0 \tag{5}$$

- Gas continuity equation

$$A\frac{\partial(\alpha_g\rho_g)}{\partial t}+\frac{\partial[(\alpha_g\rho_g A)\ v_m]}{\partial z}=0 \tag{6}$$

- Liquid continuity equation

$$A\frac{\partial[(1-\alpha_g)\rho_l]}{\partial t}+\frac{\partial[[(1-\alpha_g)\rho_l A]\ v_m]}{\partial z}=0 \tag{7}$$

3.4 Slug Flow

Momentum equations

$$\frac{1}{\rho_g}\frac{dp}{dz}+\left[\left[\frac{c_{wgf}}{\alpha_{gf}}v_{gf}^2\right]+\left[\frac{c_{if}}{\langle\alpha_{gf}\rangle}(v_{gf}-v_{lf})^2\right]+[gsin\Theta]\right]\frac{l_f}{l}$$

$$+\frac{\rho_s}{\rho_g}\left[\frac{f_{ws}}{2D}v_s^2+g\,\mathrm{sen}\,\Theta\right](1-\frac{l_f}{l})=0 \tag{8}$$

$$\left[\left[\rho_g\frac{c_{wgf}}{\alpha_{gf}}v_{gf}^2\right]+\left[\rho_g\frac{c_{if}^n}{\alpha_{gf}^n(1-\alpha_{gf})}(v_{gf}-v_{lf})^2\right]+[\rho_g gsin\Theta]\right]$$

$$-\left[\left[\rho_l\frac{c_{wlf}}{(1-\alpha_{gf})}v_{lf}^2\right]+[\rho_l g\,\mathrm{sen}\,\Theta]\right]=0 \tag{9}$$

where,

$$v_s=\alpha_{gf}v_{gf}+(1-\alpha_{gf})v_{lf} \tag{10}$$

$$\frac{l_f}{l}=\frac{(1-\alpha_g)-(1-\alpha_{gs})}{(1-\alpha_{gf})-(1-\alpha_{gs})} \tag{11}$$

$$v_l=\frac{v_s-\alpha_g v_g}{1-\alpha_g} \tag{12}$$

and

$$v_g=\frac{1}{\alpha_g}\left[v_s\left[1+\alpha_{ls}(C_o-1)-(1-\alpha_g)C_o\right]-v_d(1-\alpha_g-\alpha_{ls})\right] \tag{13}$$

The continuity equations are the same as those for annular or stratified flow.

3.5 Shear Coefficients

The shear coefficients are calculated from friction factors given by the Swamee explicit equation for turbulent flow (Reynolds number greater than 4000). If Reynolds number is less than 2000, the friction factor is 64/Re. Between 2000 and 4000 the friction factor is determined by linear interpolation. The gas-liquid interface friction factor is calculated by the Andritsos correlation for stratified flow and by the Hewitt correlation for annular flow.

4. MODEL SOLUTION

The simplified continuity and momentum equations are solved using a semi-implicit method. The set of finite difference equations is non-linear, therefore to solve them at each time step it is necessary to use an iterative method, based on Newton's Method. Starting with some estimated value for the independent variables at the new time step, the derivatives of the equations with respect to those variables are used to give the next best estimates-based on linear extrapolation from the last value - continuing until the latest estimates are equal (within prescribed tolerance) to the previous ones.

At a time step n+1, given an initial guess of the independent variables (in this case p and α), the values of the next iteration (variables without primes) are assumed to be related to those at the last (with primes) by the relations:

$$p^{n+1} = p'^{n+1} + \delta p \tag{14}$$

$$\alpha_g^{n+1} = \alpha_g'^{n+1} + \delta\alpha_g \tag{15}$$

Since the finite-difference equations are functions of p and α_g, a Taylor expansion about the last iteration's value, retaining only the terms linear in δp and $\delta\alpha_g$ *gives:*

$$\begin{aligned} f(p,\alpha_g) &= f(p' + \delta p, \alpha_g' + \delta\alpha_g) \\ &= f(p', \alpha_g') + \left.\frac{\partial f}{\partial p}\right|_{p=p'} \delta p + \left.\frac{\partial f}{\partial \alpha_g}\right|_{\alpha_g=\alpha_g'} \delta\alpha_g + \textit{higher order terms} \end{aligned} \tag{16}$$

In practice, an expansion is performed only on the gas and liquid continuity equations. v_g^{n+1} and v_l^{n+1} are calculated from the momentum equations as a function of the differential pressure of the last iteration. The solution strategy used in the code is shown in the Fig. 4.

5. PIGGING MODEL

To simulate the pig motion in the pipe, the pipeline is divided into two sections. The first section is from the inlet to the pig, and the second from the pig to the outlet of the pipeline. The velocity of the pig is given by the velocity of the mixture pushing the pig in the previous time step. The pig is assumed to run one cell in each pig time step. As the mass of the cell in front of the pig is known and also the time to run it, we know the mass flow rate entering the section in front of the pig. This mass flow rate will be the boundary condition for the transient calculation from the pig to the outlet. This transient calculation gives the pressure in the pig which will be the boundary condition for the transient calculation from the inlet to the pig. Thus the transient calculation throughout the pipeline in this pig time step is finished. This procedure is repeated until the pig reaches the end of the pipeline. After that the normal transient calculation can be done. The pigging model described above makes unnecessary to track the position of the slug front. Fig. 5 shows the pigging model flow chart.

6. RESULTS

6.1 Experimental Data and Comparisons with the Predictions of the Model

Two-phase transient flow data collected by Minami[12] are compared with the results of the model. The test facility used by Minami is a 420.12 m long, 0.0779 m diameter, horizontal steel pipe that uses air and kerosene as working fluids. Station 1 was installed at 63.7 m and Station 4 at 398.0 m from the inlet. Several types of transient flow behaviour for the different time dependent boundary conditions imposed at the pipeline inlet are compared. Different pig launching schedules and different initial conditions are also compared. Table 1 summarises the transient flow conditions of each run. Fig. 6 to Fig. 13 show the results.

Despite the good agreement with experimental data, the quasi-steady state gas flow assumption of Minami's model make it not suitable for most of gas-liquid transient flows in deepwater pipeline-riser system and for long pipelines. The developed model predictions also matched the lab data for most of the transient operations. However, the presented model has a wider range of application than Minamis's model. Gas flow rate change, liquid flow rate change, liquid build-up and liquid blow down are reasonably predicted.

6.2 Field Operation

Fig. 14 shows the results of a conventional pigging operation for wax control in the subsea flowline connecting RJS403 well to a semi-submersible platform PXX, offshore Brazil. The oil density is 940 kg/m^3 and gas-oil ratio is $60 sm^3/m^3$. The flowline is 4.3 km long and the riser is 620m height. The water depth of the well head is 547m. The average flow rate is 0.00463 m^3/s and the well was producing under severe slugging conditions. The pig arrived at the platform after 9802 s from the launching. The pigging model predicts 9822 s from the launching.

6.3 Hypothetical Operation

A proposed method of liquid transportation by Lima[9] is simulated. The method consists of two main processes. The first process comprises the filling of the horizontal section of a pipeline-riser system. In the second process high gas flow rate, with or without a pig, is injected in order to transport the liquid, accumulated in the horizontal section, to the process facilities at the platform. The processes are related to changing the liquid and the gas flow rate at the inlet of the flowline. The process facilities pressure is kept constant.
The developed code was used to simulate the process described above, with and without a pig, applied to a 5000 m long pipeline connected to a 1000 m height riser. The simulation fluids are air and oil at 20°C. The liquid viscosity is 100 Kg/m-s and the liquid density is 900 Kg/m^3. Results with a pig are shown in Fig. 15 and results without a pig are shown if Fig. 16.

Table 2 gives the flow rates of gas and liquid at the inlet of the flowline in each time of the cycle. The boundary condition at the process facilities is a constant pressure of 10 MPa.
The model predicts high instantaneous flow rates of liquid and gas arriving at the process facilities, which is the main disadvantage of this artificial lift system. The speed of depressurization of the system after the pig arrival is extremely high. The actual systems are not designed to cope with this sort of transient operation.

7. CONCLUSIONS

A more realistic transient two-phase model based on a flow regime dependent set of equations were developed. The slug flow is also mechanistically treated by a two-fluid model.

A new pigging model was developed. It predicted the time of the pig arrival at the platform extremely well, despite the severe slugging condition of the well.

The prediction of the gas pump transient operation is quite consistent, showing the full removal of liquid using pig and some remaining liquid without using a pig (10% average liquid hold-up).

Comparison with transient and pigging operations are good, giving confidence for further development. An experimental programme is currently underway at Cranfield University to collect data for gas pump and pigging of risers.

ACKNOWLEDGMENTS

The authors thank Petrobras for permission to publish this article, and all Petrobras personnel who contributed to and supported this development.

Table 1 - Summary of the transient flow conditions

Run No.	Initial Condition	Operation	Figure No.
17	Steady State two-phase flow. $q_g = 0.086$ sm³/s $q_l = 0.0015$ m³/s	Increase of the liquid flow rate to $q_l = 0.00062$ sm³/s at 301 s.	6
18	Steady State single-phase gas flow. $q_g = 0.064$ sm³/s	Start liquid flow rate of $q_l = 0.0018$ sm³/s at 275 s.	7
19	Steady State two-phase flow. $q_g = 0.067$ sm³/s $q_l = 0.00119$ m³/s	Increase of the liquid flow rate to $q_l = 0.00240$ sm³/s at 313 s.	8
22	Steady State two-phase flow. $q_g = 0.049$ sm³/s $q_l = 0.00085$ m³/s	Increase of the gas flow rate to $q_g = 0.089$ sm³/s at 241 s.	9
3	Steady State two-phase flow. $q_g = 0.065$ sm³/s $q_l = 0.00085$ m³/s	Increase of the gas flow rate to $q_g = 0.105$ sm³/s at 87 s.	10
8	Steady State two-phase flow. $q_g = 0.079$ sm³/s $q_l = 0.00147$ m³/s	Stop liquid flow at 92.1 s.	11
52	Steady State two-phase flow. $q_g = 0.085$ sm³/s $q_l = 0.00042$ m³/s	Pig launching at 66.4 s.	12
56	Steady State two-phase flow. $q_g = 0.077$ sm³/s $q_l = 0.0015$ m³/s	Pig launching at 60 s.	13

Table 2 - Flow rate change at the inlet of the pipeline-riser system

Time (s)		Flow rate of gas (sm³/s)	Flow rate of liquid (m³/s)
with pig	without pig		
0	0	1.16	0.0
10	10	0.0	0.0232
3600	3600	0.0	0.0232
3660 (pig)	4000	8.16	0.0
4400	4600	8.16	0.0
4460	5000	1.16	0.0
7200	7200	1.16	0.0

REFERENCES

No.	Author	Title	Source
1.	AEA Technology	PLAC - An advanced computer program for the analysis of transient multiphase hydrocarbon flows	Technical Manual AEAT -0096, 1996
2.	Beggs, H. D. Brill, J. P.	A study of a two-phase flow in inclined pipes	J. Pet. Technology pp. 607-617, 1973
3.	Bendiksen, K. H. Brandt, I. Fuchs, P. Linga, H. Malnes, D. Moe, R.	Two-phase flow research at SINTEF and IFE: some experimental results and a demonstration of the dynamic two-phase flow simulator OLGA	Offshore Northern Seas Conference, 1986
4.	Bendiksen, K. H. Malnes, D. Moe, R. Nuland, S.	The dynamic two-fluid model OLGA: theory and applications	SPE Production Engng. J. pp. 171-180, 1991
5.	Brill, J. P. Doerr, T. C. Brown, K. E.	An analytical description of liquid slug flow is small diameter vertical conduits	J. Pet. Technology pp. 419-432, 1967
6.	Brown, K. E. Jessen, F. W.	Evaluation of valve port size, surface chokes and fluid fall-back in intermittent gas-lift installations	J. Pet. Technology pp. 315-322, 1962
7.	Lima, P.C.R.	Liquid Removal Process in Pipelines Through a Moving Piston	U.S. Patent 5,389,155, 1995
8.	Lima, P. C. R.	Pig Lift	J. Pet. Technology pp.930-931, 1996
9.	Lima, P. C. R	Method and equipment for offshore oil production by intermittent gas injection	Brazilian Patent Application PI 9602747-9, 1996
10.	Lima, P.C.R., Alves, S.J.N.	Application of Low Density Foam Pig Offshore Brazil	OTC 7864 May, 1995.
11.	Liao, T.	Mechanistic modeling of intermittent gas lift	PhD Thesis U. of Tulsa, 1991

No.	Author	Title	Source
12.	Minami, K.	Transient flow and pigging dynamics in two-phase pipelines	PhD Thesis U. of Tulsa, 1991
13.	Neely, A. B. Montgomery, J. W. Vogel, J. W.	A field test and analytical study of intermittent gas lift	Soc. Pet. Eng. J. pp. 502-512, 1974
15.	Pauchon, C. Dhulesia, H. Lopez, D Fabre, J	A comprehensive mechanistic model for two-phase flow	6th Int. Conference on Multiphase Production, 1993
14.	Scoggins Jr., M. W.	A numerical simulation model for transient two-phase flow in a pipeline	PhD Thesis, U. of Tulsa, 1977
16.	Taitel, Y. Shoham, O. Brill, J. P.	Simplified transient solution and simulation of two-phase flow in pipelines	Chem. Eng. Sci. Vol. 44, No. 6, pp. 1353-1359, 1989
17.	Wallis, G. B.	One-dimensional two-phase flow	McGraw-Hill, 1969

Figure 1 - 200 m^3 of wax removed from a multiphase offshore pipeline

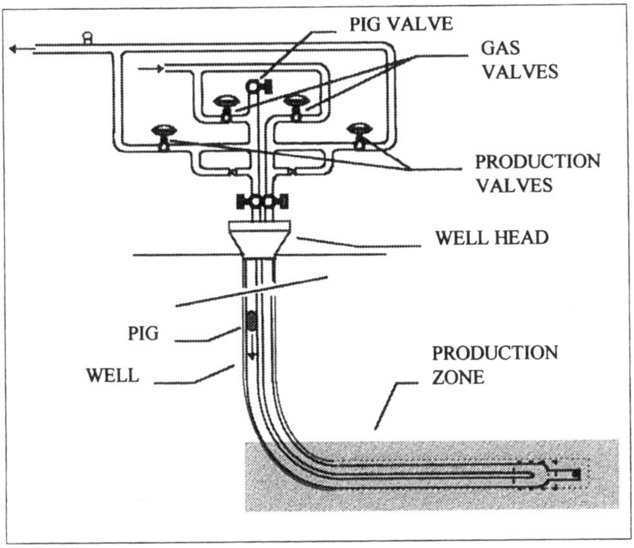

Figure 2 - Pig lift system in horizontal well

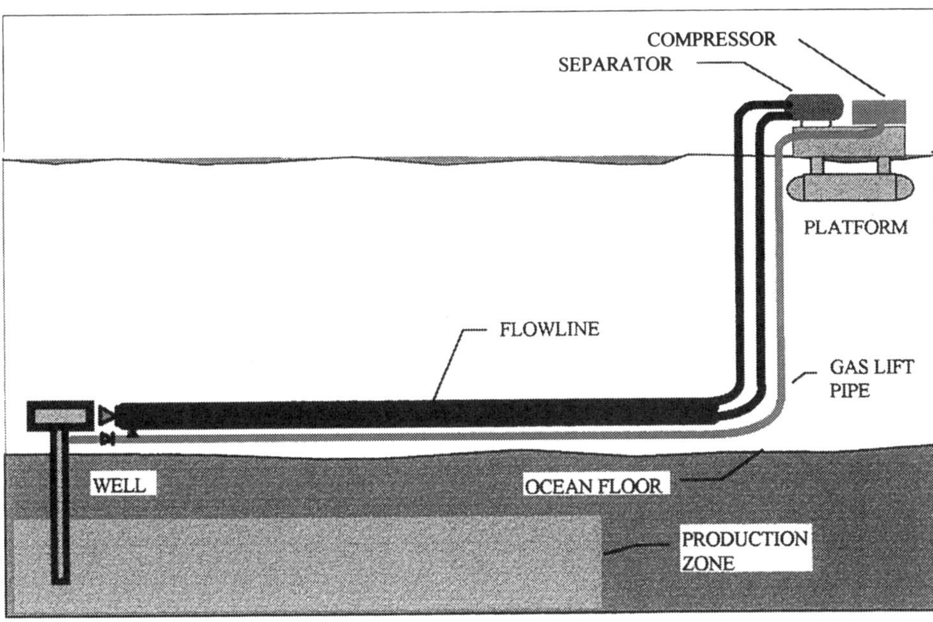

Figure 3 - Gas pump system

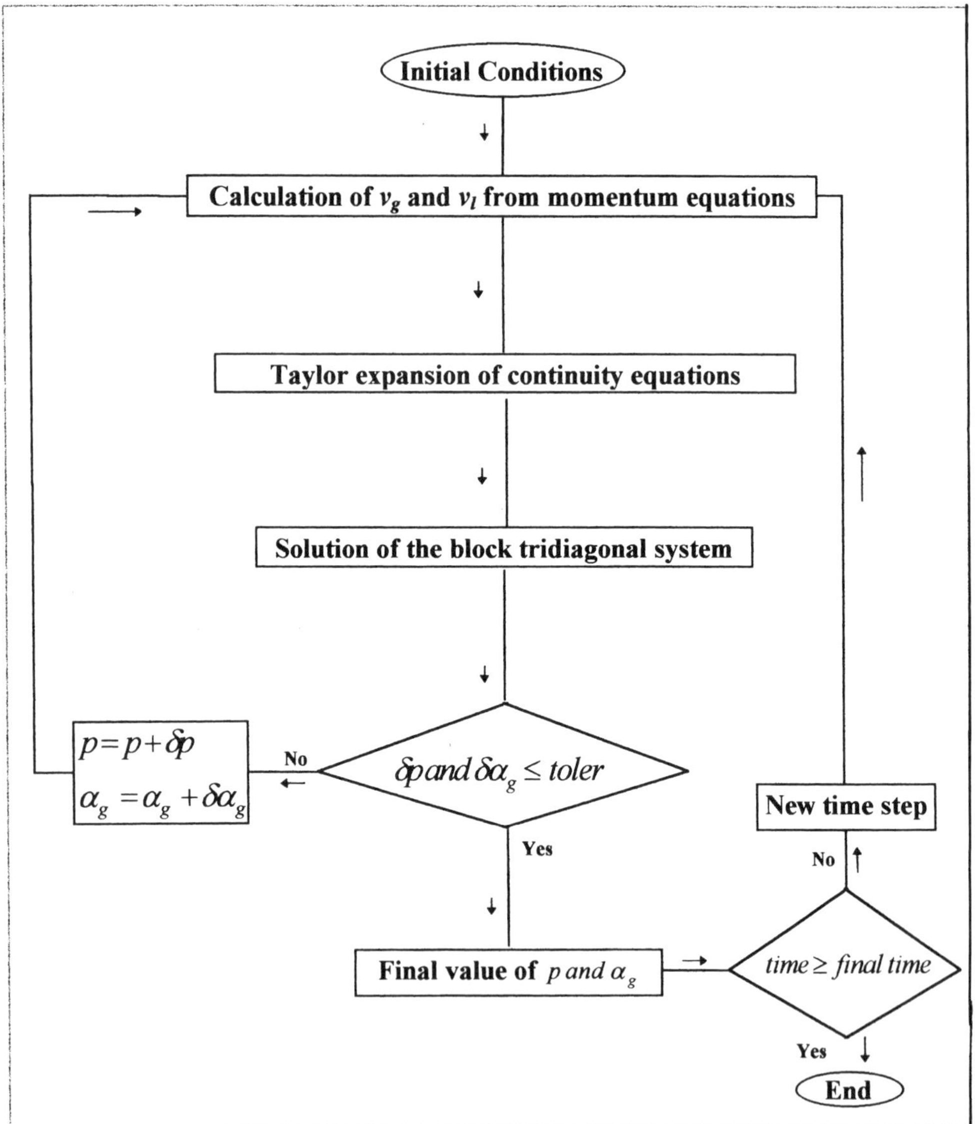

Figure 4 - Code flow chart

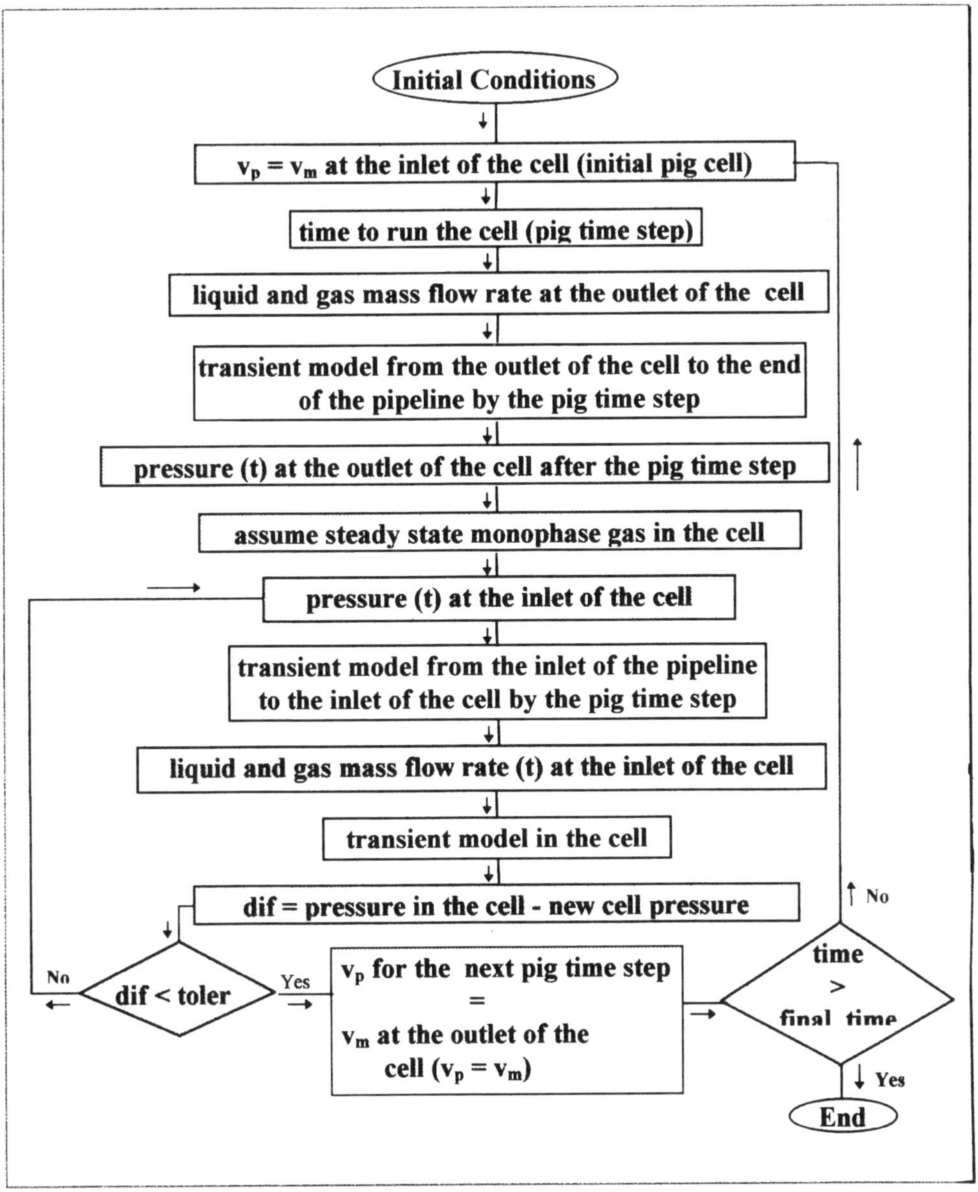

Fig. 5 - Pigging Model Flow Chart

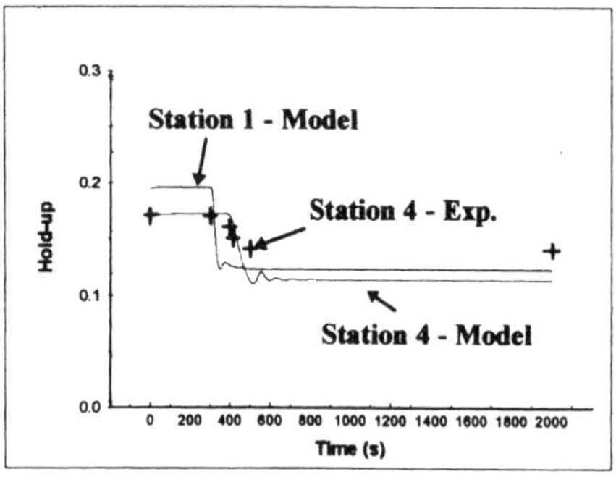

Figure 6a - Run 17 (hold-up)

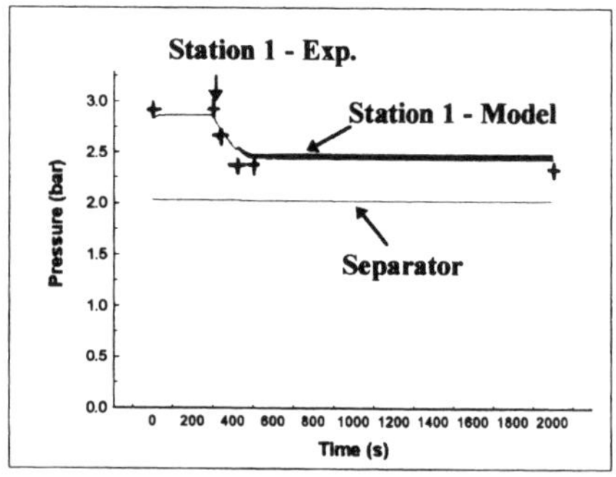

Figure 6b - Run 17 (pressure)

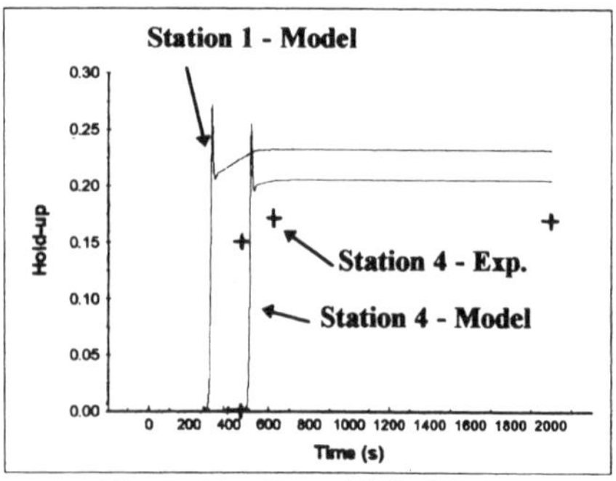

Figure 7a - Run 18 (hold-up)

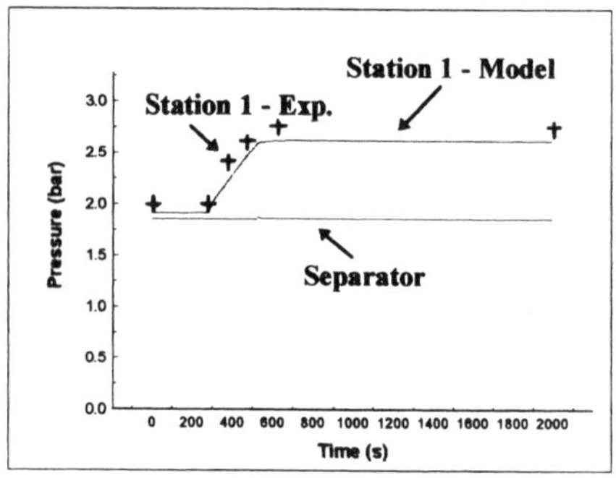

Figure 7b - Run 18 (pressure)

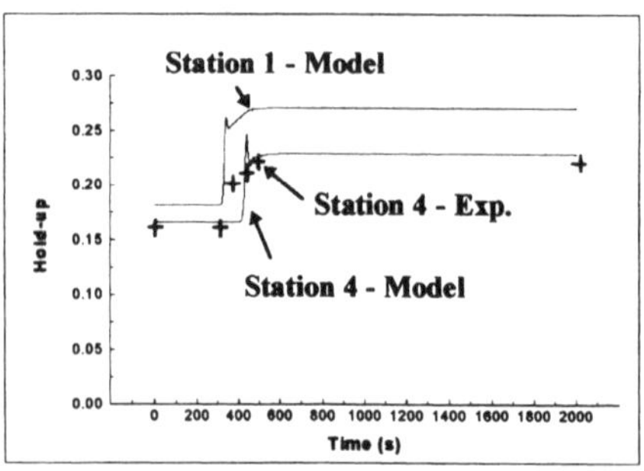

Figure 8a - Run 19 (hold-up)

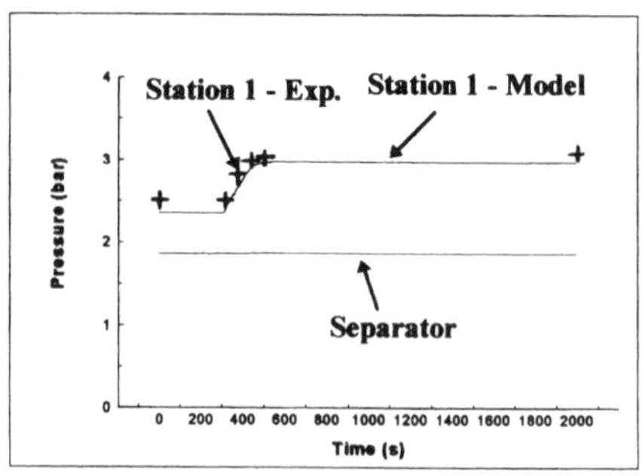

Figure 8b - Run 19 (pressure)

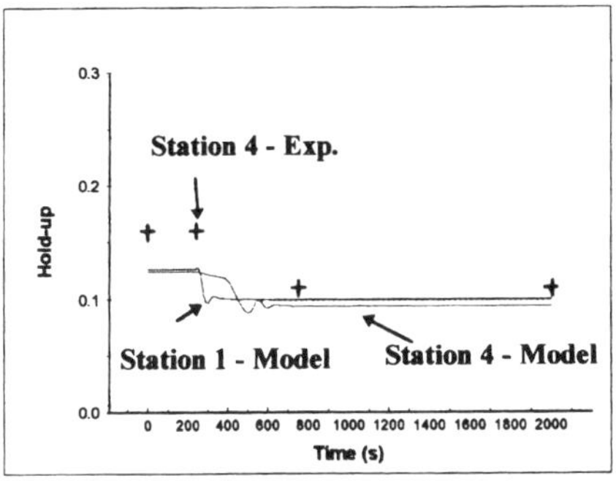

Figure 9a - Run 22 (hold-up)

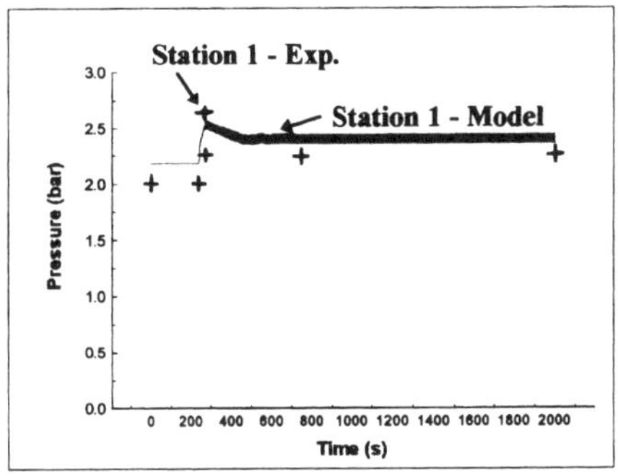

Figure 9b - Run 22 (pressure)

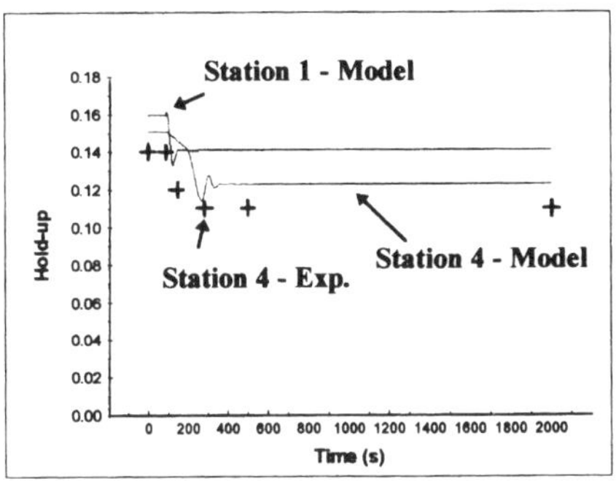

Figure 10a - Run 3 (hold-up)

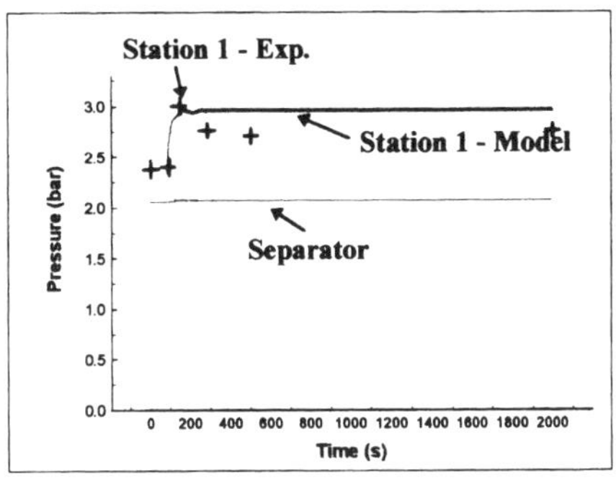

Figure 10b - Run 3 (pressure)

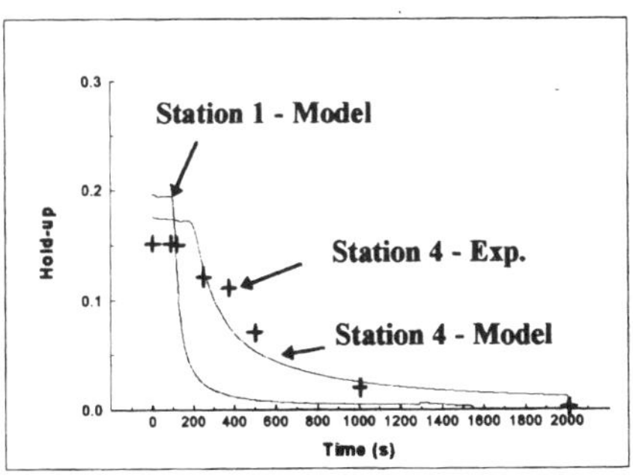

Figure 11a - Run 8 (hold-up)

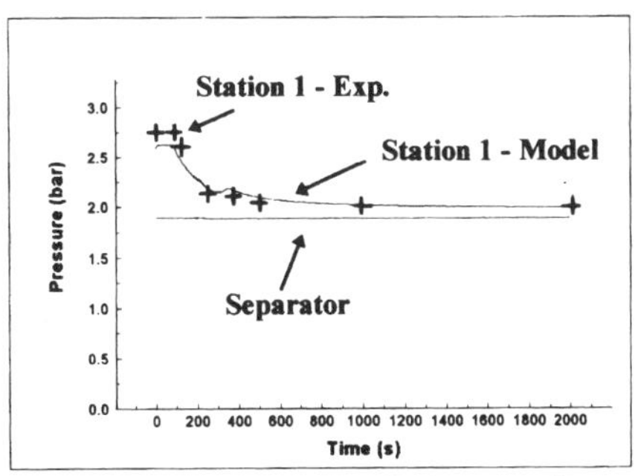

Figure 11b - Run 8 (pressure)

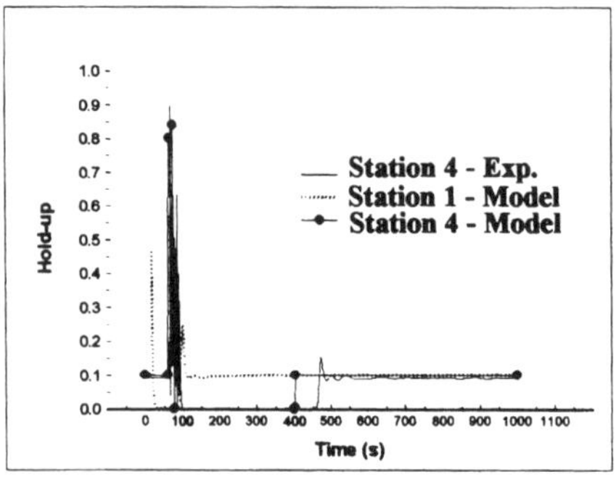

Figure 12a - Run 52 (hold-up)

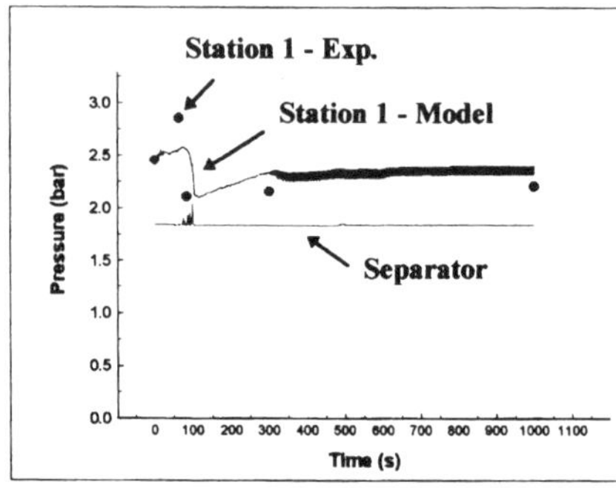

Figure 12b - Run 52 (pressure)

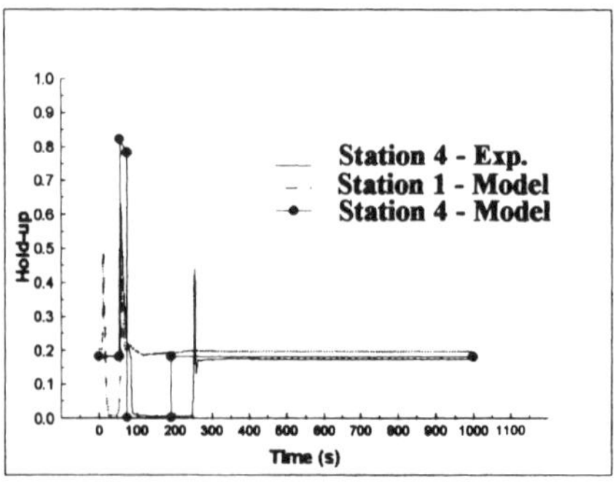

Figure 13a - Run 56 (hold-

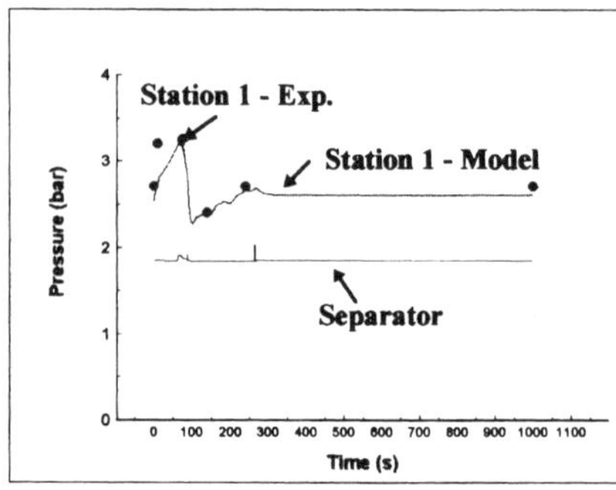

Figure 13b - Run 56 (pressure)

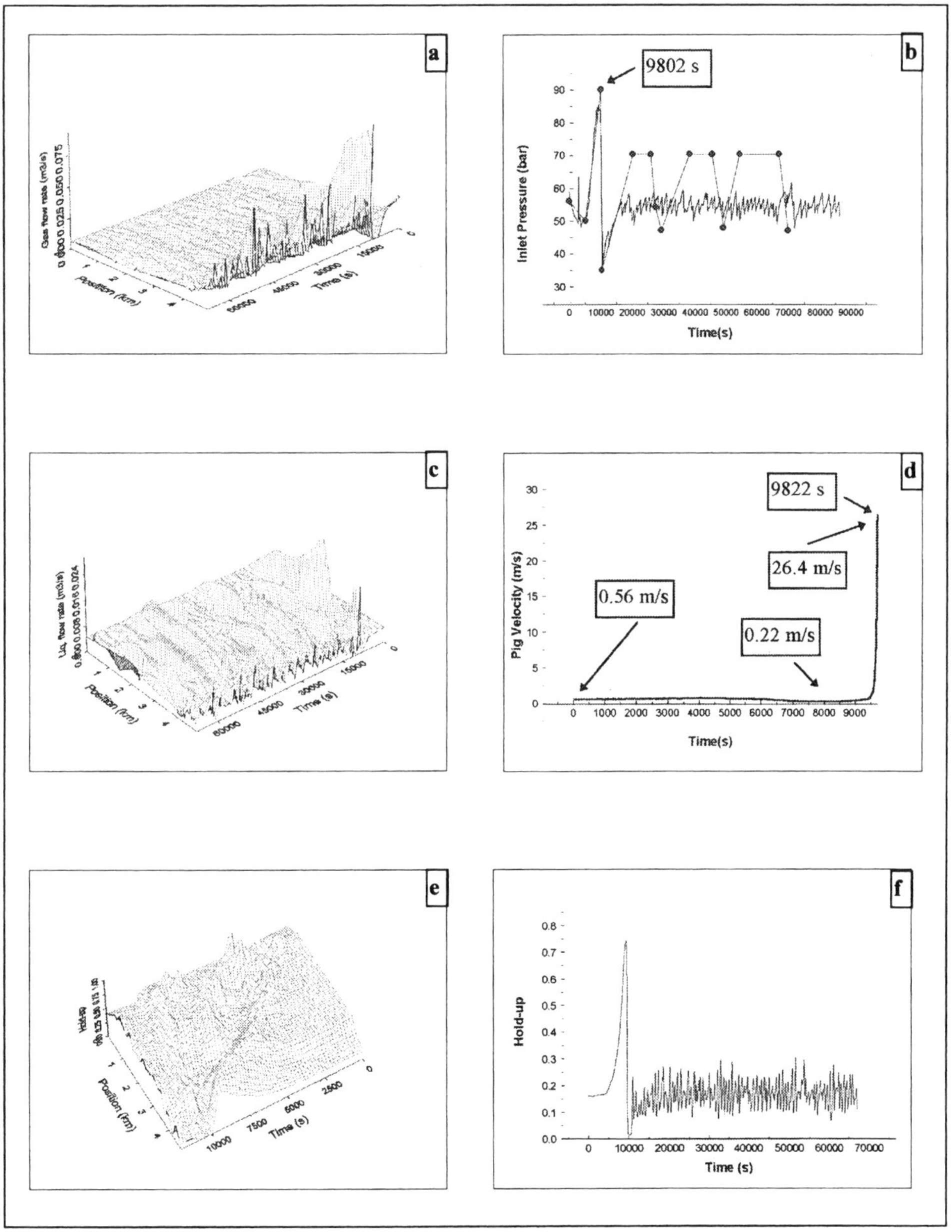

Figure 14 - Pigging of RJS403

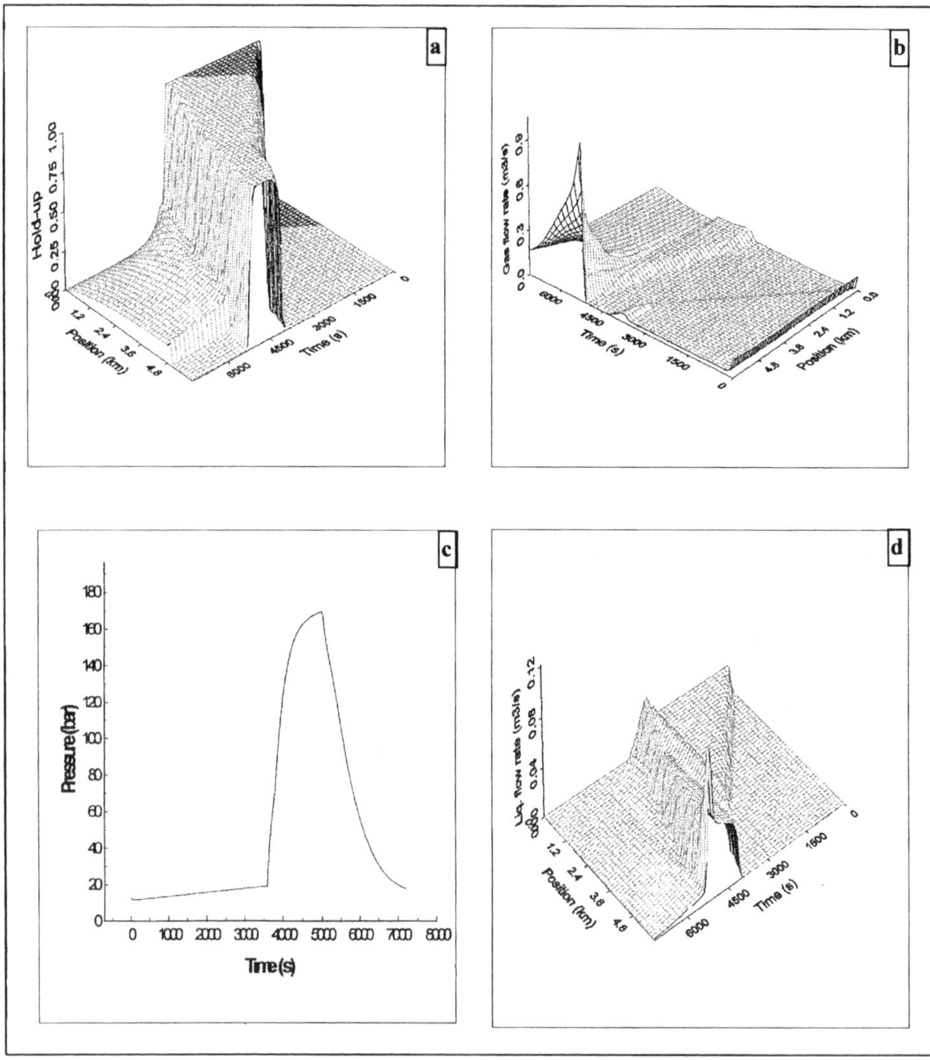

Figure 15 - Gas Pump without Pig

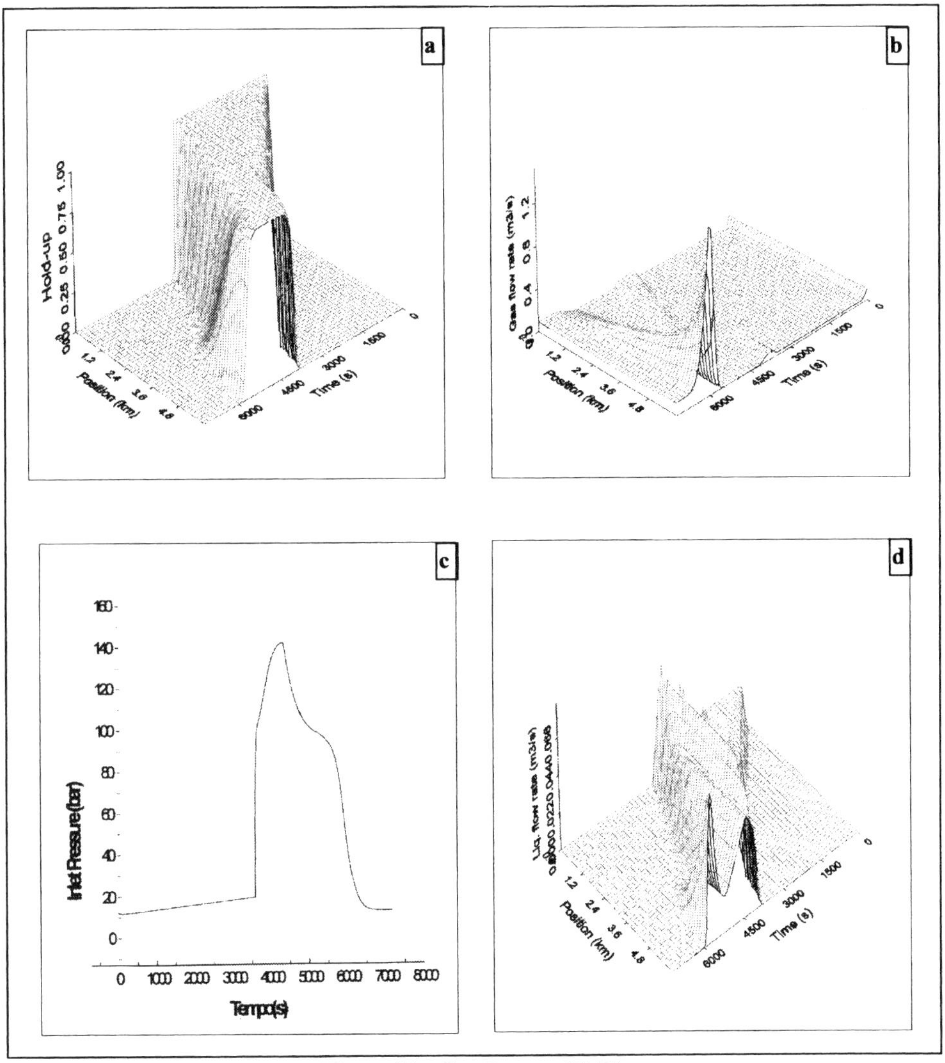

Figure 16 - Gas Pump with Pig

Slugs

Novel design concept and operational enhancements of vessel-type slug catchers for multiphase transfer at the Theddlethorpe gas terminal

R PRICE
Conoco (UK)/Kværner Process Alliance, UK
C THOMAS
Conoco Inc., USA

1 INTRODUCTION

The development concept for the CONOCO(UK) Southern North Sea gas basin involves multiphase transfer through various large-bore pipelines over extended distances. The concept has been developed and refined over a 25 year period which commenced with production from the Viking Fields in 1972.

The Viking infrastructure includes a 138 km long, 28" pipeline to shore transporting fluids of relatively low Condensate to Gas Ratios (CGR), typically in the range of 5-10 M^3/MSCM (1-2 bbls/MMSCF). The water to gas ratio has also been consistently low. Subsequent developments - the Lincolnshire Offshore Gas Gathering System (LOGGS), the Pickerill system and the 188 km pipeline system for Caister-Murdoch - have typically contained greater liquid production both in terms of hydrocarbon condensate cut and potential aqueous phase.

These parallel systems converge on a central gas processing plant at Theddlethorpe St Helens in Lincolnshire, England. The development and operation of this infrastructure has in the past and continues to present a number of technical and logistical challenges. Multiphase transfer challenges include tracking and controlling liquid inventories, particularly in systems with high water cuts, and corrosion control/management which is of particular concern in CO_2-rich gas systems in contact with a free water phase.

This paper outlines key pipeline liquid control mechanisms and focuses on the evolutionary design of Theddlethorpe Gas Terminal Slug Catchers over the last 12 years.

2 C(UK)L SOUTHERN NORTH SEA LAYOUT AND FACILITIES

CONOCO's Southern Gas Basin infrastructure is dispersed over a wide area off the coast of Lincolnshire, Eastern England. See Figures 1 & 2. The infrastructure comprises four central complexes - Viking, LOGGS, Pickerill and Caister-Murdoch (CMS) which are hubs for various fields. Typically, fluids are fiscally metered and re-pressurised where necessary for onwards transmission to shore at the Theddlethorpe Terminal. No offshore processing is undertaken and all fluids are transferred to shore. The characteristics of the four pipelines connecting the central complexes to the Terminal are as follows:

	Size(NB)	Length	Start-Up
Viking	28"	138 km	1972
LOGGS	36"	119 km	1988
Pickerill	24"	66 km	1992
CMS	26"	188 km	1993

The overall system contains more than 90 wells and currently can flow in excess of 56 MSCMD (2 BSCFD) of gas with the potential to increase transportation volumes to over 84 MSCMD (3 BSCFD).

The Viking, LOGGS and Pickerill fields are typically of Rotliegendes quality which is relatively sweet and lean - less then 0.5 mol% carbon dioxide and 90-95 mol% methane. However the CMS system is typically of Carboniferous or Bunter quality with CO_2 levels up to 3.5 mol%, Nitrogen content ranging up to 14 mol% and methane levels between 80-85 mol%. CMS fields also tend to have a greater heavy hydrocarbon content which generates condensate to gas ratios in the 50-100 M^3/MSCM range (10-20 bbls/MMSCF). The commercial transportation agreements contractually allow all fields to produce up to 3 bbls/MMSCF of associated water without penalty. Consequently, large volumes of methanol are provided for hydrate inhibition adding to the multiphase transportation effect of the system.

This concoction of field qualities, liquid generation and wet gas transportation produces a unique array of challenges for CONOCO's Southern North Sea operation necessitating close monitoring and management.

2.1 Blending

The mixture of high and low quality systems converging on Theddlethorpe enables poorer quality fields to be homogeneously mixed with Rotliegendes fields to produce a combined Sales Gas acceptable for the UK gas distribution system. This philosophy has allowed lower volume, low specification gas to be economically recovered from the Southern Gas Basin without recourse to further, expensive gas treatment (e.g. CO_2 amine stripping or cryogenic nitrogen removal). Blending Management at Theddlethorpe is a key process that can affect CONOCO from a safety, commercial and financial standpoint. The on-line monitoring systems have a level of redundancy that prevents any low specification gas reaching the gas distribution system from the Terminal. It has spawned a successful pipeline integrity monitoring system (PIMS) for multiphase pipeline operation which tracks 'packets' of gas and associated liquid as it approaches Theddlethorpe for blending. This on-line system has been adapted to facilitate liquid and slug management issues as well.

2.2 Corrosion Control

Wet gas pipelines containing carbon dioxide are prone to carbonic acid attack and can readily be compromised without a carefully controlled corrosion management system.

A corrosion inhibitor is injected into the transportation system along with the methanol. This adheres to the pipeline walls and is replenished either by regular sphering of the line, by the natural slugging effect of the gas flow, or by deliberately ramping-up of field flowrates. As field conditions change, the philosophy for each segment of the SNS is reassessed. For example, as the LOGGS system has generated higher aqueous phase fractions in recent years, the corrosion inhibitor has been changed from a hydrocarbon soluble chemical to a water-based inhibitor.

The need to sphere for corrosion control significantly affects the liquid management strategy for long multiphase trunklines.

2.3 Gas Management and Substitution

All gas fields converging at the Theddlethorpe Terminal have their own unique Sales Gas Agreements with single or multiple buyers via the UK Network Code System. However, the contracts must abide by a generic Allocation Agreement that defines the parameters for the gas processing arrangements for redelivery of Theddlethorpe gas. This contract is called the Theddlethorpe Allocation and Commingling Agreement (TACA), which identifies all gas arriving at Theddlethorpe as constituting part of a 'Pool' of gas to be allocated and attributed dependent upon the field Nominations. This provides a powerful tool for managing individual fields as well as entire pipelines in that Substitution between fields is possible to alleviate any short-term outages without affecting the ability to meet the field Nomination. For example, Field A can be experiencing an operating problem and be shut-in whilst the Theddlethorpe 'Pool' can over-produce elsewhere in order to cover Field A's Nomination preventing under-deliveries. Later on Field A can produce at a rate greater than its Nomination to pay back into the Pool. Alternatively, fields can be managed in campaign modes irrespective of whether they are nominated or not. Whilst there are limits to the 'Lend-Borrow' philosophy the ability to optimise gas production can substantially reduce operating costs, wear and tear on systems and minimise the financial risks to individual fields.

Gas management can also be used to control the impact of multiphase transfer by minimising ramp-up rates or by operating systems close to their sweep velocities irrespective of the gas deliverability requirements. However, there are occasions where this system can be detrimental. For example, one major trunkline could suffer an emergency shutdown requiring the other lines to compensate and hence ramp-up quickly generating uncontrolled slugs which can, if unchecked, shut down other segments of the infrastructure.

2.4 Liquid Management

The basic methods of system liquid management can be summarised as Primary and Secondary. These are as follows:-

Primary

- Sphering (or pigging)
- Maintaining 'Sweep Velocity'
- Flowrate Ramp-ups

Secondary

- Slug Catcher size and design; and
- Downstream process capabilities and rates

Primary methods are day-to-day mechanisms that can generally be varied to optimise the liquid handling of the system whereas secondary methods are design related that can significantly widen the operating envelope of the primary methods. For example, an undersized slug catcher will severely impact on either sphere launch frequencies or ability to vary the pipeline flowrate away from a "Sweeping" zone.

Each pipeline to the Terminal has unique characteristics requiring individual liquid management strategies. The LOGGS line is over-sized for current throughputs and contains a high proportion of aqueous phase - over 80-85% of the liquid production. Consequently rapid deposition of liquids can occur generating substantial slug sizes when the line is sphered. As the aqueous cut increases, the gas velocity required to keep the line swept of liquids without exceeding the LOGGS Slug Catcher available liquid volume increases rapidly such that daily sphering is required to maintain an acceptable operating regime.

Conversely, the Pickerill line has a reducing operating pressure and maintains a relatively flat gas sales profile in a pipeline sized to meet the demands of the field profile. Consequently, even though gas sales volumes may fall, the sweep velocity is maintained in the line over a wide variation of Sales Contract conditions. Pipeline hold-up is minimised and no sphering is required.

The Viking line is adequately sized to meet demands and has a potential winter to summer gas rate turndown ratio of 4:1 but lower liquid loading than the LOGGS line. The offshore installation will shortly become Remotely-Operated from Theddlethorpe and hence the logistics of sphering will become problematic and expensive. Consequently, this line is hydraulically controlled by flowrate ramping and an adequately sized Slug Catcher.

The CMS line is the most problematic line in that it has a high liquid loading (predominately hydrocarbon condensate), a long trunkline (excess of 180 km) and an installed linesize designed for future developments resulting in current gas velocities that may be inadequate for liquid sweeping. Currently a mixed strategy is adopted for this line alternating between non-sphering in winter (high gas flows) and a 3 to 5 day sphering cycle in shoulder/summer periods and utilising the maximum capacity of the 1670 M^3 (10500 bbls) Slug Catcher.

3 THEDDLETHORPE GAS TERMINAL PROCESSING

All pipelines have dedicated Reception facilities comprising of sphere receipt and slug catching equipment. The Slug Catchers have various forms and sizes (discussed in Sections 4 & 7). The Pickerill and CMS fluids are fiscally metered onshore for allocation purposes and Pickerill gas is compressed to attain gas export pressure. All other fields are compressed offshore to attain the contractual BG Transco minimum pressure.

Pipeline pressure is reduced to plant operating pressure of approximately 73 barg (1060 psig) and the resulting Joule-Thomson cooling effect is supplemented by mechanical refrigeration using propane or Freon-22 to chill the gas to -26^{0}C (-15^{0}F). Hydrocarbon condensate, water and associated methanol condense from the saturated gas. Subsequent separation and gas heating provides a Sales Gas quality within BG Transco specifications of 1^{0}C (33^{0}F) and a cricondentherm of -2^{0}C (29^{0}F).

Hydrocarbon condensate from each system is commingled and distilled to generate a stabilised condensate suitable for on-site storage, metering and subsequent export via pipeline to the Humber Oil Refinery 40 km from the Terminal.

The aqueous phase is similarly collected and distilled to recover the methanol for re-use. The waste water stream which is predominantly condensed water with varying proportions of reservoir produced water is further processed to remove hydrocarbons via a Tilted Plate Separator, monitored and then discharged to the North Sea via a 3/4 mile 6" effluent outfall line.

4 THEDDLETHORPE SLUG CATCHERS

4.1 Historic Perspective

The development concept for the original Viking Gas Terminal in 1972 was to utilise standard technology of a pipe-type slug catcher even though the infrastructure operated in a three-phase region containing the potential for higher water cuts. Pipeline hydraulic simulation capability in the late-60s was in its infancy and consequently the onshore slug catcher was designed for worst-case expectations and focused on early-life capability when gas velocities were high. A Slug Catcher volume of 500 M^3 (3000 bbls) was installed. It contained a standard design of parallel, inclined pipe for gas-liquid separation and storage but incorporated an additional lower ring of pipework to enable liquid-liquid separation and aqueous removal at the bottom. The hydrocarbon condensate withdrawal occurs approximately one-third from the pipework base and aqueous phase or gas breakthrough to the condensate processing facility is prevented by phase interface level control. The aqueous withdrawal rate only designed to accommodate approximately 163 m^3/day (30 gpm).

The volume of the catcher and the associated liquid handling equipment reflected the low condensate to gas ratio and water production rates expected and the relatively flat terrain for the pipeline route - there is no incline greater than 1 degree. Consequently, the liquids management philosophy was to sphere regularly (once or twice a week dependant upon field

Nominations) and enable the two liquid phases within the slug catcher to separate over a number of hours before withdrawing. During the liquid processing period, gas would continue to flow although no liquid would typically arrive at the reception facilities. Once the optimum sphering frequencies had been determined through trial and error early on, this philosophy proved very successful for many years through into the early 1980's.

In 1985, CONOCO(UK) commenced design of the LOGGS infrastructure and closely analysed the anticipated hydraulic nature of the proposed system. The assessment relied heavily upon the operating experience from 14 years of Viking trunkline and Slug Catcher data. Whilst this approach was a success for the Viking system, a number of issues indicated potential problems for the LOGGS system:-

- The LOGGS system contained condensate-to-gas ratios three to four times greater than Viking. Hydraulic steady-state would be reached relatively quickly reducing the time between slug arrivals and/or increasing the potential risk of interference to liquid-liquid separation from natural slugs.

- LOGGS was anticipated to generate higher volumes of associated water later in life altering pipeline liquid hold-up characteristics, slug sizes, and the required Slug Catcher withdrawal rates to handle larger aqueous slugs.

- The 36" pipeline would be over-sized for initial operation to enable future expansion in the LOGGS catchment area. To date, the proposed expansion has not occurred. Over-sizing of the line has exacerbated hold-up volumes and slug volumes formed on gas rate ramp-up.

- A Slug Catcher volume in excess of 800 m^3 (5000 bbls) was identified. The Viking unit required a large area of land which was at a premium during the LOGGS development. A more compact arrangement would be necessary to accommodate the greater LOGGS volume reflecting land availability and still provide potential to expand in the future.

- On one occasion, a Viking well produced larger volumes of water and solids. The resulting higher water/methanol rates reduced sphering frequencies and produced higher proportions of aqueous phase in the sphered slugs. The low Slug Catcher liquid withdrawal rate extended processing time and began impinging on the arrival time of the next slug. The associated solid production settled in the pipes of the Slug Catcher slowly reducing the effective volume of the unit and adversely affecting the liquid-liquid separation and could not readily be extracted from the pipe-type Slug Catcher. With the possibility of later solids production from LOGGS fields, a suitable design to allow maintenance and cleaning of the unit would be critical.

- If there had not been regular sphering on the Viking system the potential 4:1 turndown from winter to summer caused natural slugging problems as the gas rates were ramped up during the Fall campaign. With the LOGGS system characteristics and a similar ramp-up capability this problem required to be addressed.

Consequently, a radical re-design was required for the LOGGS reception facility if it was to mirror the successful Viking design concept of "All-Fluids-to-shore".

4.2 Vessel-type Slug Catcher

A compact, versatile but very simple arrangement of interconnected vessels was designed to maximise storage volume in the minimum of layout area whilst maximising the ability to provide undisturbed liquid-liquid separation in either a batch or natural slugging regime. The arrangement is schematically shown in Figure 4.

The top vessel (3 m diameter by 30 m tan/tan) is a standard gas/liquid separator designed to operate with very low liquid accumulation and to separate all liquid droplets above 600 microns. See Figure 5.

Two downcomers are sized to accommodate swift liquid removal from the top vessel during slug receipt. Typically, the gas flowrate behind a sphere is not reduced as the slug enters the catcher and hence the downcomers range from 36" up to 50" depending upon volumetric transfer to the lower vessels.

The two side vessels (3 m diameter by 30 m tan/tan) are primarily liquid surge vessels with volumes of approximately 300 m^3 (1500 bbls) for slug receipt. See Figure 6. However, they also provide a buffering and stilling effect for the liquid-liquid separation and withdrawal function which occurs in the central vessel. No liquid removal from the unit occurs in the side vessels. All liquids transfer to the central vessel via an interconnecting 36" line beneath the vessels. There is a pressure balancing line between all three vessels so, in effect, the 36" lower line raises and lowers the liquid levels simultaneously in each vessel.

The central vessel is key to the design and operation of the unit. See Figure 7. It is the same tan-to-tan length as the other vessels (30 m) and the same initial diameter (3 m) but has a bulbous end section with an increasing diameter up to 4 m. At this point the bulbous end lies below the elevation of the two side vessels, allowing the liquid phases to segregate in this low point. Accurate level measurement and withdrawal control can be achieved.

Immediately after arrival of a large slug the three bottom vessels contain a turbulent mixture of hydrocarbon condensate and raw methanol. In condensate fields like the UK's Southern North Sea, the differential between the hydrocarbon condensate (s.g. = 0.6) and raw methanol specific gravities (s.g. range = 0.9 - 1.1) is considerable and rapid, coarse gravitational separation occurs in the vessels. Typically this separation can occur within 10-15 minutes. However, the Operators will wait between 1 to 2 hours before commencing processing of the liquid phases to enable maximum settling. This minimises the effects of cross-contamination in downstream liquid processing systems. Hydrocarbon carry-over into lower pressure water handling systems produces excess flash gas whilst water contamination in the condensate stabilisation system can produce excessive corrosion and water "boiling" in the fractionation columns.

When good separation is evident via a defined liquid interface level reading, the aqueous phase is removed at a steady rate from the low point of the bulbous section of the central vessel and routed to an intermediate storage facility. The rate is controlled to prevent possible coning effects drawing excess condensate into the raw methanol handling system. At this stage, no condensate is removed from the system. Raw methanol extraction continues until the interface level falls to a minimum level. This level is low in the bulbous section to provide unrestricted condensate contact between the volumes in the side vessels and the central vessel.

When the aqueous phase has been removed, the condensate phase can be accessed for removal via the withdrawal connection which is approximately 1m above the base of the bulbous section. Condensate is routed direct to a distillation processing facility with minimal intermediate hold-up. Consequently the Slug Catcher is utilised as a feed-tank to regulate a steady flow to the downstream unit(s). Dependant upon the anticipated timing of the next slug arrival the withdrawal rate can be modulated by slowly ramping up downstream column rates or starting up another column. Clearly, these ramp-ups cannot be achieved instantaneously and hence a realistic knowledge of slug arrival times and frequency is critical to optimal operation of the Slug Catcher.

4.3 Natural Slugging Systems

The basic Slug Catcher operation indicated above empties the system within a 24-hour period. For batch operation systems, where regular sphering is the main factor in liquid management, this cycle time is acceptable. However, some of the Theddlethorpe pipelines operate either in a non-sphering mode or contain significant liquids such that a pipeline hydraulic steady-state is reached many hours before a planned sphere 'run'.

In both these scenarios, three-phase natural slugging will occur in a generally haphazard manner. Although it is possible to predict the generic slugging characteristic of a pipeline or when a flow perturbation is introduced into the system, the specific impact on the Reception facilities is unpredictable until the natural slug arrives. It is under these conditions that the vessel arrangement provides significant benefit.

A single vessel, or two vessels one above the other, attempting to provide three-phase separation, requires adequate residence time to provide liquid-liquid separation and then a period of time to empty the vessel(s). During these periods, any liquid receipt disturbs the settling process suspending withdrawal until adequate separation can be re-established. For a pipeline system at hydraulic steady-state the recovery of phase separation is extremely unlikely. The outcome would typically be regular interruption of gas flow through high level trips or a need to gas throttle to prevent natural slugs for a period of processing. Neither option is acceptable within the commercial framework of gas sales in the UK.

The Theddlethorpe Slug Catchers provide a simple solution to this issue as all liquid phase fluids are routed only to the two side vessels. There is no downcomer direct to the central vessel. This provides a buffering effect to the separation and removal functions being undertaken in the central vessel. The two outer vessels absorb the turbulent effect of uncontrolled natural slugs and although it would be unacceptable to withdraw liquids from the side vessels, the phase separation achieved in the central vessel is virtually unaffected. Consequently, it is possible to continue withdrawing liquids from the Slug Catcher whilst simultaneously receiving a slug. This capability prevents unnecessary accumulation in the Slug Catcher, minimising risk of interruption of gas flow and provides a stable processing rate for downstream equipment. Of course this arrangement cannot buffer ALL scenarios - receipt of a very large slug in a short period of time would clearly transmit turbulent conditions in to the central vessel. However, natural slugs are significantly smaller in volume than sphered slugs and generally do not impact the operation of a unit which is always sized to accept larger sphered slug volumes.

Theddlethorpe Slug Catchers operate across the entire spectrum of possible scenarios from batch-wise sphered operation and non-sphering with regular ramp-ups, to high liquid content lines requiring regular sphering but also generating frequent natural slugs in between sphere runs. The latter scenario is prevalent for the CMS system and is the most onerous operating case for this unit design. The CMS system has been highly successful since start-up in 1993 and only fails when the regular sphering frequency is interrupted by weather conditions restricting re-load of remote-operated sphere launchers.

Even under these conditions it is possible to operationally recover the situation by flow ramping and generating frequent dynamic slugs which the catcher can "buffer" and process.

5 BENEFITS

There are numerous benefits of this design over the pipe-type units, some of which have only become evident since system start-up.

5.1 Separation Capability

Liquid/Liquid gravity separation in a vessel is significantly faster than in a long inclined pipe due to the reduced settling path. The comparison between the Viking pipe-type unit and the LOGGS vessel-type has been very marked. This was a major factor in determining the need to replace the Viking pipe-type with a vessel-type when CONOCO(UK) developed the Phoenix complex into the Viking infrastructure adding significantly wetter fields which affect the hydraulics of the entire system.

5.2 Plot Area

A vessel-type unit typically utilises only a third of the plot area of an equivalent volume pipe-type unit. The land purchase or rental cost in the UK can yield substantial cost reduction. Additionally, knock-on effects involve more compact plant layout for ancillary and downstream equipment contributing savings of pipe runs, pipe-racks, cables, ducting, underground services, etc.

5.3 Maintenance

Pressure vessels can clearly be designed for entry and are readily maintainable and available for internal inspection, remediation and removal of deposits. Pipe-type units are notorious for lack of access either offering cramped working environments, or requiring remote-controlled inspection tools. Additionally solid deposit removal and any rectification work is problematic and time-consuming.

5.4 Robustness of Operation

5.4.1 Water Cut

Typically, operation of the unit is unaffected by the percentage of aqueous phase in the produced fluids. Over the past five years, the LOGGS Slug Catcher has witnessed a dramatic increase in water cut up to 80% of the total liquid feed. Separation timing is unaffected and although condensate withdrawal is prevented until the interface level drops below the condensate withdrawal nozzle the aqueous phase outflow rate can be increased significantly when there is a high interface level. Consequently as condensate and aqueous removal are sequential processes, the overall time to empty the unit is virtually unaffected by condensate/aqueous phase ratio.

5.4.2 Solids

Solid deposition does not adversely affect the primary objectives of the unit. The central vessel contains a weir and 'settling area' just prior to the eccentric expander section of the vessel. The combined effect of turbulence during slug receipt and the solids 'trap' in the central vessel generally confines deposition to the mid-section of the central vessel or the outer vessels where there are no critical process operations occurring (e.g. liquid withdrawal, level indication , etc.). Additionally, removal of solids is comparatively easy during maintenance periods as access is unrestricted.

5.4.3 Natural Slugging

As outlined in the previous section, settling and subsequent processing through the unit is insensitive to natural slugging events.

5.5 Expandability

This vessel-type philosophy is readily expandable to support major alterations to the pipeline hydraulics characteristics. Events that could alter the hydraulics include:-

- addition of new developments with higher CGR's
- watering out of reservoirs
- altering operating pressures (e.g. reservoir 'suck-down')
- reducing production rates
- cessation of regular sphering capability (e.g. demanning of offshore locations)

This vessel arrangement functionally divorces the liquid-liquid separation from the storage volume issue. Consequently, alteration to pipeline hydraulics only affect storage volume requirements. This can simply be accommodated by addition of extra vessel(s) by the side. Blind flanges are provided for all the balance lines and space is usually allocated at the outset of a project to allow an additional one or two vessels. These new vessels do not need to have identical dimensions to the original unit - a recent addition to the CMS Slug Catcher had a large diameter but a shorter tan-to-tan length with an overall volumetric increase of 66% over the original vessels.

Expandability potential has proved an invaluable asset in securing new developments and combating changing processing conditions.

6 DRAWBACKS

Unfortunately, the system is not all 'positives' as there are a number of drawbacks:-

6.1 Capex

The cost for a pressurised vessel of these volumes is higher than a equivalent volume pipe-type unit. A Theddlethorpe Slug Catcher unit (4 vessels) with a 600# design rating will have an installed cost in excess of £4MM ($6MM). However, allowing for land cost savings and elimination of downstream storage volumes, the system typically will prove financially sound. If opportunity costs are included in the financial analysis as a result of reduced risk of shutdowns due to hydraulic slugging then the financial case is usually secured.

6.2 Manufacture

The vessel walls are over 100 mm (4") thick and the size and number of the units limits the number of potential manufacturers. Theddlethorpe Slug Catchers are now fabricated outwith the UK and CONOCO(UK) have experienced increasing difficulty in securing acceptable bids with companies that have the capability, track record, yard ullage, logistical/transportation understanding and accessibility to the UK. Additionally, fabrication schedules ensure that Slug Catchers are usually the longest lead-item in a project.

6.3 Transportation

Theddlethorpe is in rural Lincolnshire where the road network is particularly limiting. Transportation of these vessels through picturesque, English hamlets and over weight-restricted bridges is especially awkward. The sleepy village of Alford has now had 15 vessels pass within a few inches of the 13th century church spire.

6.4 Safety Systems

A pressurised vessel containing a large volume of gas and/or hydrocarbon condensate within a limited plot area requires careful consideration from a safety perspective. The potential for a BLEVE is significantly greater than for a pipe system and consequently the safety systems must conform to rigorous blowdown criteria and/or additional passive fire protection measures.

7 ENHANCEMENTS TO THE DESIGN

Theddlethorpe expansion over the last 10 years has generated a unique opportunity to innovate a novel design, monitor its success, identify weaknesses and modify subsequent units. Modifications can be categorised into two types - firstly improvements to the inherent mechanical design of the unit and secondly upgrades to the overall system from wellhead to delivery point which enhance the operation of the Slug Catcher.

7.1 Liquid downcomers

The original LOGGS design incorporated 2 off 36" downcomers from the top vessel to the lower side vessels. Although calculations indicated that this would be adequate for routing liquids to the lower vessels, the practice showed that at peak sphering velocities insufficient liquid volumes were draining away. The level in the top vessel would slowly rise initiating a high level trip. This required the Operators to slow the approach of the sphere as a slug arrived. On subsequent projects this 'pinching' approach was deemed undesirable and so the downcomer size was increased to 50"

7.2 Vortex Breakers

Operation of the LOGGS unit caused a higher level of condensate carry-over in the aqueous phase than originally anticipated limiting the aqueous withdrawal rate. Subsequent designs installed vortex breakers to reduce this 'suck down' effect.

7.3 Improved Re-Run Capability

Over the years, 'rules of thumb' to predict slug sizes were generated by operators. These were directly correlated to trunkline flowrates and which fields were on-line. There were regular errors of judgement and some sphered slugs were too large for the Slug Catcher.

Over-sized slugs are a major concern due to the extended period to process the excess fluids during which time no gas is delivered. The original concept allowed for an emergency liquid removal, routing turbulent liquids from the top vessel to a low pressure flash drum prior to entering liquid hold up facilities in the Slops Tank system.

It was found that withdrawal from the top vessel did not significantly aid processing an over-sized slugs as the liquid volume in the top vessel is deliberately low with an undefined liquid-gas interface. Regular 'gassing' to the downstream flash drum limited the reprocessing rate. By the time a clear interface was determined the bottom vessels were full and the system was close to shut-down. Early, pro-active intervention is necessary to reduce the risk of shut-down. This was achieved on subsequent designs by connecting the re-run line to the balance line underneath the lower vessels. Consequently, an early decision on the need to re-run can be made as there is always guaranteed liquid head above the off-take location. Although the volume of flash gas generated limits the withdrawal rate due to downstream PSV restrictions and environmental discharge consents this consistent rate is a significant advantage in preventing unnecessary shut-downs due to large slugs.

7.4 Pipeline Virtual Operator (VO) System

The most important system modification has been the introduction on the CMS system of a Pipeline Integrity Monitoring System (PIMS) incorporating "Intelligent" functions with respect to wellhead control/choke positioning. The wellhead control function provides valuable on-line control for meeting Daily Nominations, blending constraints, tubing head pressure restrictions and minimising erosional effects across the integrated system which incorporates various Third Party equity fields. However from a liquids management perspective, the most critical aspects are the inventory tracking, liquid hold-up and slugging prediction elements.

The Virtual Operator tracks 'packets' of fluid (gas and liquid) from the individual wellheads and intra-field lines, through the mixing manifolds and into the main trunkline to shore. Liquid hold-up and ramp-up characteristics for the system over a wide range of potential operating scenarios have been formulated using a refined version of WOLGA. Simple algorithms for these systems have been generated and run on-line with the fluid tracking function. The on-line databank is completed with level information supplied by the onshore slug catcher indicating available liquid volume. Diagnostics are provided to the Operator to indicate the need for sphering, liquid hold-up volumes, condensate-aqueous phase splits and whether the system has attained steady-state conditions and is in natural slugging mode.

The Virtual Operator algorithms are regularly tuned dependent upon changes in operating conditions and particularly for alteration in aqueous cuts. Over the past three years the VO has proved very accurate and a significant aid in controlling pipeline liquid management.

8 CONCLUSIONS

The CONOCO(UK) Southern North Sea gas-field infrastructure has successfully operated multiphase pipelines for over 25 years. A key element to that success has been the design and refinement of an innovative vessel-type Slug Catcher providing:-

- Robust operation insensitive to widely varying water-to-condensate ratios and operating conditions;
- The ability to accept natural slugging whilst continuing to withdraw liquids;
- Enhanced liquid-liquid separation compared to pipe-type units;
- Reduced plot area compared to pipe-type units;
- Improved maintainability and clean-out capability; and
- Flexibility and easy expansion as further fields are added to the infrastructure.

The addition of a fast-response, on-line Virtual Operator to the CMS Pipeline based upon WOLGA-generated algorithms has proved particularly successful in controlling pipeline operations containing large volumetric throughputs of associated liquids.

Acknowledgement

This paper is published with the permission of CONOCO(UK) and Kvaerner Process.

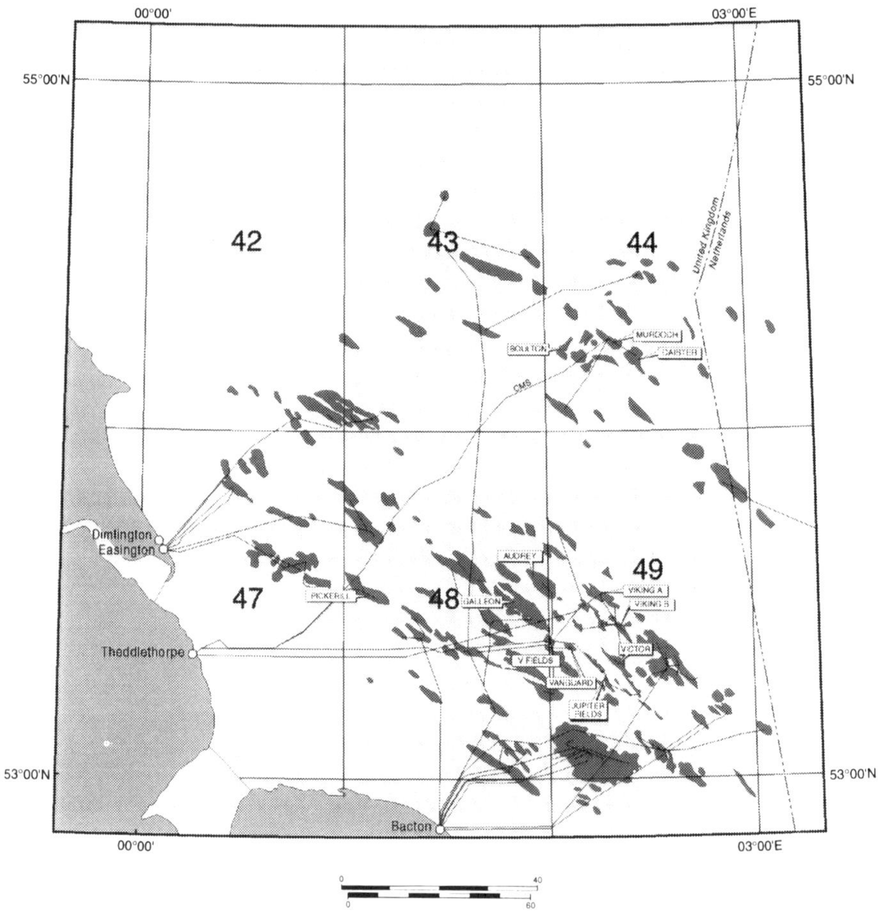

Figure 1 Southern North Sea Location

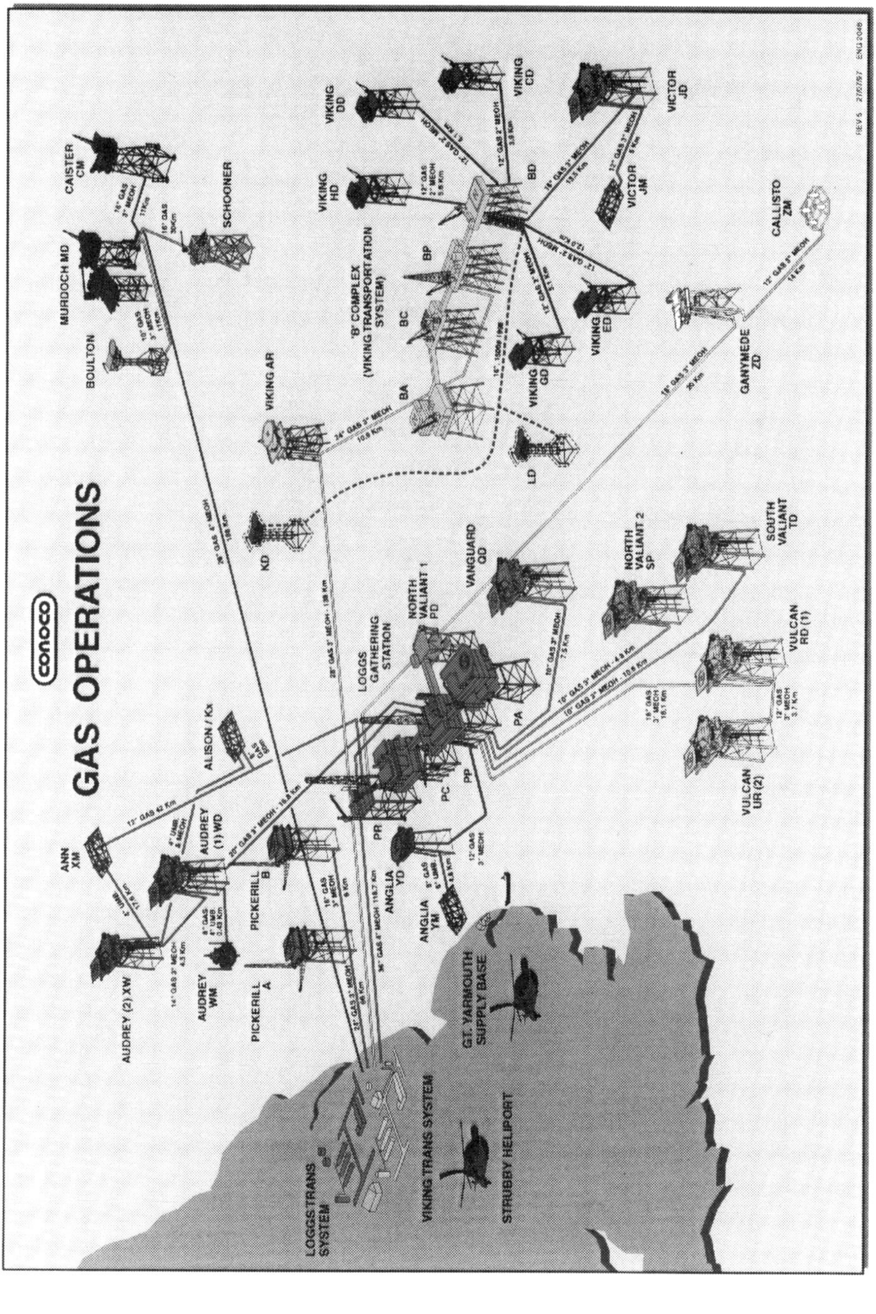

Figure 2 Southern North Sea Infrastructure

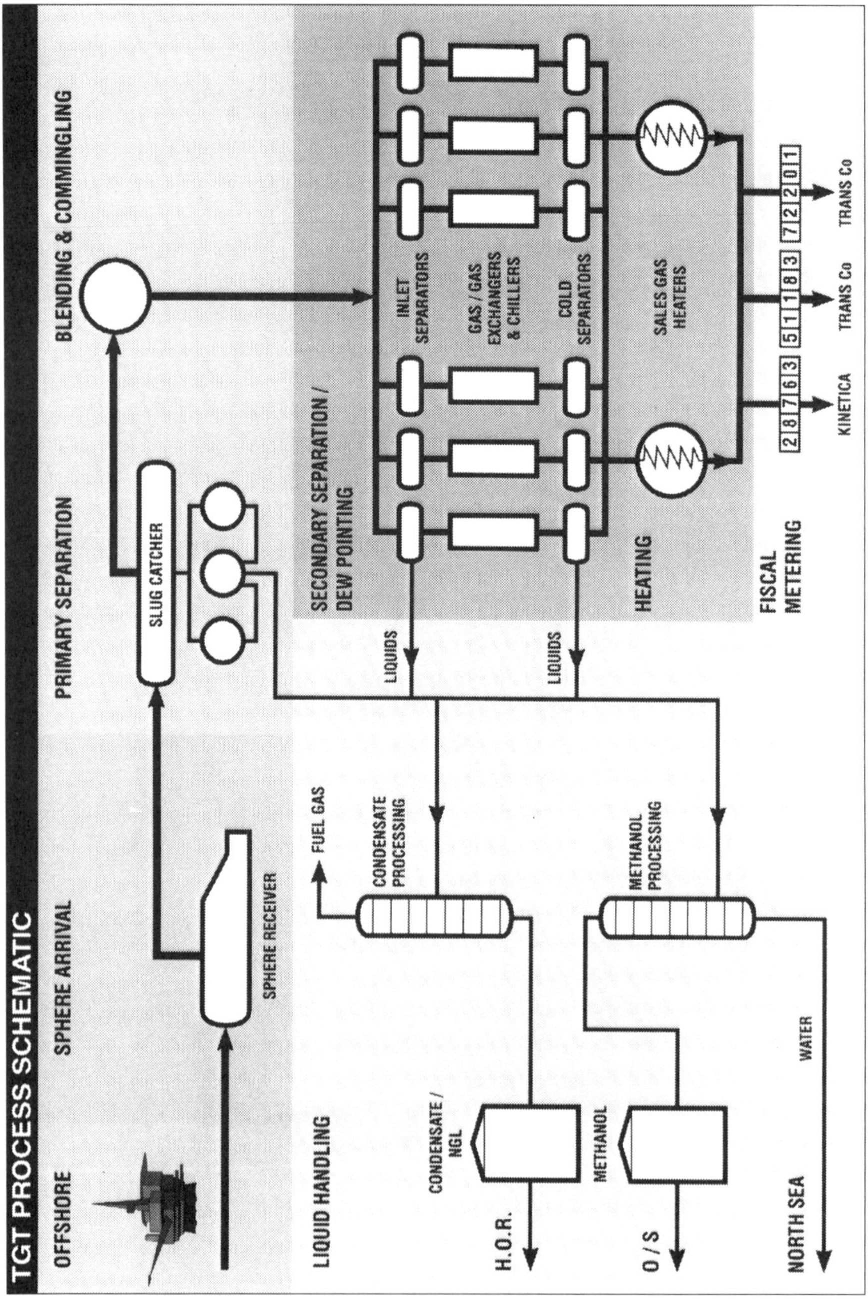

Figure 3 Theddlethorpe Operation

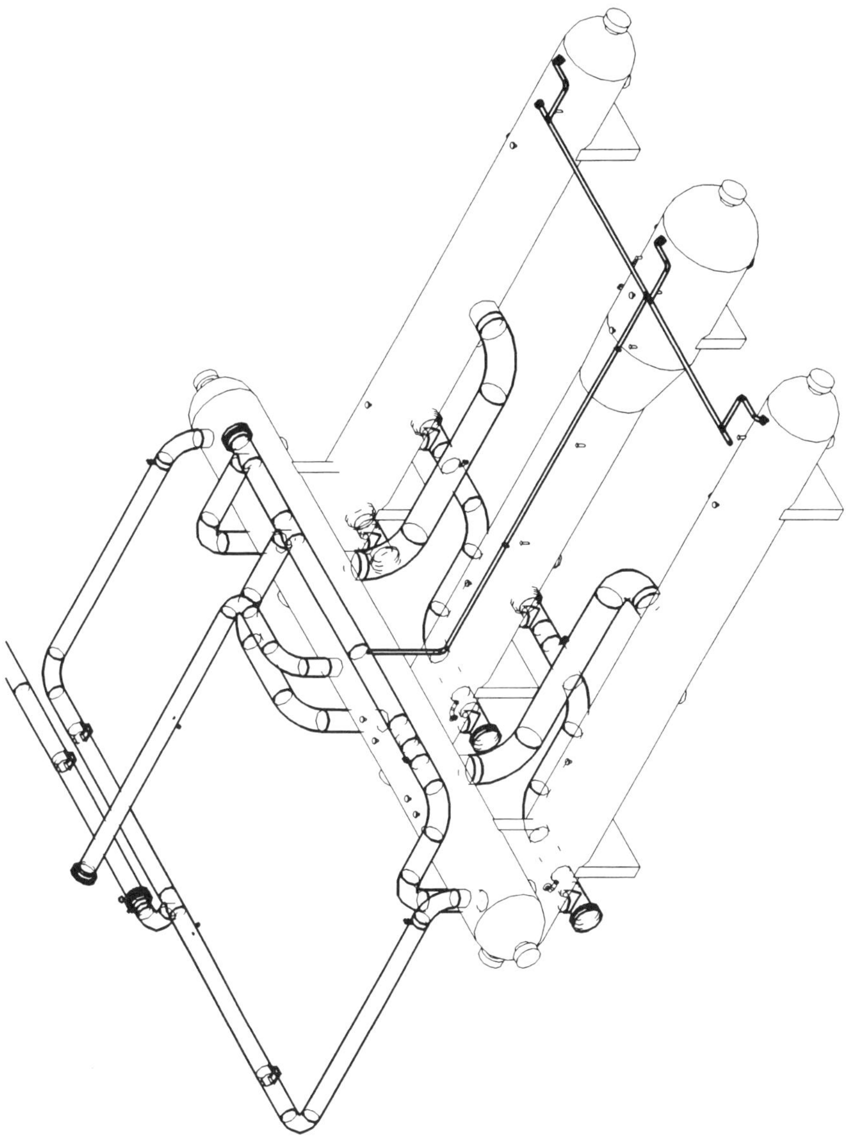

Figure 4 Theddlethorpe Slug Catcher

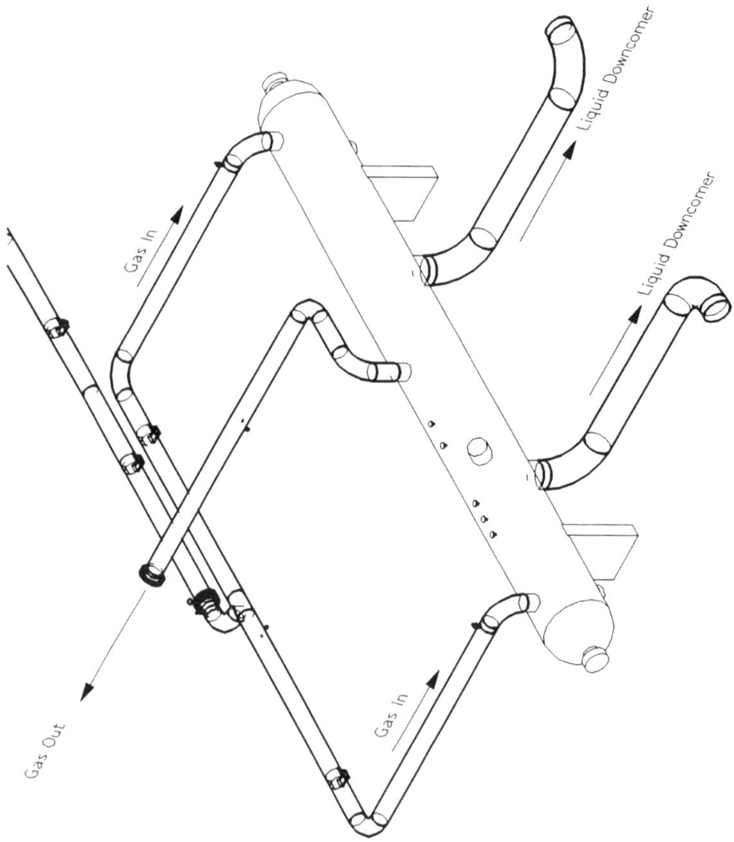

Figure 5. Top Vessel

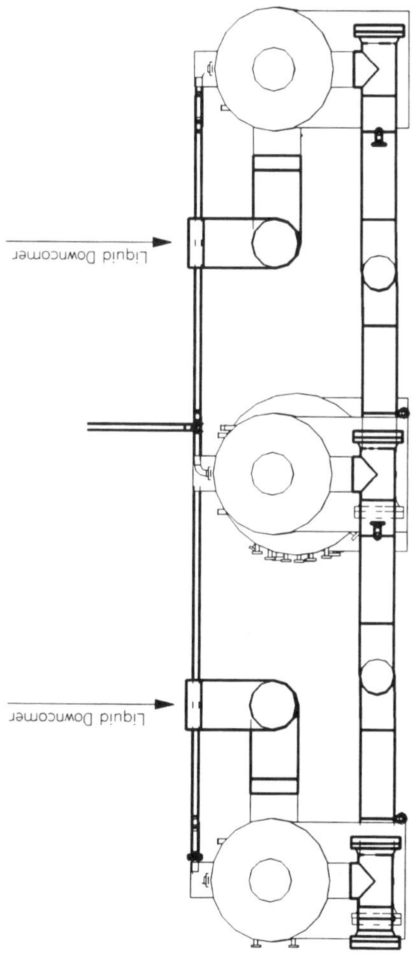

Figure 6. Storage Vessels

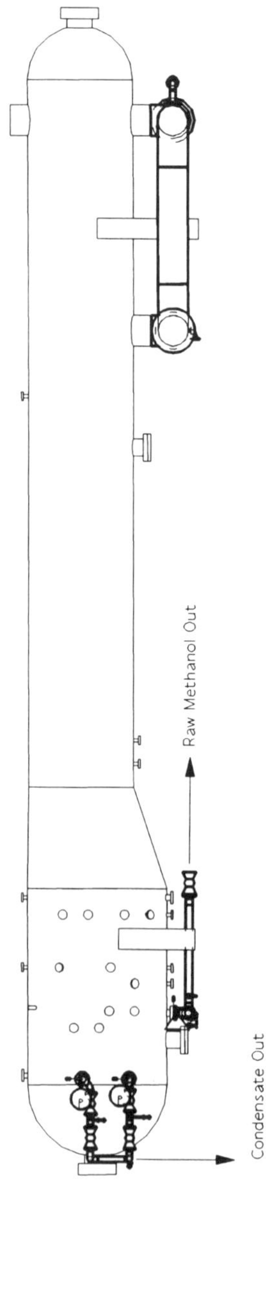

Figure 7. Central Vessel

Understanding and development of a slug hysteresis model for hilly terrain pipelines

N BARRATT, T J HILL, and **M KING**
BP Exploration, UK

ABSTRACT

In the basic design of a pipeline, one of the most critical parameters to know is the steady state pressure drop through the system as it affects line size, operability and sizing of process plant. In recent years, it has been highlighted that the current steady state pressure drop methods, whether mechanistic or empirical, fail to accurately predict the measured pressure drop over hilly terrain. This is due to the downhill flow regime being incorrectly predicted because of slug hysteresis effects.

One such set of field data is used here to highlight these inaccuracies and a new model capable of capturing some of the hysteresis that slugs undergo is proposed. Comparisons to experimental data show that inclusion of these effects results in a distinct improvement over currently used techniques.

1. INTRODUCTION

Arguably the most common flow regime encountered in the transportation of oil and gas is slug flow. Slug flow can occur for a variety of different reasons, but however formed it will always fall into one of the following categories - hydrodynamic, transient, terrain induced or severe slugging. Slug flow is an unstable type of flow regime, characterised by a succession of pistons of liquid moving at a velocity relating to that of the mixture. Each slug is bounded by a stratified layer before and after it. This layer is engulfed in the slug front and shed away at the slug tail, resulting in a net bulk forward movement of fluid and areas of high mixture. This flow always gives rise to time dependent behaviour, where the pressure, mass flow and velocity past a point vary with time.

Slug flow can occur in horizontal, vertical or inclined flowlines. In particular, the inclination of flowlines in the upward direction tends to favour the onset of slug flow;

consequently the area of the slugging regime on a flow pattern map increases. This is shown when using a flow pattern map such as that proposed by Taitel & Dukler (1). Conversely when pipelines are inclined downwards, slug flow becomes less common.

Flow regime maps are based on steady state considerations and are constructed only for a single point in a system. They do not take account of any transients that may cause a change in flow regime (2). Neither do they consider any hysteresis of the system, which can commonly occur from changes in inclination, often leading to the onset of terrain induced slugging.

2. EXISTING WORK ON SLUG FLOW

Extensive work has been conducted on the prediction of slug flow, as well as on the characteristics that it exhibits. Some of the areas of interest are the liquid holdup in a slug, the frequency at which they occur and the shedding behaviour/film profile of the back of the slug.

Slug flow is very chaotic in nature and can occur at many different frequencies. Consequently much of the modelling work performed by Dukler & Hubbard (3), Crowley (4) and Taitel & Barnea (5) is based on an average slug length. They consider a unit cell characterising all the slugs along a given section of pipe. Integration is performed along the film to obtain its detailed profile. These techniques require the input of some additional parameters in addition to the normal input parameters (i.e. gas velocity, liquid velocity, etc.). These are closure relationships which provide values for slug body holdup and average frequency and are typically empirical correlations.

Although mechanistic in nature, these models are all dependent on the prediction of slug flow. Slugs are not tracked and so slugs cannot pass into downhill sections where slug flow is not predicted.

More recent work, driven by the need to follow slugs through the pipeline, was performed by Zheng et al. (6). They proposed the concept of sinks and sources to create/destroy slugs in dips and hills. This has full hysteresis but the model is too rigorous for quick analysis of a pipeline and would over complicate current steady state design methods. However all steady state designs should be checked using a transient code to check the operability of the system.

3. HILLY TERRAIN CHALLENGES

The term 'hilly terrain' refers to an extreme topography over which a pipeline may be laid. Typical topographies previously encountered have involved sections of pipe inclined at ±1 or 2° to the horizontal. Current design methods have therefore been developed to model these systems.

Recently the oil industry have begun the development of oil fields in more challenging environments, where the inclination of pipelines may be 20° or more from horizontal. These hilly terrain systems can occur both on land over hills and mountains or subsea where trenches, canyons and sand waves cause extreme topographies. A good example of this type of oil field is that of Cusiana oilfield in Colombia. This produces about 180,000 bpd through a complex network of hilly terrain pipelines. Some of the inclination angles of some of these lines are in excess of 35°.

A description of the Cusiana field and a description of the some of the multiphase lines and their operability is available in (7). An important observation was made for the Cusiana field during a long term test (LTT) - the measured well head flowing pressure (WHFP) was considerably different to that predicted using current design methods.

3.1 Design Methods

Data was collected from the LTT between 1993 and 1995 to assess slug catcher operation and gauge current design methods in hilly terrain. If the topography of the LTT line is plotted (Figure 1), the large variations in inclination can be seen and it becomes apparent that this topography will pose a challenge for current design methods which are based on small inclination changes.

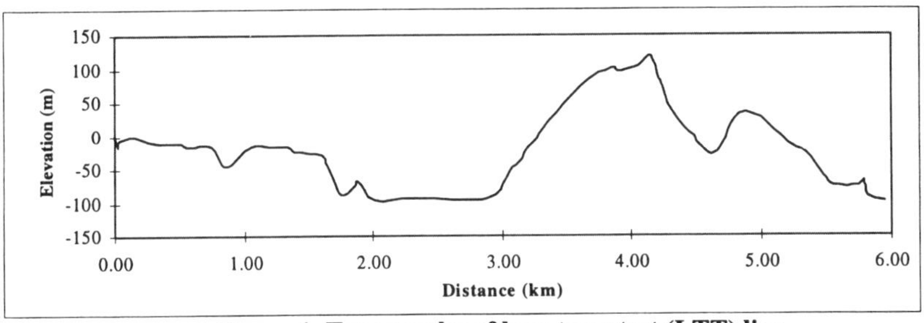

Figure 1. Topography of long term test (LTT) line.

If a comparison between some current design methods and actual data recorded from the LTT is made, then inaccuracies are clearly seen. Experimental values reported are shown in Table 1.

Table 1. Data collected from LLT taken from Hill et al. (7)

WHFP (psia)	WHFT (°F)	LTT Pressure (psia)	LTT Temp (°F)	Oil Flowrate (bopd)	Gas Flowrate (MMScfd)	Mixture V (m/sec)	GOR (scf/stb)
1171	210	1013	165	11361	21.18	4.08	1865
1162	211	1015	165	11249	21.23	4.03	1888
1786	209	1646	144	7666	12.63	1.58	1648
1486	184	1351	152	9923	16.45	2.45	1658
1419	184	1282	150	10493	17.33	2.7	1651
1458	199	1300	149	10092	16.65	2.72	1650
1830	191	1683	142	8061	12.89	1.6	1599
1854	180	1697	135	6957	11.3	1.38	1624
1860	180	1706	135	6975	11.43	1.39	1639

The fluid was characterised by a gas SG of 0.761, an oil API of 35.2° and a water cut of 0.05% and the Glaso correlations were used to estimate the physical properties. The topography is shown in Figure 1 and the line size was 7.625" internal diameter with 0.5" thick walls. The line was buried to 4ft, with an ambient temperature of 81°F.

3.2 Comparison of the Data using Current Methods

Several models were compared to the data collected above; the model was provided with the inlet pressure (WHFP) and the outlet pressure was predicted. The physical properties were the same for all the comparisons. Figure 2 shows the error between the predicted and measured values, according equation (1).

$$Err(\%) = \left(\frac{\Delta P_{Predicted} - \Delta P_{Measured}}{\Delta P_{Measured}} \right) \times 100 \quad \text{.......(1)}$$

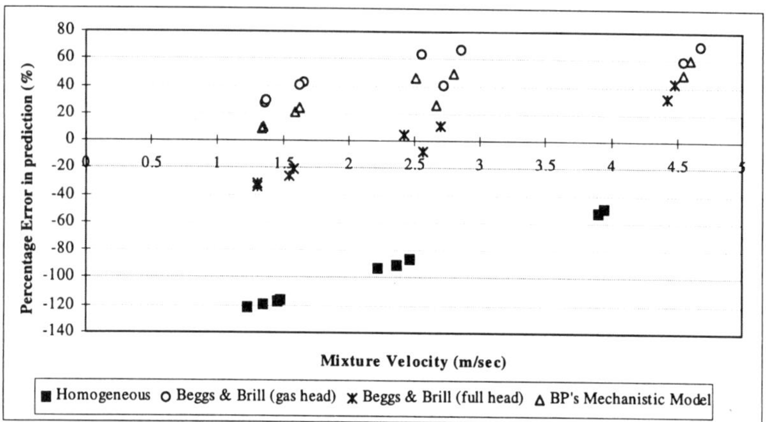

Figure 2. Comparison of different models against collected data.

3.2.1 Homogeneous

The simplest model used the homogeneous model. This uses the no slip liquid holdup and an equivalent single phase pressure drop method. It under-predicts the pressure drop at all velocities. At the highest velocities the error is lower but still under-predicts by 50%. The reason for this consistent under prediction is that there is a lot of head recovery in all downhill sections in the pipeline.

3.2.2 Beggs & Brill (8)

This correlation is more complex than the homogeneous model but still simple and widely used. Two versions of this correlation were tested; one where gas head only recovery in the down hill sections was used and one where full liquid head recovery was allowed. These were both coupled with to slight modification of the friction factor, believed to enhance the frictional pressure drop. On study of Figure 2, the difference between full head and gas head recovery leads to a large variation in accuracy at low and intermediate velocities. It is also worth noting that the Beggs & Brill correlation consistently predicts slug flow to occur in both the uphill and downhill sections of the pipe. Full head recovery appears appropriate at intermediate velocities.

3.2.3 BP's Mechanistic Model

This is the most complex of the models compared. It utilises the Taitel & Dukler flow pattern map (1) and selects an appropriate pressure drop model for each flow regime predicted. The methods used with in the mechanistic model seem to be accurate at low velocities. This is perhaps due to better prediction of the frictional pressure gradient coupled with only gas head recovery in the downhill sections. The accuracy is lost at progressively higher velocities to as frictional losses increase and the hydrostatic head term is still modelled using gas head recovery alone.

3.3 Discussion of the Results

It is thought that the cause of these errors is the way in which most models use a flow pattern map based on the local inclination and insitu properties, ignoring any hysteresis. Hysteresis plays an important part in the prediction of pressure drop in hilly terrain as slugs do not simply vanish at the end of a section. Instead they can persist for a finite distance into the next section. This means that liquid head recovery is possible in the next section; the pressure is therefore higher than would be predicted by only gas head recovery in that section, but lower than only liquid head recovery as slugs decay before reaching the section end.

Persistence is dependent on the velocity at which the slugs arrive into the next section and also other parameters such as predicted flow regime and inclination of the next section. Figures 2-4 summarise the behaviour of slugs as they pass through a change in inclination.

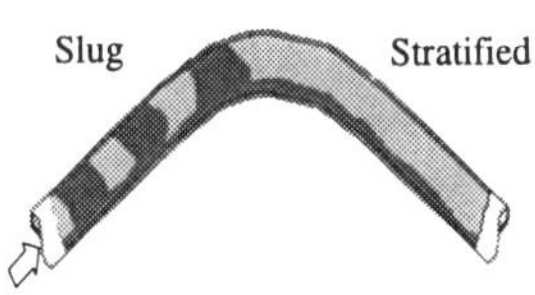

Figure 2. Low flowrate flow regimes, slug flow uphill and stratified downhill.

Figure 3. Intermediate flowrate flow regimes, slug flow uphill and slugs decaying downhill.

Figure 4. High flowrate flow regimes, slug flow uphill and slug flow downhill.

This behaviour has been observed in the laboratory by Zheng et al. (6) and Yang (9) but has not been quantified as yet. This type of flow behaviour can be modelled by either modifying existing models or by developing a new model which incorporates slug hysteresis.

4. MODEL DEVELOPMENT

As reported in (7) and again re-iterated here, the driving force for a new model is that current design methods do not model the additional liquid head recovery, that exists when slugs move from an upwardly inclined section exhibiting slug flow, to a horizontal or downward section where slug flow is not predicted. A new method is required which will add a simplified model for slug decay to the system. This is best achieved by superimposing a new condition on traditional design methods. This method needs to be general in application, so that modifications can be made to other existing mechanistic models, rather than a completely new model unable to be placed inside an existing framework. All the approaches are therefore viewed from a general point of view.

4.1 Proposed Application

As seen in Figures 2 & 4, one extreme of slug decay is when decay occurs at an infinite rate and the flow instantaneously becomes stratified in the downhill section. Conversely, an infinitely small rate of decay, leaves slugs unaffected and for all purposes the downhill section is in fully developed slug flow. The real case is that depicted in Figure 3. It would be very difficult to try and include a new flow regime into a flow regime map as it would involve additional information about the flow history in the pipe. It is proposed to superimpose a fully consistent decaying slugs regime on the incorrectly predicted stratified flow as shown in Figures 5&6.

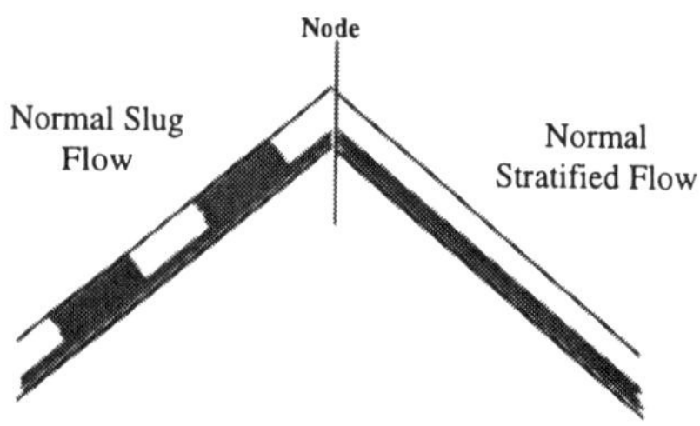

Figure 5. Node reached: Flow regime recalculated, stratified flow is incorrectly predicted.

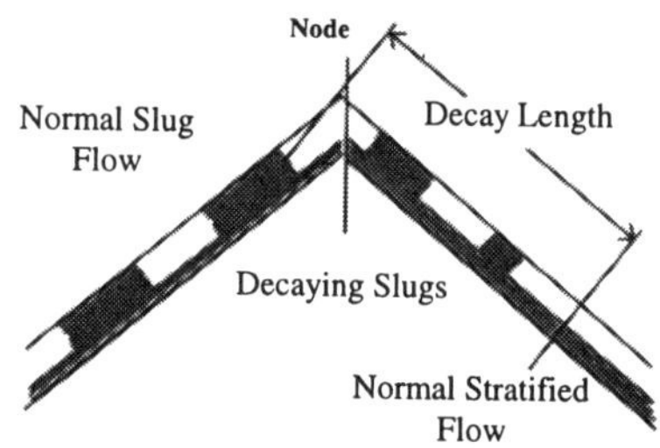

Figure 6. Node reached: Decaying Slug Flow is superimposed over incorrectly predicted stratified flow.

The model should begin with a slug being passed into the downhill section. A slug will then decay at a rate dependent on its position and the film in front of it until after a distance ('the decay length'), the slug ceases to exist any more. *The resulting film must therefore be equivalent to the stratified layer that would exist if slugs had not been superimposed.*

4.2 Steady State Approach

Slug flow is an unsteady flow pattern in which the key parameters (mass flowrate, holdup, etc.) vary with time. It may however be approximated by a stratified flow of the same mass flowrate as a time-averaged slug flow for any particular section (see Figure 7).

Figure 7. Time averaged mass over each section of pipe.

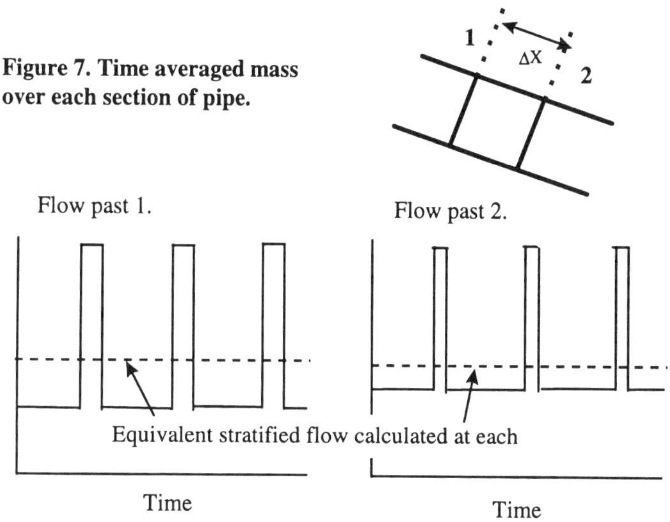

As a slug passes through this point, the section will see a varying liquid mass flowrate which depends on the ratio of the slug to bubble length. It has already been argued that the mass flowrate on average must be the same as for a stratified flow and therefore a mass balance can be taken over any section of pipe of length ΔX can be equated to a mass balance for a equivalent stratified flow.

4.3 Mass Balance over Unit Slug

If a long section of pipe is examined, typically that between two nodes then the section must be divided into small sections (step lengths), commonly referred to as ΔX, the purpose of these small sub-divisions is provide an accurate solution to the momentum equation which is integrated over ΔX. The increments of ΔX are chosen so that any reduction in ΔX has a negligible effect on the resulting solution of pressure. Note that many different criteria are used to determine the maximum stable length of ΔX, they are not discussed here but are commonly used within computer code to achieve the stability mentioned above. All criteria are based on making comparisons of the pressure gradient predicted by two different ΔX's, different in-situ conditions and flow regimes thus have different stable step lengths.

If a section of pipeline which sees decaying slugs is taken, for example the entire decay length, then a pictorial representation of what each ΔX contains is shown below in Figure 8, note also that ΔX can be reduced slightly to give an exact multiple of the decay length.

Parameters that describe the ratio of slug and bubble per increment of pipe can now be defined. The ratio is calculated by considering the average length of slug that would pass through a particular section. This length is the length of slug entering this section minus the amount that has decayed.

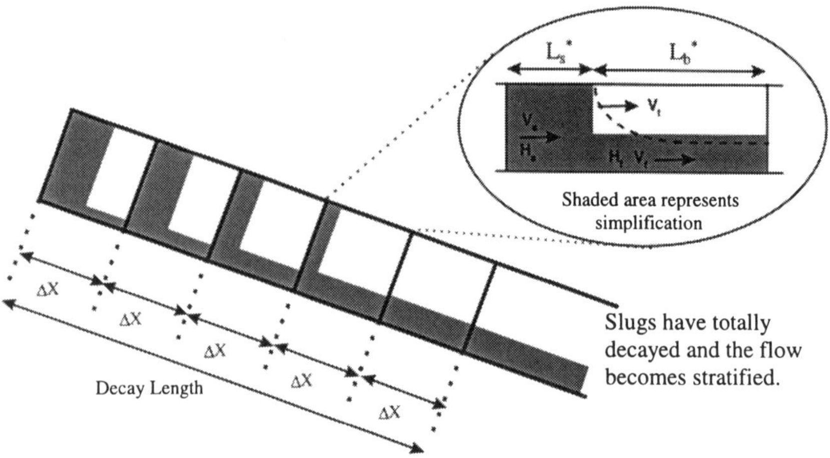

Figure 8. Varying ratio of slug to bubble seen in each section of pipe as distance increases up to the decay length.

If we take a general point in the system, then the following equations can be written:

$$L_s^* = \frac{L_s^n}{L_s^n + L_b^n} \Delta X \qquad (2a)$$

and

$$L_b^* = \Delta X - L_s^* \qquad (2b)$$

Where L_s^n is the slug length in section n, L_b^n the bubble length in section n and ΔX the length of section n. If a mass balance is taken over this increment and equated to an equivalent stratified flow, then the following is obtained:

$$H_f = \frac{\Delta X\,(\rho_L H_{LE} + \rho_G(1 - H_{LE})) - L_s^*\,(\rho_L H_s + \rho_G(1 - H_s))}{L_b^*\,(\rho_L - \rho_G)} - \frac{\rho_G}{\rho_L - \rho_G} \qquad (3)$$

Where H_f is the liquid holdup in the film region, H_{le} the liquid holdup of an equilibrium stratified flow and H_s is the liquid holdup of the slug body. ρ_G and ρ_L are the densities of the gas and liquid phases respectively. Equation (3) which reduces to,

$$H_f = \frac{H_{LE}\Delta X - L_S^* H_S}{\Delta X - L_S^*} \qquad (4)$$

Equation (4) is an expression for finding the film height in a section of pipe and is dependent only on the input lengths. These are calculated from the slug length at the end of the previous section.

4.5 Slug Decay Rate

As mentioned previously, slugs take a finite distance to decay upon entering a new section of flowline. The rate at which this decay or indeed growth occurs is often referred to as the growth rate (G). If the growth rate is considered to be zero, then amount of liquid that a slug pick-up as it moves over a film, must be equal to the amount that is shed in the film behind the slug. This is due to the back of the slug moving faster than the bulk fluid. There are two basic cases of the decay rate - a minimum and a maximum - both are based on the principle that a growth rate can be defined as:

$$G = \frac{dL_S}{dx} = \frac{dL_S}{dt} \cdot \frac{dt}{dx} \qquad \text{.......(5)}$$

Where L_s is slug length and x & t are distance along the pipe and time respectively. If it is assumed that:

$$\frac{dL_S}{dt} = V_F - V_B \qquad \text{.......(6)}$$

and

$$\frac{dt}{dx} = \frac{1}{V_F} \qquad \text{.......(7)}$$

Where V_F and V_B are the slug front and back velocities. Then by utilising a simple mass balance over the front of the slug and substituting (6) & (7) into (5), the expression below is obtained:

$$G = \frac{dL_S}{dx} = 1 - \frac{V_B(H_S - H_F)}{H_S V_S - H_F V_{LF}} \qquad \text{.......(8)}$$

Where V_S is liquid in the slug velocity and H_S is the slug body hold up, similarly V_{LF} is liquid in the film velocity and H_F is the film hold up in front of the slug. Simple examination of equation (8) shows that decay rate can be unbounded when the film height and velocity are equal to that of the slug body holdup and the slug liquid velocity respectively. A limit must be imposed on the growth rate of -1. This simply means that when the growth rate is predicted to be below -1, a slug would decay instantaneously. Conversely if the growth rate is positive then slugs will grow and so the applicability of the model must be questioned.

There are several unknowns in equation (8), some which can be obtained from established closure relationships for slug body holdup and slug translational velocity. H_F can be found from the mass balance, but V_{LF} is still unknown. One additional useful parameter can be specified, decay length (D_L), which is the length over which the slug is reduced to zero by decay.

4.6 Liquid film Velocity (V_{LF})

V_{GF} can be related to V_{LF} as the film ahead of a slug is considered to be a normal stratified flow. Thus at steady state, the pressure drop over a short section of the liquid film must be the same as that in the bubble existing over the film. We can write the momentum equation for each phase (Taitel & Dukler, 1) as:

$$- A_G\left(\frac{dP}{dx}\right) - \tau_{WG}S_G - \tau_i S_i + \rho_G A_G g \sin\alpha = 0 \quad (9)$$

$$- A_L\left(\frac{dP}{dx}\right) - \tau_{WL}S_L - \tau_i S_i + \rho_L A_L g \sin\alpha = 0 \quad (10)$$

Where A_G and A_L are the cross sectional areas occupied by the gas and liquid phases. S_G, S_L and S_i are the gas-wall, the liquid-wall and interfacial perimeters and τ_{wl}, τ_{wg} and τ_i are the respective shear stresses. α is the pipe inclination angle.

If the pressure gradients are eliminated the system can be solved for the equilibrium liquid holdup (H_{LE}). Conversely if H_F and V_{GF} are known, then V_{LF} can be calculated as the pressure gradient is the same in each phase.

V_{GF} is obtained by considering the flow of gas past a point and equating the mass flow rate of gas for a slug and bubble pair to that for a stratified layer.

$$V_g A(1 - H_{LE})\rho_G = \frac{L_S}{L_U}(1 - H_S)V_S A\rho_G + \frac{1}{L_U}\int_0^{L_f}(1 - H_F)AV_{GF}\rho_G \, dx \quad (11)$$

Where V_g is the velocity of the gas phase in a stratified flow, L_u is the length of a slug unit and V_s is the velocity of the fluid in a slug. Assuming a constant area duct and a locally incompressible fluids, then (11) reduces to:

$$V_{gf} = \frac{V_g(1 - H_{LE}) - (L_S / L_U)(1 - H_S)V_S}{(L_F / L_U)(1 - H_F)} \quad (12)$$

with

$$\beta = \frac{L_s}{L_u} \quad (13)$$

We can express the slug length to unit length by defining a parameter β which can vary between 0 and 1. $\beta = 1$ corresponds to liquid/bubbly flow and $\beta = 0$ to stratified flow. Equation (12) with the substitution of (13) now gives the final relationship required to calculate the decay rate at a given point.

4.7 Pressure Drop over ΔX.

The pressure drop over a small increment of pipe is made up of four components. These are hydrostatic pressure drop in the slug and bubble regions and the frictional pressure drop in both regions.

4.8 Pressure drop in the Slug

The velocity of the bulk fluid in the slug is a function of the local superficial velocities (taking account of the compressibility of the gas phase). The slug body holdup (H_s) is also a function of the slug liquid velocity. The slug properties are therefore defined as:

$$\rho_{SLUG} = \rho_L H_S + \rho_G (1 - H_S) \quad \text{and} \quad \frac{1}{\mu_{SLUG}} = \frac{X}{\mu_G} + \frac{(1-X)}{\mu_L}$$

Where X is the mass quality. Using the above properties, the pressure drop in the slug can be calculated as:

$$\left(\frac{dP}{dx}\right)_{Friction} = f_{SLUG} \frac{\rho_{SLUG} V_S^{2}}{2} \quad \text{and} \quad \left(\frac{dP}{dx}\right)_{Hydrostatic} = \rho_{SLUG} g \sin\alpha$$

Where f_{SLUG} is the friction factor obtained using a mixture Reynolds number and a normal friction factor correlation. The overall pressure drop over the slug is therefore defined as:

$$\Delta P_{SLUG} = \left(\left(\frac{dp}{dx}\right)_{Friction} + \left(\frac{dp}{dx}\right)_{Hydrostatic}\right) L_S^{\bullet} \quad \text{.......... (14)}$$

4.9 Pressure drop in Bubble

After V_{LF} has been found, the pressure gradient in the bubble region can be calculated. However in the process of obtaining V_{LF}, the pressure gradient over the bubble was computed and can simply be applied over the length of the bubble.

$$\Delta P_{BUBBLE} = \left(\frac{dP}{dx}\right)_{From\ ()} L_b^{\bullet} \quad \text{.......... (15)}$$

4.10 Total Pressure Drop

The overall pressure drop over an increment ΔX is the sum of that over the slug and bubble, the outlet pressure of the increment is simply calculated as follows.

$P_{OUT} = P_{IN} - \Delta P_{TOTAL}$ .... (16) where $\Delta P_{TOTAL} = \Delta P_{SLUG} + \Delta P_{BUBBLE}$

So for a particular inlet pressure, the outlet pressure can be calculated using (16).

5. CALCULATION PROCEDURE

A flow diagram is presented below (in Figure 9) which gives the procedure for finding the pressure drop for decaying slug flow. A simple check is firstly performed to identify that the previous section of pipeline is in slug flow and that the section that is being entered is in stratified flow. The validity of applying this procedure for annular flow is discussed later.

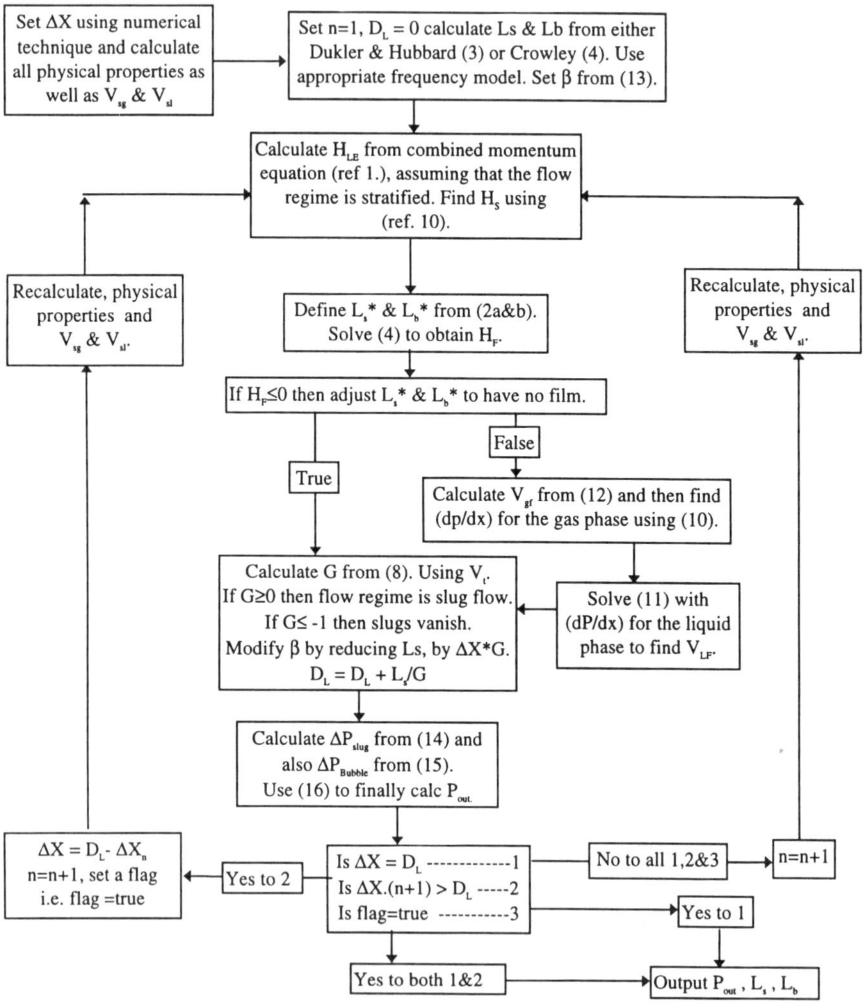

Figure 9. Flow Diagram showing the procedure for calculating the pressure drop and resulting slug characteristic for a section of pipe in a pipeline.

6. VALIDATION

The construction of this model has been modular and therefore various different closure relationships can be used. This is to take advantage of better correlations and models as they become available.

Figure 10 shows a comparison the model to the model of Beggs & Brill with full head recovery. The following closure relationships were used in applying the decaying slug model:

Slug Frequency Correlation: Gregory & Scott (11)[†]
Slug Length Correlation: Crowley & Rothe (4)[†]
Slug Body Holdup Correlation: Gregory et al. (10)

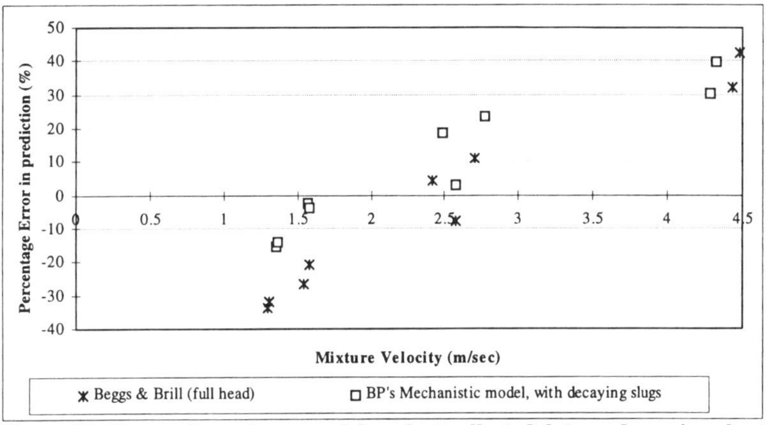

Figure 10. Comparison of new model against collected data and previous best prediction.

It can be seen that from Figure 10, the ability to model decaying slugs greatly enhances the capacity of the model to accurately predict pressure gradient. The model is definitely an improvement at both low and intermediate velocities and at higher velocities is comparable to the full head recovery version of the Beggs & Brill correlation.

7. CONCLUSIONS

The model for decaying slugs proposed here is an improvement on a general mechanistic modelling approach. It demonstrates that slugs continuing through a system give rise to a substantial head recovery that is a significant component of the actual pressure drop.

It is believed that the inclusion of decaying slug model acts as an improvement to all models where slugs exist in uphill sections. The construction of the model is entirely consistent with current modelling practice of stratified, stratified wavy and annular type flows. The slugs therefore decay into one of these flow regimes.

In the results shown above some of the cases predict a holdup of zero for the film region. This reflects the case where only a very thin film of fluid exists and the calculation preferentially puts all the liquid into the slug.

At higher velocities annular flow tends to be predicted in both uphill and downhill sections, i.e. it is almost independent of pipe inclination. Therefore discrepancies at high velocities are thought to be a combination of the transition to annular flow (less head loss uphill, therefore

[†] Both taken from (4), where it is recommended to use an increased frequency to maintain continuity at a variety of flowrates.

there is less to regain) and that slug frequency methods break down. The slug-bubble ratio at high velocities is incorrectly driven by the equilibrium superficial velocities. The ratio is therefore under-predicted and since head recovery is directly proportional to the slug-bubble ratio, an under-prediction of this ratio will lead to an under-prediction of the head recovery in the downhill section and hence an overprediction in pressure gradient.

REFERENCES

(1) Taitel, Y. & Dukler, A.E. 1976, "A Model for Predicting Flow Regime Transitions in Horizontal and Near Horizontal Gas-Liquid Flow" *AIChE Journal, Vol 22, 47-55.*

(2) King, M.J.S, Hale, C.P., Roberts, I.F., Fisher, S.A., Lawrence, C.J., Mendes-Tatsis, M.A. & Hewitt, G.F. 1997 "Experimental Investigations of Flowrate Transients in Horizontal Pipes" *Multiphase '97, Cannes, France, June 18-20.*

(3) Dukler, A.E. & Hubbard, M.G. 1975 "A Model for Gas-Liquid Slug Flow in Horizontal and Near Horizontal Tubes" *Ind Eng Chem Fundam,* ***14****, 337-347.*

(4) Crowley, C.J. & Rothe, P.H. 1986 "State of the Art Report on Multiphase Methods for Gas and Oil Pipelines volume 2 Guide to Computerised Calculations" *Creare Inc. report TN-409 vol. 2 (prepared for Project PR-172-609, American Gas Association, Catalog No. L51527).*

(5) Taitel, Y. & Barnea, D. 1990 "Two-Phase Slug Flow" *Advances in Heat Transfer,* ***20****, 83-132.*

(6) Zheng, G., Brill, J.P. & Taitel, Y. 1994 "Two-Phase Slug Flow in Hilly Terrain Pipelines" *Int J Multiphase Flow,* ***20****, 63-79.*

(7) Hill, T.J., Fairhurst, C.P., Nelson, C.J., Becerra, H. & Bailey, R.S. (1996) "Multiphase Production through Hilly Terrain Pipelines in Cusiana Oilfield, Columbia" *SPE Annual Technical Conference and Exhibition, Denver, CO, 6-9 October.*

(8) Beggs, H.D. & Brill, J.P. 1973 "A Study of Two-Phase Flow in Inclined Pipes" *Trans J Pet Tech,* ***May 1973****, 607-617.*

(9) Yang, J. 1996 "A Study of Intermittent Flow in Downward Inclined Pipes" *PhD Dissertation, University of Tulsa, OK.*

(10) Gregory, G.A., Nicholson, M.K. & Aziz, K. (1978) "Correlation of Liquid Volume Fraction in the Slug for Horizontal Gas-Liquid Slug Flow" *Int J Multiphase Flow,* ***4****, 33-39.*

(11) Gregory, G.A. & Scott, D.S. (1969) "Correlation of Liquid Slug Velocity and Frequency in Horizontal Cocurrent Gas-Liquid Slug Flow" *AIChE J,* ***15****, 933-935.*

Development of a mechanistic model for the prediction of slug length in horizontal multiphase flow

M GOPAL
Corrosion in Multiphase Systems Center, Ohio University, Athens, USA

Synopsis

A mechanistic model for the prediction of slug length in multiphase flow is presented based on a unique concept involving the Froude number. It is shown that the Froude number in the liquid film ahead of the slug is greater than unity. It decreases to values less than unity inside the mixing zone of the slug and then gradually increases within the body of the slug. The slug tail occurs as the Froude number tends to unity once more. Agreement with experimental data is good. The model also closely predicts the data of other researchers in large diameters pipes.

Notation

a_{GF}	=	area occupied by the gas above the stratified layer of liquid
a_{LF}	=	area occupied by the stratified layer of liquid
S_L	=	perimeter of liquid contact with wall over which shear stress acts.
S_G	=	perimeter of gas contact with wall over which gas phase shear stress acts.
S_i	=	width of gas-liquid interface.
τ_{WL}, τ_{WG}	=	wall shear stress for liquid and gas respectively
τ_i	=	shear stress at gas-liquid interface.
ρ_L, ρ_G	=	density of liquid and gas respectively
θ	=	pipe inclination (small values, close to horizontal).
f_i	=	interfacial friction factor
f_G	=	gas phase friction factor
v_t	=	translational velocity of the slug.
v_{LF}	=	liquid film velocity
h_{EF}	=	effective film height ahead of the slug

u_L = average velocity of the liquid
p_{LF}, p_S = pressures in the liquid film and the slug
a_{LF}, a_{LS} = areas of the liquid film and the liquid in the slug
x_o, x_1 = points in the liquid film and slug respectively
ρ_L = density of liquid

$\bar{a}_{LF}$, $\bar{a}_{LS}$ = fraction of area occupied by the liquid in film and slug
v_{LF}, v_{LS} = average liquid velocity in film and slug

$\bar{a}_{GF}$ = fraction of area occupied by the gas over the liquid film.
p_{GF} = pressure in gas pocket over liquid film
F_D = drag force exerted by the gas bubbles on the liquid.
p'_o = $p_{LF} - \rho_G \zeta'_o$, and ζ'_o is the depth from the surface of the center of pressure in the gas over which p'_o acts.
v_{GS}, v_{GF} = gas velocity in the slug and gas pocket over liquid fiilm

$\bar{a}_{GS}$, $\bar{a}_{GF}$ = fraction of area occupied by gas in slug and by gas pocket over liquid film
h_F, h_j = hydrostatic head corresponding to the pressures p_o, p_1
v_{LS} = the effective average liquid velocity in the slug at any point
h_j = the effective "height" of the jump.
q_l = volumetric flow rate of liquid as part of slug
$v(y)$ = velocity at distance y from the pipe wall
v_O = centerline velocity
δ = thickness of the boundary layer

$\bar{h}$ = nondimensional height from the bottom of the pipe
a,b = nonlinear regression coefficients for void fraction distribution
α_{avg} = average void fraction over a cross sectional area
l_m = The length of the mixing zone
m, c = linear regression coefficients for the mixing zone
u_o = centerline velocity in a moving coordinate system = $v_t - v_O$
f = the Fanning friction factor
C = constant in friction factor calculation = 0.046
n = exponent in friction factor calculation = - 0.2
C_D = drag coefficient
d_b = diameter of bubble
v_r = measured rise velocity of bubbles in the slug body
Re_b = bubble Reynolds Number

1 INTRODUCTION

Multiphase flow has become a technology central to oil and gas production today. With the development of "full well stream transportation" systems (Fairhurst, 1988), both from remote sites, such as Alaska, and in offshore production, the emphasis is on the transport of all components of the produced fluids from the well over long distances to central processing facilities. An accurate prediction of multiphase flow characteristics is essential for the design, and economical operation of these systems.

Several flow regimes occur in multiphase flows, including, stratified, slug, and annular flows. The production rates from the wells are such that these multiphase flow pipelines are expected to be in slug flow at some time in their lives. This is a highly turbulent flow regime, leading to increased pipe damage from internal corrosion and mechanical impacts. This is related to the slug length. It is therefore, important to obtain a detailed, mechanistic understanding of slug flow characteristics, and to determine the overall slug length.

Mathematical models have been developed that describe the relationship between different variables as knowledge of slug flow features have increased over the last decade. However, a detailed understanding of the motion of gas within the slug, and the distribution of phases in the different zones of the slug is not known. This is essential information that may be used to develop a complete mathematical model to predict slug length.

This paper describes a mathematical model that has been developed to predict slug length utilizing the phase distribution and velocity profiles within the slug. The experimental techniques have been described elsewhere (Gopal and Jepson, 1997, 1998a,b).

2 BACKGROUND

Figure 1 shows the profile of a slug. Waves form on the liquid film, that grow to bridge the pipe. This causes the liquid to be accelerated by the gas. As the slug front moves through the pipe, it overruns the slow moving liquid film ahead of it and accelerates it to the velocity of the slug. A mixing vortex is created in this process. This leads to a scouring mechanism on the pipe wall with high rates of shear. Also, as the liquid is assimilated by the slug, a considerable amount of gas is entrained (Jepson, 1987). This leads to the creation of a highly frothy, turbulent region behind the slug front called the mixing zone.

Beyond the mixing region of the slug, the level of turbulence is reduced, and buoyancy forces move the gas towards the top of the pipe. The cross sectional area available for liquid flow increases, a boundary layer develops, and the liquid velocity decreases. This is the slug body. Eventually a point is reached where the liquid velocity is no longer sufficient to sustain the bridging of the pipe, and the slug body is curtailed. This is called the slug tail. The liquid velocity decreases in the liquid film, its height rebuilds with waves forming on its surface, and the next slug is initiated.

Dukler and Hubbard (1975) published the first realistic mechanistic model for slug flow characteristics. They established fundamental equations that could predict several slug flow characteristics. The agreement with experimental data was good and a better understanding of the mechanisms was achieved. However, many parameters, such as slug frequency and the void fraction within the slug, were required to complete the calculation. Also, their definition of the mixing zone is not adequate (Gopal and Jepson, 1998b). The slug lengths in their studies varied from 12 to 25 pipe diameters.

Nicholson et al. (1978) found a non-zero gravity induced drift velocity, even for horizontal pipes, and determined that this velocity needed to be incorporated in the calculation of the slug translational velocity. They modified and extended the model of Dukler and Hubbard to apply to the entire intermittent flow regime. The slug lengths in their case varied from 10 to 60 pipe diameters. It is to be noted that the model ignores any slip between gas and liquid within the slug. It has been found (Jepson, 1987) that for large diameter pipes, this is not true for higher slug velocities.

Maron et al. (1982) derived a model for slug flow based on periodic distortion of the hydrodynamic boundary layer followed by a recovery process. As the slug front overruns the liquid film, the boundary layer is destroyed by the mixing eddy. At the end of the mixing zone, this boundary layer begins to redevelop. The model considered two separate types of slug, one with aeration and the other without. In the first case, the entrained gas bubbles in the slug leave the boundary layer region due to buoyancy effects and tend to agglomerate in the upper portion of the pipe. The slug length is the distance required for complete separation of the gas from the liquid. In the second case, the slug length is given by the distance required for the boundary layer to fully develop and reach the center of the pipe. They developed their boundary layer analysis in a coordinate system moving with the slug front and introduced a one-seventh power law model to describe the velocity profile within the boundary layer. They showed that the model could be applied to predict pressure drop for a wide quasi-steady frequency range of slugs. However, stability analysis indicated that the slug pattern stabilized over a narrow frequency range corresponding to a minimum pressure drop. It is to be noted that no information about the pipe diameter or working fluids were given.

Dukler et.al. (1985) applied the concepts developed by Maron et al. (1982) and formulated a generalized model for the prediction of the minimum stable slug length for horizontal and vertical slug flow. The model utilized the velocity profile developed by Maron et al. (1982) in the boundary layer, and combined this with an inviscid potential core. An assumption was made that a flat velocity profile resulted at the end of the mixing zone. The velocity at the center was allowed to decrease due to frictional effects predicted by a Blasius-type equation. The slug length was predicted by the distance required for the complete development of the boundary layer. The results of the model were applied to 5 cm I.D. vertical and horizontal, and 3.8 cm horizontal pipes, with air and water as the working fluids. It was found that the experimental results were bounded between the value predicted by the model and twice that value. It should be noted that the model ignored the contribution of gas to the slug characteristics, and hence, to the slug length. Their slug length could have been the mixing length of the slug (Gopal and Jepson, 1998a).

Kouba (1986) formulated a model to account for both liquid and gas phase distribution as well as velocity distribution within the slug. By considering a mass balance between the slug and the liquid film, he was able to formulate a generalised expression for liquid film velocity ahead of the slug. From observations in a 7.5 cm I.D., 418 m long pipeline, using kerosene and air as the working fluids, he concluded that the drift velocity, v_d , was significant even for horizontal slug flow. Utilising the theory of shearing flow over a wavy boundary developed by Benjamin (1968), he developed an expression for the drift velocity as a function of pipe diameter. The incorporation of the drift velocity improved slug length predictions by twenty percent. The range of slug lengths in this study were twenty-five to one hundred pipe diameters. This was due to the length of his pipeline, which was 418 m. There was an effect of gas expansion over this length resulting in an acceleration of the slugs and an increase in slug length.

Figure 2 describes the physical model for stratified flow developed by Taitel and Dukler (1976). The gas and liquid flow in stratified layers within the pipe. They used this model along with other parameters to determine the transition to other flow regimes. This model is used as the foundation for predicting the liquid film height in slug flow in this study. From a consideration of a momentum balance in the liquid and gas phase the following equation is obtained:

$$\tau_{WG} \frac{S_G}{a_{GF}} - \tau_{WL} \frac{S_L}{a_{LF}} + \tau_i S_i \left(\frac{1}{a_{LF}} + \frac{1}{a_{GF}}\right) + (\rho_L - \rho_G) \qquad (1)$$

Equation (1) can be used to predict the liquid film height in slug flow as well. Details are given elsewhere (Gopal and Jepson, 1998b).

Jepson (1989) presented a physical model for the prediction of transition to slug flow. The model assumes that the slug is formed as a result of a hydraulic jump propagating along the pipe. He defined a dimensionless Froude Number for the film ahead of the slug using a comparison between slugs and hydraulic jumps. Using equations of continuity and momentum conservation at a condition where the jump just touches the top of the wall of the pipe, he provided the necessary conditions for the existence of slugs. This theory forms the foundation of the proposed model for slug length in this paper.

Several investigations on slug flow have been conducted for large diameter pipes. Crowley et al. (1986) conducted slug flow studies in 17 cm pipes, with water and glycerine (viscosity 400 cP) for the liquid phase and Freon for the gas phase at densities one to twenty times that of air. They found that the translational velocity of the slug was predicted for all cases using the drift flux model. There was a negligible effect of the gas density on the slug velocity. However, there was a large effect of liquid viscosity. The slug velocities for glycerine were fifty percent geater than water for the same conditions. It was found that the interfacial friction factor was about ten times the gas phase friction factor in this case. The slug lengths ranged from 0.5 m to about 3.5 m.

Jepson and Taylor (1993) found that there was an increase in slug length with an increase in pipe diameter. Below a gas superficial velocity of 5 m/s, the slugs appeared to be growing. This indicates that there may be gas expansion in large diameter pipes, or a significant drift velocity causing the the slug to grow.

Scott et.al. (1986) developed correlations for predicting slug length in large diameter pipes using data from 30, 40, 50, and 60 cm I.D. pipes in oil fields in Prudoe Bay, Alaska. They found that in these pipelines there was an additional factor in slug length analysis. They termed this factor the "long term growth". This was related to gas expansion within the pipeline due to pressure changes.

Fairhurst (1988) discussed several important issues related to slug flow in large diameter pipes. From an analysis of the data gathered in the 30, 40, 50, and 60 cm pipes in Alaska, he concluded that generally available design methods for slug flow failed to predict the behavior of real oil and gas pipelines, and that future work should be tested with field data.

Recently statistical characterizations have been developed to describe the variations of slug lengths in pipes (Nydal et al., 1992, Barnea and Taitel, 1993). A log-normal distribution of slug lengths was found to be applicable by Nydal et al. (1992), while Barnea and Taitel (1993) found that the mean slug length was 1.5 times the minimum stable slug length and the maximum length was about 3 times the minimum stable length.

3 MATHEMATICAL MODEL

A mathematical model is developed to relate the variables involved in slug flow. Figure 4 shows a schematic flow chart of the model development.

Step 1 involves the calculation of the Froude number in the liquid film ahead of the slug. Knowing the input superficial velocities of liquid and gas, the pipe diameter, and the fluid properties, the liquid film height, h_{LF}, and the average velocity of the liquid in the film, v_{LF}, are calculated. Using the superficial liquid and gas velocity, the slug translational velocity, v_t, is also calculated. These three variables, v_t, v_{LF}, and h_{LF}, are then used to calculate a film Froude number.

Step 2 involves the calculation of the Froude Number in the slug. Once the film Froude number is known, the length of the mixing zone in the slug, l_m, and the effective height of the slug, h_j, are calculated. Next, knowing the superficial gas and liquid velocities, the slug velocity, v_S, is computed. Then, using the void fraction distribution and the velocity profile equations, the effective average liquid velocity in the slug, v_{LS}, is calculated. Finally, using v_t, v_{LS}, and h_j, the Froude Number in the slug is determined.

In step 3, the decrease of the liquid velocity in the slug is calculated using pressure drop relations. The decrease in the effective average liquid velocity in the slug is then related to the increase in Froude number in the slug. The point where the Froude Number equals unity is found, and the slug length is calculated.

3.2 Liquid Film height

As mentioned previously, the liquid film height is predicted using a modified Taitel and Dukler (1976) model. The modification of the interfacial friction factor as given by Equation (1) is insufficient to predict the liquid film height in slug flow. It has been found that the interfacial friction factor needs to be increased over those predicted by Equation (7), due to the presence of large three dimensional waves on the liquid film. Details of this explanation can be found elsewhere (Gopal and Jepson, 1998b). Hence it was estimated as a constant multiplied by the gas phase friction factor as follows:

$$f_i = k_1 f_G \tag{2}$$

where, $k_1 = 50$

3.2 Liquid Film Velocity

It is very difficult to estimate the liquid film velocity, since it is not constant between slugs. The liquid drains from the rear of the slug and mixes with new incoming liquid. Further, at high gas velocities, roll waves are formed on the liquid surface that can also affect the liquid film velocity. In this study the liquid film velocity was estimated to be, on average, equal to the liquid superficial velocity. The details are given elsewhere (Gopal and Jepson, 1998b). This model sets:

$$v_{LF} \approx v_{SL} \tag{3}$$

This is used in the model for slug length prediction

3.3 Film Froude Number

It has been shown that slugs are hydraulic jumps (Kouba and Jepson, 1987) and their strengths may be determined by the Froude Number ahead of the slug. The Film Froude Number, Fr_f, is defined as follows:

$$Fr_f = \frac{v_t - v_{LF}}{\sqrt{g\, h_{EF}}} \tag{4}$$

The details of the Film Froude number derivation are given in a previous paper (Gopal and Jepson, 1998b).

Pressure Relationship in Slugs

In a coordinate system moving with the slug front, the momentum equation becomes the same as that for a hydraulic jump (Jepson, 1989). This is given for channel flow by Stoker (1957) as:

$$\frac{d}{dt}\int_{x_o}^{x_1} \rho_L\, a_L\, u_L\, dx = \int_{a_{LF}} p_{LF}\, dA - \int_{a_{LS}} p_s\, dA \tag{5}$$

In a coordinate system moving with the slug front, the relative velocities of the fluids in the slug and in the film are described by (v_t - v_S) and (v_t - v_{LF}) respectively, and the slug front itself becomes stationary. Under such a condition, following Stoker (1957), Equation (5) can be integrated to give:

$$_L \left[\bar{a}_{LS}\,(v_t - v_{LS})^2 - \bar{a}_{LF}\,(v_t - v_{LF})^2\right] = p_{LF}\,\bar{a}_{LF} - p \tag{6}$$

A momentum equation for the gas phase can be similarly written:

$$_G \left[\bar{a}_{GS}\,(v_t - v_{GS})^2 - \bar{a}_{GF}\,(v_t - v_{GF})^2\right] = p'_o\,\bar{a}_{GF} - \tag{7}$$

Adding equations (6) and (7) and neglecting the terms involving ρ_G, since $\rho_G << \rho_L$ we get:

$$\rho_L \left(\bar{a}_{LS} (v_t - v_{LS})^2 - \bar{a}_{LF} (v_t - v_{LF})^2\right) = p_{LF} \qquad (8)$$

Next, dividing throughput by $\rho_L g$, the following expression is obtained:

$$\frac{(v_t - v_{LS})^2}{g} \bar{a}_{LS} - \frac{(v_t - v_{LF})^2}{g} \bar{a}_{LF} = \frac{p_{LF} - p_{LS}}{\rho_L g} = h \qquad (9)$$

This can then be rearranged, after some manipulation, to give:

$$\frac{(v_t - v_{LS})^2}{g h_j} \frac{h_j}{h_F} \bar{a}_{LS} - \frac{(v_t - v_{LF})^2}{g h_F} \bar{a}_{LF} = 1 - \frac{h_j}{h_F} \qquad (10)$$

The second term in Equation (10) provides a basis for the definition of a Froude number in the slug. The term $(v_t - v_{LF}) / \sqrt{(g h_F)}$, is similar to the film Froude number defined in Equation (4).

In open channel flow, the film Froude Number is defined as:

$$Fr_f = \frac{v_t - v_{LF}}{\sqrt{g\, h_F}} \qquad (11)$$

In pipe flow, the definition of the film Froude Number must be modified to account for the geometry. In this case, h_F is the height of the liquid film at the center of the pipe. The definition of the film Froude Number is then written as in Equation (4). Equation (11) is also based on a channel geometry. To use it for a pipe, h_F is replaced by h_{EF}. The term $(v_t - v_{LF}) / \sqrt{(g h_F)}$ would then be replaced by the term $(v_t - v_{LF}) / \sqrt{(g h_{EF})}$. This would then be the Froude Number in the film, as defined in Equation (4). The term $(v_t - v_{LS}) / \sqrt{\{g h_j\}}$, may then be said to be the Froude Number in the slug.

3.5 Slug Froude Number Definition

As is indicated from Equation (10), a Froude number in the slug, Fr_s, may be defined as:

$$Fr_s = \frac{v_t - v_{LS}}{\sqrt{g h_j}} \qquad (11)$$

It should be remembered, that v_{LS} is not constant in the slug. Farther into the body of the slug, momentum losses occur and, the fraction of the total pipe cross sectional area available for liquid flow, increases with distance into the slug body. Here the gas is pushed towards the top of the pipe decreasing the overall void fraction. This leads to a decrease in the slug liquid velocity, v_{LS}, and consequently, gives an increase in the slug Froude number.

3.6 Ratio of Heights in Slugs

A relationship to predict h_j as a function of the Film Froude number is now needed. Chow (1959) gives a correlation between h_F and h_j for channel flow:

$$\frac{h_j}{h_F} = \frac{1}{2}\left[\sqrt{1+8Fr_f^2}+1\right] \tag{12}$$

where, Fr_f = Film Froude Number in open channel flow as given by Equation (10).

This equation also needs to be modified for the geometry of the pipe. Figure 5 shows the variation of the ratio h_j/h_{EF} and h_j/h_F for a stationary slug (Kouba and Jepson, 1987) where, h_{EF} is the effective height of the liquid film and h_F is the height of the film at the center. The data shown in Figure 5 were obtained from single point pressure measurements. Table 1 shows the variations of h_j/h_{EF} as a function of distance into the slug at various Film Froude Numbers for water-carbon dioxide stationary slug systems. It is seen that beyond a distance of 45 cm from the slug front, the ratio h_j / h_{EF} varies similar to the values predicted by Equation (12), with h_{EF} used to estimate h_j.

It is seen that Equation (12) provides a reasonable estimate for h_j for slug flow in pipes, if the effective film height, h_{EF}, is used. This is due to the curved geometry of pipes. In channels, the depth is uniform across the width. Hence, a single height is sufficient to describe the liquid film. Using the value of h_{EF} for the liquid film height in Equation (12) gives a good estimate of the effective height of the slug, h_j.

3.7 Average Liquid Velocity in Slug

The volumetric flow rate of liquid in the slug is given by:

$$q_l = \int_0^A v\,(1 - \alpha)\,dA \tag{13}$$

The local velocity, v, and the void fraction, α, within the differential area in the slug body are given by the following equations:

$$v(y) = v_o\left(\frac{y}{\delta}\right)^{1/7} \tag{14}$$

and,

$$\alpha(\bar{h}) = \frac{a\,\bar{h}}{(1 - b\,\bar{h})} \tag{15}$$

In Equation (15), the values of a and b are taken as 0.1 and 0.085. The details of Equations (14) and (15) are given elsewhere (Gopal and Jepson, 1997).

Once q_l is known, the effective average velocity in the slug is calculated as:

$$v_{LS} = \frac{q_l}{a_{LS}} = \frac{q_l}{A\,(1 - \alpha_{avg})} \tag{16}$$

where, α_{avg} = average void fraction over a cross sectional area.

Once v_{LS} is known, the Froude Number at any distance in the slug can be calculated using Equation (10).

3.8 Slug Length Model

The slug front is a propagating hydraulic jump (Jepson, 1989) and represents a transition from subcritical flow to supercritical flow when viewed in a coordinate system moving with he slug front. Figure 6 shows how the Froude Number varies throughout the film and the slug in this case. When a moving slug is observed in a stationary coordinate system, the reverse occurs and the flow is subcritical in the film ahead of the slug and supercritical in the slug.

The above analysis implies that in a coordinate system moving with the slug front, the Froude Number in the film is always greater than unity, indicating supercritical flow, and in the slug is always less than unity, indicating subcritical flow. The Froude Number decreases rapidly at the slug front and subsequently, gradually rises back up to unity within the slug body. The point where it tends to unity corresponds to the end of the slug and the formation of a new liquid film. The increase in Froude Number, according to Equation (10), may be used to find the point where the Froude Number in the slug equals unity, and hence the slug length. The slug is broadly composed of a mixing zone, and the slug body. The total slug length is then a sum of the individual lengths of each of these zones.

3.9 Length of Mixing Zone

The mixing length was defined as the minimum distance into the slug required for the void fraction profile to reach a quasi-steady state described by Equation (15). The mixing length was estimated as an empirical function of the film Froude number. The dependence is given by:

$$l_m = m.Fr_f + c \tag{17}$$

The values of the regression coefficients are found to be m=0.13, and c = -0.31, to calculate the length of the mixing zone in metres.

3.10 Slug Body Length

In the slug body, the void fraction distribution is obtained by using equation (15) and the velocity profile may be obtained by Equation (14). The effective average liquid velocity may be calculated using Equations (13) and (16). The liquid velocity decreases due to momentum losses. This is expressed by the following relation:

$$u_o \frac{du_o}{dx} = -\frac{1}{\rho_L}\frac{dp}{dx} \tag{18}$$

It is noted that the velocity profile in the slug body is similar to that given by the one-seventh power law, with v_O as the centerline velocity. However, unlike a fully developed flow in a pipe, the center line velocity decreases in a slug due to momentum losses. The profile, however, has the same shape throughout the body of the slug (Gopal and Jepson, 1997).

The pressure gradient, dp/dx, in the slug body is given by the sum of two components, the pressure drop due to friction alone, dp/dx_f, and, an excess pressure drop due to agitation by the gas bubbles in the slug dp/dx_e.

The pressure drop due to friction is given by the well known equation:

$$\frac{dp}{dx_f} = \frac{4f}{D}\frac{\rho_L v_{LS}^2}{2} \tag{19}$$

where f is calculated using a Blasius-type equation:

$$f = C\left(\frac{D\, v_{LS}\, \rho_L}{\mu_L}\right)^n \tag{20}$$

The pressure drop due to agitation by bubbles is more difficult to evaluate. This is given by the expression:

$$\frac{dp}{dx_e} = \frac{3C_D \alpha_{avg}}{2d_b}\frac{\rho_L v_r^2}{2} \tag{21}$$

The drag coefficient C_D, is a complicated function of several dimensionless groups. Wallis (1969) modified the derivation of Peebles and Garber (1953) and used the following:

$$Re_b = \frac{\rho_L d_b v_r}{\mu_L} \tag{22}$$

Also, a dimensionless group G_1, is defined as follows:

$$G_1 = \frac{g\mu_L^4}{\rho_L \sigma^3} \tag{23}$$

Then a table for the drag coefficient may be used as shown in Table 2. Hence, the total pressure gradient can be calculated. Starting at the end of the mixing zone, the centerline velocity, u_o, can be updated using Euler's method in Equation (19). v_{LS} can then be found as before. The slug Froude number can then be checked using Equation (12) and a value of unity yields the length of the slug body.

The sum of the mixing length and the slug body length would then yield a measure of total slug length.

4 RESULTS AND DISCUSSION

The detailed results of the slug translational velocity, mixing length, and the detailed profiles of velocity and voids within the slug are given elsewhere (Gopal and Jepson, 1997, 1998a,b). Here, the emphasis is on Froude number results, and slug length modeling. Froude numbers in the slug were calculated for water-carbon dioxide systems.

4.1 Liquid Film Height

Figures 7a and 7b show the variation of the mean film height ahead of the slug as a function of slug velocity. It is seen that the mean height of the film, in both cases, lie between 0.3 and 0.4 pipe diameters. The mean height does not change significantly over the range of velocities studied for either cases. The modified Taitel and Dukler model used in this study, gives a reasonable value of film height prediction. The slight overprediction of the model for the case of water is due to the neglect of liquid flow rate as part of slugs and large roll waves that occur in slug flow. These do not seem to have a major effect for the ARCOPAK90™ slug systems.

4.2 Froude Number

The details of the film Froude number have been given in a separate paper (Gopal and Jepson, 1998a). Here the details of the variation of the Froude number in the slug is given.

Figures 8a to 8c show the variation of Froude number with distance in the slug, for a water-carbon dioxide slug system. Figure 8a describes the Froude number variations for a superficial liquid velocity of 0.2 m/s, and a superficial gas velocity of 1.07 m/s. The film Froude number in this case is 3.8. It is seen that the Froude Number decreases rapidly to a value of about 0.8 to 0.9 in the slug, at a distance of 20 cm from the slug front. This also corresponds to the end of the mixing zone in this case. The Froude Number then gradually rises in the slug body and tends to a value close to unity near the tail of the slug. The total length of the slug at which the Froude Number tends to unity is about 60 to 80 cm.

Figure 8b shows the Froude Number variation in the slug for a superficial liquid velocity of 0.3 m/s, and a superficial gas velocity of 1.07 m/s. The film Froude number in this case is also about 3.8. Again, it is seen that the Froude number drops to a value between 0.8 and 0.9 at 20 cm into the slug. However, there is an oscillation of the Froude Number near the end of the mixing zone at approximately 30 cm into the slug.. This is due to a release of pulses of gas bubbles observed in the mixing zone. The pulse of bubbles results in increased local void fraction, and an increase in the average velocity of the liquid. In the slug body, these effects are dissipated and the Froude Number rises again to a value close to unity, near the tail of the slug. This is at a distance of 40-50 cm into the slug.

Figure 8c shows the Froude number variation in the slug for a superficial liquid velocity of 0.4 m/s, and a superficial gas velocity of 1.43 m/s. The film Froude number is about 4.6. In this case, the increase in the Froude number value is rapid. As is seen, the Froude number drops rapidly below unity to a value of around 0.7 at a distance of 15 cm into the slug. At this Froude number, there are high levels of turbulence within the mixing zone. The Froude number can therefore be expected to oscillate rapidly in this zone. From this point, the Froude number increases rapidly and reaches a value near unity at the tail of the slug. This occurs at approximately 55 cm.

Figure 9 shows the variation of the Froude Number from the film to the slug and beyond the tail of the slug for slug velocities of 1.27 m/s, 1.37 m/s, and 1.83 m/s. It is seen that the Froude Number in the film ahead of the slug in each case is of the order of 4 to 5. This decreases rapidly to a value of about 0.7-0.9 within the slug. The Froude number then gradually increases within the body of the slug and reaches a value close to unity at the tail of the slug. Beyond the slug tail, it is seen that the Froude number begins to rise rapidly again. For slug velocities of 1.27 m/s and 1.37 m/s, the Froude number in the tail rises to 2.5, while for a slug velocity of 1.83 m/s, the Froude number beyond the tail rises back to about 4.5.

Figures 8a, b, and c, and Figure 9 strongly indicate the existence of a transition in the flow characteristics based on Froude number. It is seen that when viewed in a coordinate system moving with the front of the slug, the Froude number in the film is greater than unity. It drops to below unity in the slug and rises again to above unity beyond the slug tail. There is a transition from supercritical flow in the film, to subcritical flow in the slug, and a transition back to supercritical flow at the end of the slug. This point of transition marks the end of the slug.

4.3 Total Slug Length

Figures 10a and 10b show the variations of total slug length as a function of slug velocity for water-carbon dioxde slug and ARCOPAC90™-carbon dioxide slug system. Tables 3 and 4 list the variations for the two systems.

It is seen that the slug length is distributed about the mean value within ±2 standard deviations. In general, the total slug length increases with an increase in slug velocity, from 0.7 m at a slug velocity of 1.27 m/s to about 1.3 m at a slug velocity of 3.2 m/s.Table 3 shows the variations of total slug length as a function of slug velocity for ARCOPAK90™-carbon dioxide slug system. Again, it is seen that the slug length is within ±2 standard deviations. In general, the total slug length increases with an increase in slug velocity, from 0.6 m at a slug velocity of 1.27 m/s to about 1.16 m at a slug velocity of 5.88 m/s.

Figure 11 shows the variation of slug lengths for different systems. Data from this study is shown along with those of Nicholson et.al. (1978), Kouba (1986), Crowley et.al. (1986), and Jepson (1988). The slug lengths have been represented as a number of pipe diameters, to compare slug lengths from systems of different pipe diameters. Nicholson et.al. performed experiments in 2.58 cm pipes.

It is seen that, in general, the slug length varies from about five to twenty pipe diameters. The data from Kouba (1986) are twice as large as the rest of the data. This is ascribed to gas expansion over the length (418 m) of the pipe. The data in this study for both water and ARCOPAK90 are very similar to those of others.

Figure 12a and 12b shows a comparison of the model with average total slug length for water-carbon dioxide and ARCOPAK90™-carbon dioxide slug systems. The model agrees well with the experimental data for the mean slug length, in the case of water-carbon dioxide slug systems. The mean slug length varies from about 0.7 m at a slug velocity of 1.3 m/s, to 1.3 m at a slug velocity of 3.2 m/s. This translates from about 10 pipe diameters to about 17 pipe diameters. It should be noted that there is a distribution of the slug length of approximately two standard deviations from the mean, and the model predictions are well within the range of this distribution.

Figure 12b shows the results for slug lengths for ARCOPAK90™-carbon dioxide slug system. The mean experimental slug length varies from about 0.6 m at a slug velocity of 1.6 m/s to about 1.2 m at a slug velocity of 6 m/s. The model agrees well with the mean slug length results up to a slug velocity of 3 m/s. Beyond that velocity, it begins to over predict the slug length. This may be due to variations in the length of the mixing zone for this mixture. Again it is to be noted that there is a distribution of lengths around the mean slug length and the model predictions are within this range.

Figure 13 shows the model predictions for the data of Jepson and Taylor (1988) and Crowley et.al. (1988). The lengths vary from about 3 pipe diameters to 17 pipe diameters. It is seen that there is good agreement between the model and the experimental data.

5 CONCLUSIONS

Slug flow characteristics have been studied and modelled for two-phase gas-liquid systems. The effect of liquid properties were investigated by the use of two different liquids, water and ARCOPAK90™.

Experimental conditions maintained in this study were in the slug flow regime. The liquid superficial velocity ranged from 0.2 m/s to 0.7 m/s for water, and 0.15 m/s to 0.88 m/s for ARCOPAK90™. The superficial gas velocity ranged from 1 m/s to 5 m/s.

There was a wide variation of slug lengths for both water-carbon dioxide slugs and ARCOPAK90™-carbon dioxide slugs. The slug lengths were distributed within ±2 standard deviations. The data agrees with those of Jepson and Taylor (1988) and Crowley et.al. (1988) as well as Nicholson et.al. (1978) and Dukler and Hubbard (1975).

A slug Froude number was defined which was used to determine slug lengths. From a knowledge of the Film Froude Number, the effective height of the slug was defined. An effective average velocity of the liquid in the slug was also defined and used in the slug Froude number.

It was found that the film Froude Number was always greater than unity and the slug Froude number was always less than unity. The slug Froude number tended to unity near the tail of the slug, and this criteria was used to determine the slug length. Beyond the tail, the film Froude Number was seen to increase beyond unity once more.

A mathematical model was developed to predict slug length. Knowing the superficial gas and liquid velocities, the fluid properties, and the pipe diameter, the liquid film height ahead of the slug was predicted. The Film Froude Number was then calculated. The mixing length in the slug was then determined as a function of the Film Froude Number. A mechanistic model was developed that incorporated the void fraction distribution and the velocity profiles in the slug body to predict the variation of the slug Froude Number. The slug Froude Number variation was then computed, and the point where it equalled unity was determined. This provided the criterion to determine the slug body length. The total slug length was then estimated as the sum of the mixing length and the slug body length.

The model gives good prediction of the mean slug length for the data in this study. It also is able to predict the data of Jepson and Taylor (1988), and Crowley et.al. (1988).

Table 1: Table for h_j/h_{EF} with distance for a range of film Froude numbers

Froude Number	h_j / h_{EF} with distance into slug					
	0 cm	15 cm	30 cm	45 cm	60 cm	Eq. (19) $h_F = h_{EF}$
6.6	6.8	9.4	10.2	10.2	10.3	8.9
9	3.6	9.3	11.42	11.9	11.95	12.4
12		5.7	11.05	14.6	14.0	16.4
14			11.1		15.8	19.3

Table 2: Summary of drag coefficient equations

Region	Drag Coefficient, C_D	Range of Applicability
Region 1	$24\ Re_b^{-1}$	$Re_b \leq 2$
Region 2	$18.7\ Re_b^{-0.68}$	$2 \leq Re_b \leq 4.02\ G_1^{-0.214}$
Region 3	$0.0275\ G_1\ Re_b^4$	$4.02\ G_1^{-0.214} \leq Re_b \leq 3.10\ G_1^{-0.25}$
Region 4	$0.82\ G_1^{0.25}\ Re_b$	$3.10\ G_1^{-0.25} \leq Re_b$

Table 3: Variation of total slug length with slug velocity for water-carbon dioxide slugs

Slug velocity, m/s	Total slug length, m			
	Minimum	Maximum	Mean	St. Deviation
1.27	.44	1.04	.70	.19
1.37	.34	.82	.61	.15
1.63	.51	1.03	.68	.16
1.73	.35	.81	.54	.14
1.83	.43	.85	.61	.14
2.7	.73	1.45	1.0	.23
3.2	1.01	1.70	1.31	.22

Table 4: Variation of total slug length with slug velocity for ARCOPAC90™-carbon dioxide slug systems

Slug velocity, m/s	Total slug length, m			
	Minimum	Maximum	Mean	St. Deviation
1.6	.40	.83	.61	.17
2.3	.46	.76	.67	.14
2.47	.50	1.07	.73	.17
3.3	.50	1.10	.78	.16
4	.60	1.60	.94	.25
5.88	.98	1.76	1.16	.34

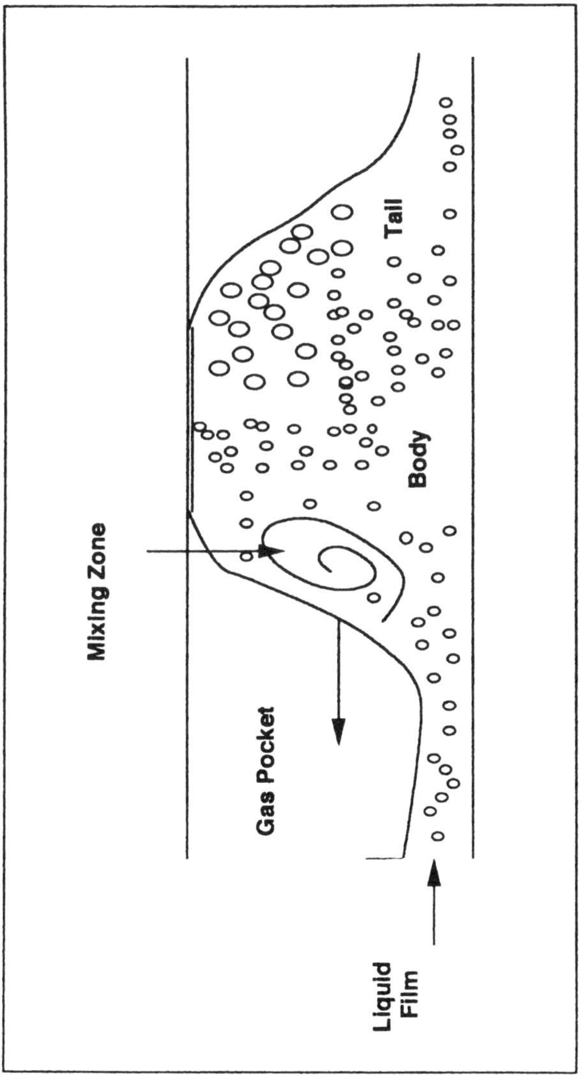

Figure 1: Profile of a slug unit

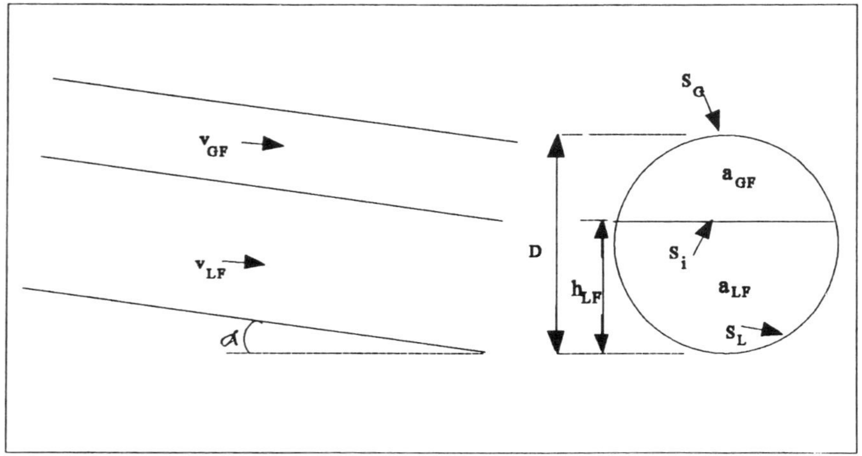

Figure 2: Physical model for stratified flow

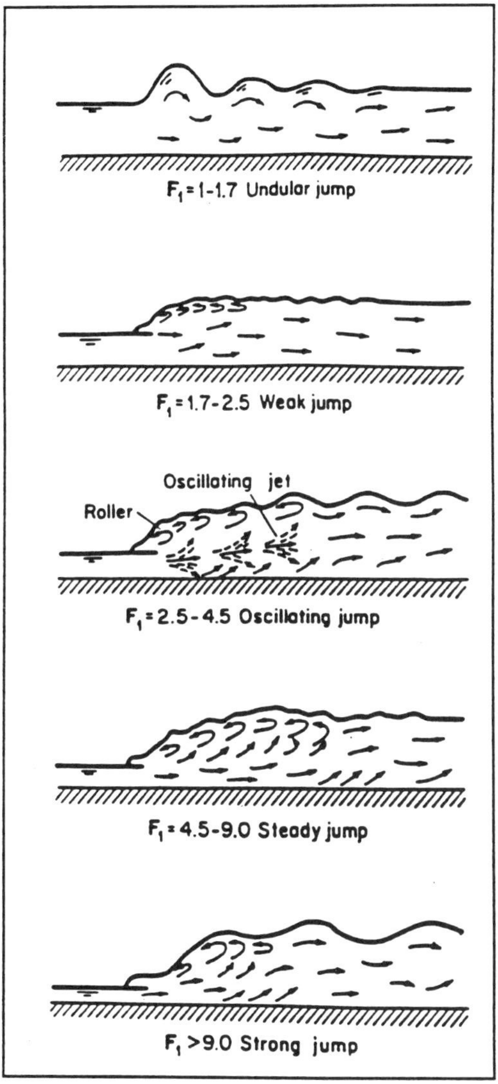

Figure 3: Various types of hydraulic jumps

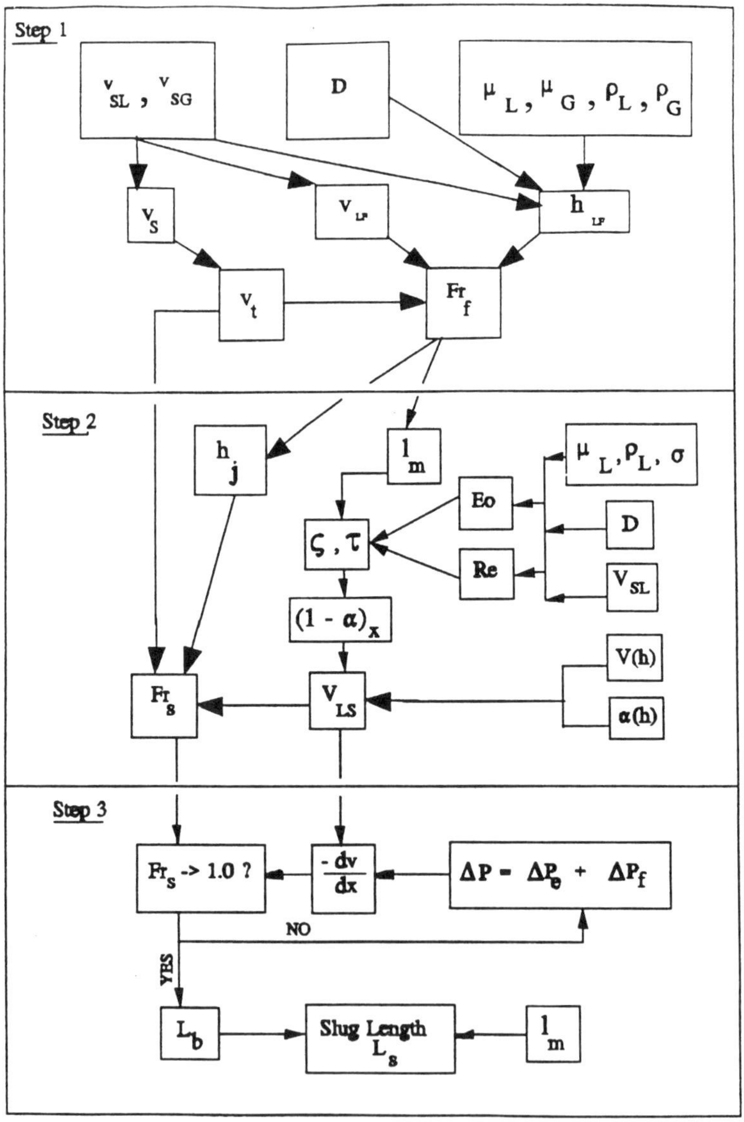

Figure 4: Schematic flow chart of slug length model

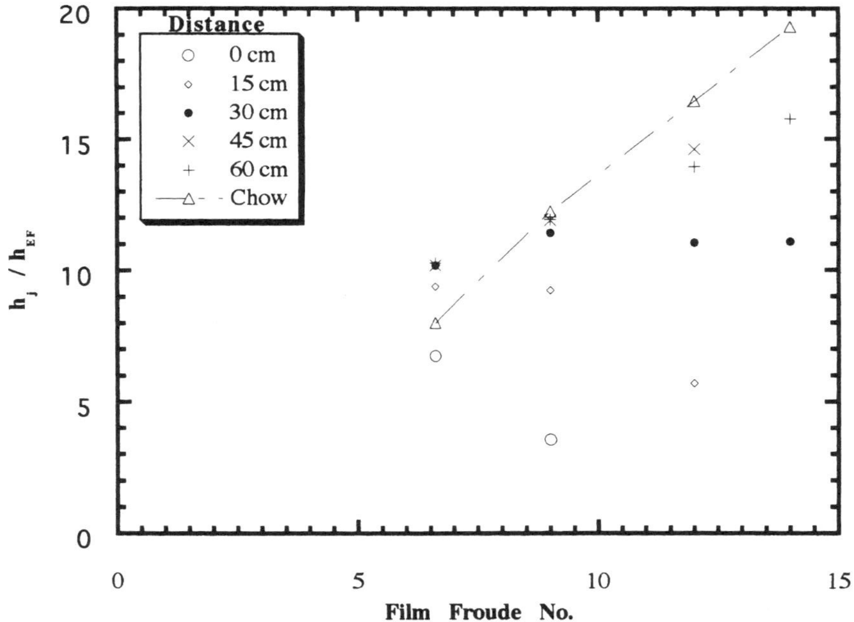

Figure 5: Variation of equivalent height ratio with distance into slug as a function of Film Froude No. for stationary slug

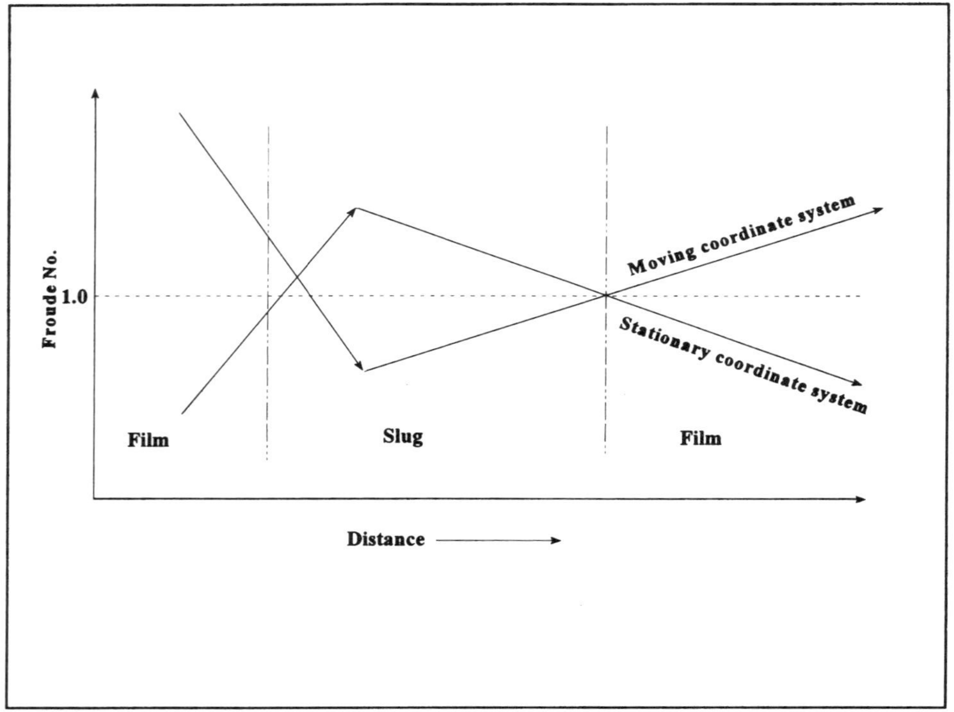

Figure 6: Schematic diagram of Froude number variation with distance into the slug

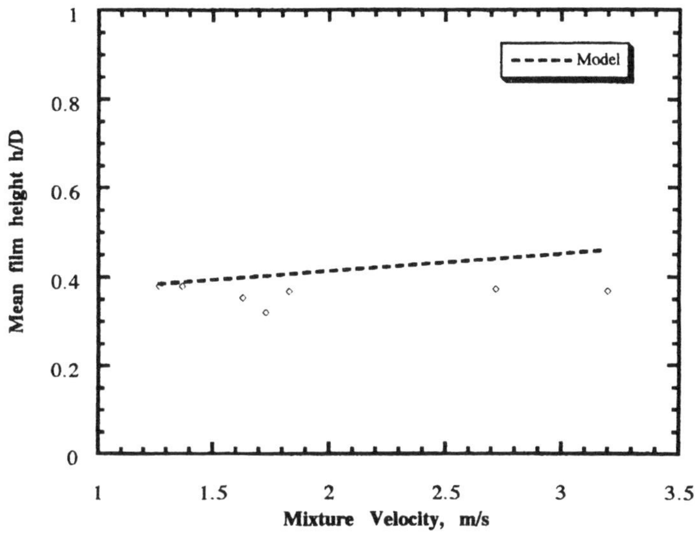

Figure 7a: Variation of mean liquid film thickness for water-carbon dioxide slug systems. Comparison with model.

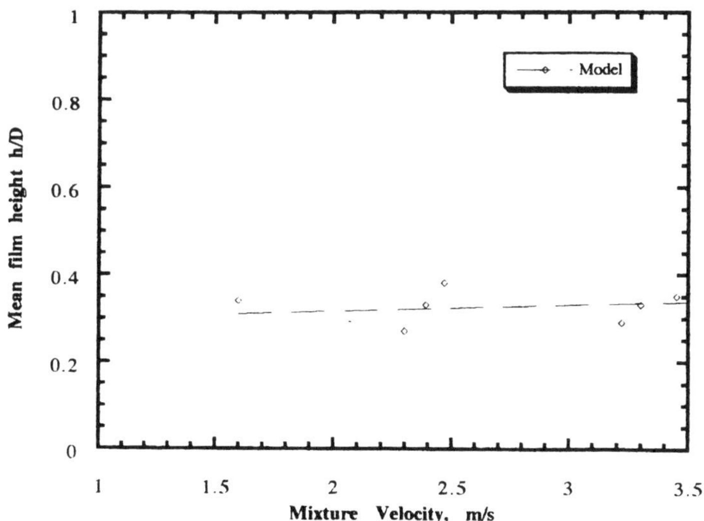

Figure 7b: Variation of mean liquid film thickness for ARCOPAK90™-carbon dioxide system. Comparison with model.

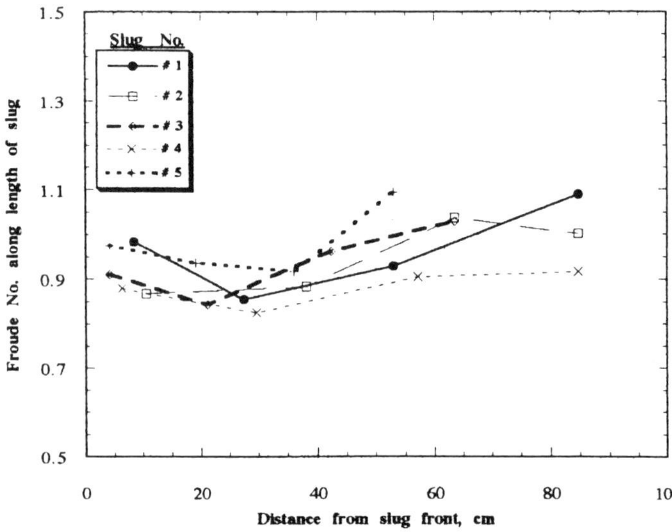

Figure 8a: Variation of Froude number with distance into the slug for water-carbon dioxide system. vsl = 0.2m/s, vsg = 1.07m/s.

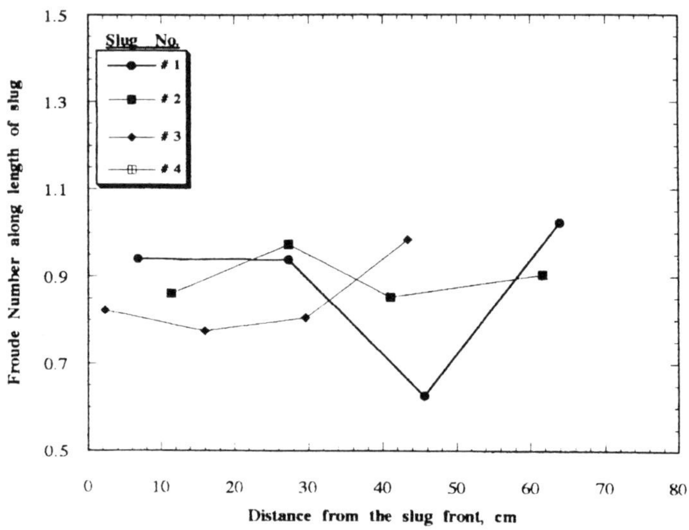

Figure 8b: Variation of Froude number with distance into the slug for water-carbon dioxide system. vsl = 0.3m/s, vsg = 1.07m/s.

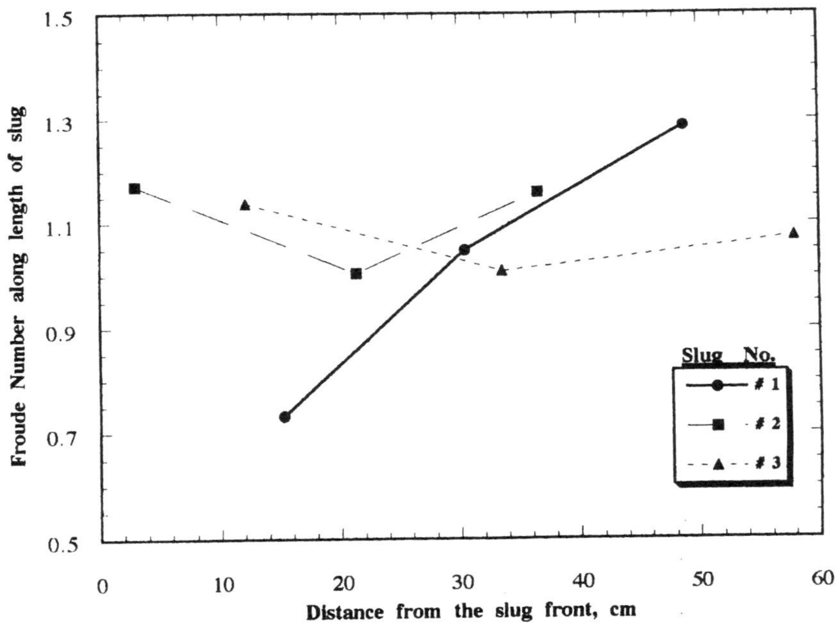

Figure 8c: Variation of Froude number with distance into the slug for water-carbon dioxide system. vsl = 0.4m/s, vsg = 1.43m/s.

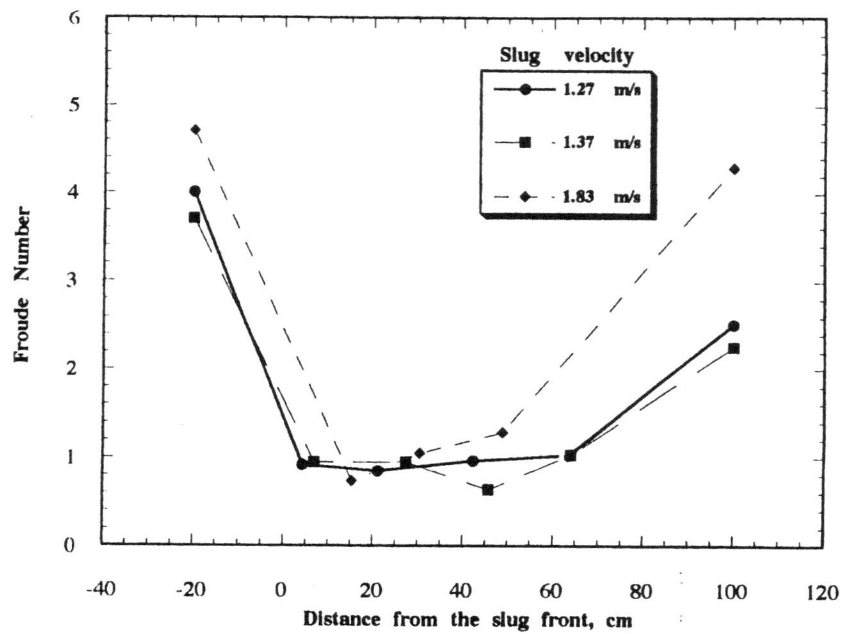

Figure 9: Variation of Froude number across a slug for different slug velocities.

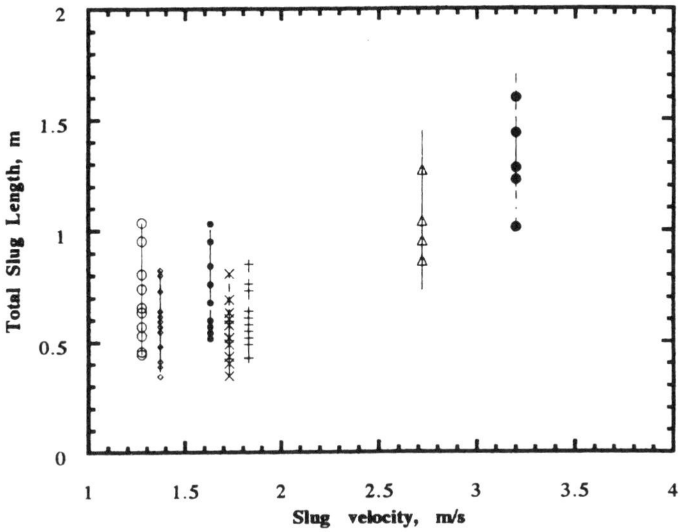

Figure 10a: Variation of total slug length with slug velocity for water-carbon dioxide system.

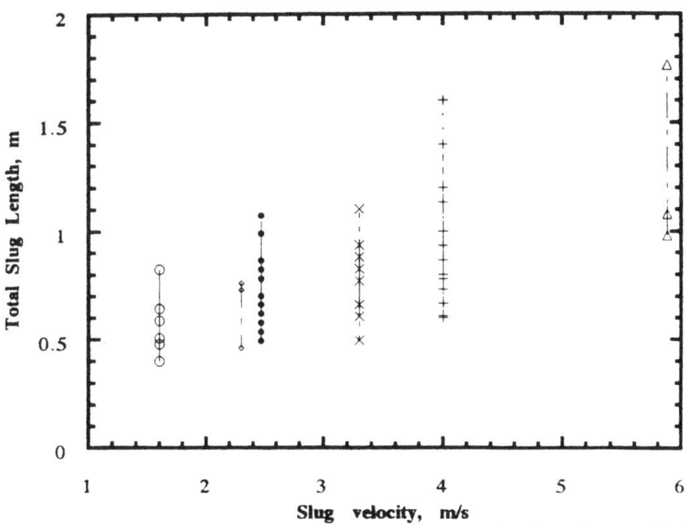

Figure 10b: Variation of total slug length with slug velocity for ARCOPAK90™-carbon dioxide system.

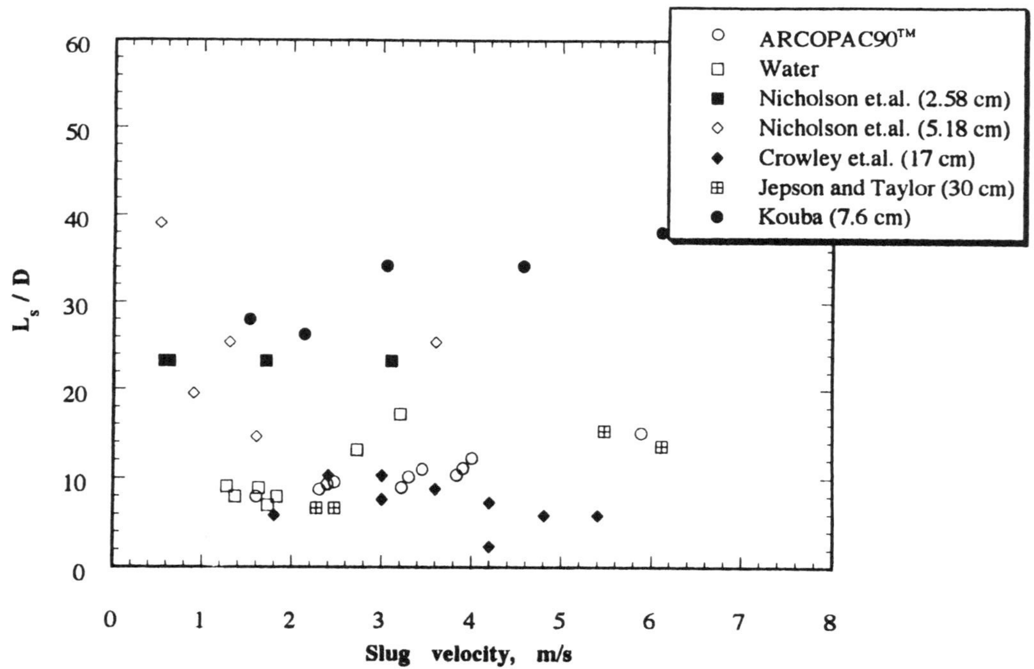

Figure 11: Comparison of total slug length from different systems.

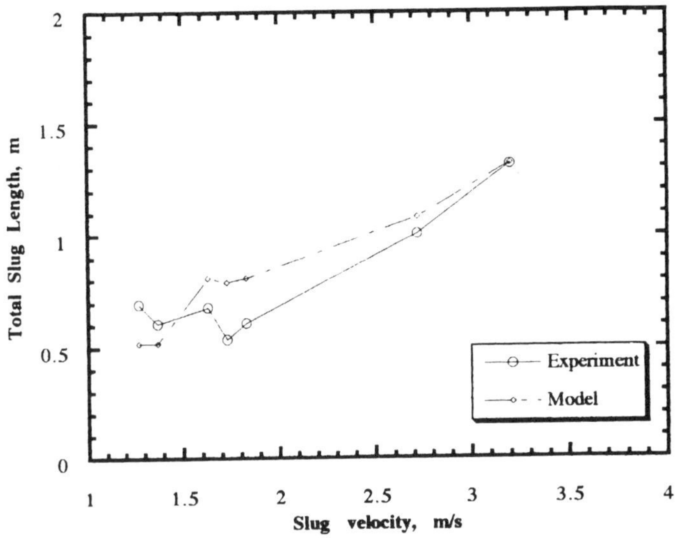

Figure 12a: Comparison of slug length model with average slug length for water-carbon dioxide system.

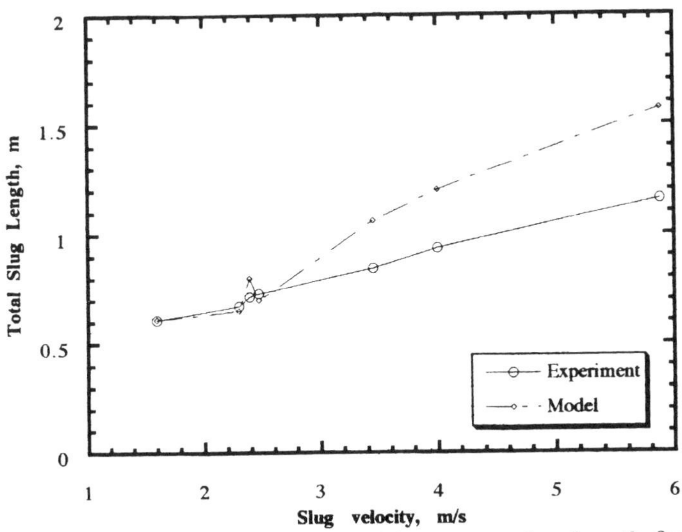

Figure 12b: Comparison of slug length model with average slug length for ARCOPAK90™ carbon dioxide system.

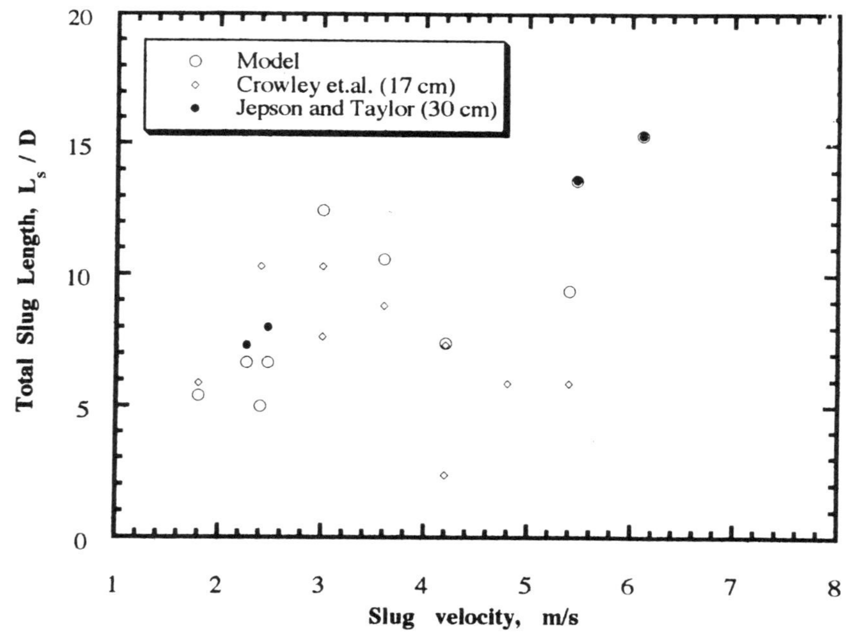

Figure 13: Comparison of model predictions with slug length from other systems.

References

Agarawal, S.S., G.A. Gregory, and G.W. Govier, " An Analysis of Horizontal Two-Phase Flow in Pipes", *Can. J. Chem. Eng.*, **51**, 280, (1973).

Andritsos, N., and Hanratty, T. J., "Influence of Interfacial Waves in Stratified Gas-Liquid Flows", AIChE Journal, **33**, 3, 444, (1987).

Barnea, D., and Taitel, Y., "A Model for Slug Length Distribution in Gas-Liquid Slug Flow", Int. J. Multiphase Flow, **19**, 5, pp. 829, (1993).

Benjamin, T. B.,"Gravity Currents and Related Phenomena", J. Fl. Mech., **31**, II, pp. 209, (1968).

Brill, J. P., Schmidt, Z., Coberly, W. A., Herring, J. D., and Moore, D. W., "Analysis of Two-Phase Tests in Large-Diameter Flow Lines in Prudhoe Bay Filed", SPE J., 363, (June 1981).

Chow, V. Te, "Open-Channel Hydraulics", McGraw-Hill, (1959).

Crowley, C. J., Sam, R. G., Wallis, G. B., and Mehta, D. C., Slug Flow in a Large Diameter Pipe: I. Effect of Fluid Properties", AIChE Annual Meeting, San Francisco, CA, (1984).

Dukler, A. E., and Hubbard, M. G., " A Model for gas-Liquid Slug Flow in Horizontal and Near Horizontal Tubes", Ind. Eng. Chem. Fundam., **14**, 4, 337, (1975).

Dukler, A. E., Maron, D. M., and Brauner, N., "A Physical Model for Predicting the Minimum Stable Slug Length", Chem. Eng. Sci., **40**, 8, 1379, (1985).

Fairhurst, C. P., "Slug-Flow Behaviour Clarified in Large Diameter Pipeline Study", O & G J., 49, (Oct. 3, 1988).

Gopal, M., and Jepson, W. P., "The Study of Dynamic Slug Flow Characteristics using Digital Image Analysis", Part I & II, JERT, Vol. 120, June 1998 (a, b, to be published).

Gopal, M., and Jepson, W. P., "Development of Digital Image Analysis Techniques for the Study of Velocity and Void Profiles in Slug Flow", Int. J. Multiphase Flow, Vol. 23, No.5, 945-965, 1997.

Green, A. S., Johnson, B. V., and Choi, H. J., "Flow Related Corrosion in Large Diameter Multiphase Flowlines", SPE paper 20685, (Houston, TX: SPE 1989).

Jepson, W. P., " The Flow Characteristics in Horizontal Slug Flow", 3rd Int. Conf. Multiphase Flow, Paper F2, 187, (1987).

Jepson, W. P., "Modelling the Transition to Slug Flow in Horizontal Conduit", Canad. J. Chem. Engg., **67**, 731, (1989).

Jepson, W.P. and Taylor. R.E., "Slug Flow and its Transitions in Large Diameter, Horizontal Pipes," Int. J. Multiphase Flow, **19**, 411, (1993).

Kordyban, E. S., and Ranov, R. R., "Mechanism of Slug Formation in Horizontal Two-Phase Flow", J. Basic Engg., **92**, Series D, 857, (1970).

Kouba, G. E., "Horizontal Slug Flow Modelling and Metering", Ph.D. Thesis, University of Tulsa, 1986.

Kouba, G. E., and Jepson, W. P., "The Flow of Slugs in Horizontal, Two Phase Pipelines", Pipeline Eng. Symp., 37, (1989).

Lin, P. Y., and Hanratty, T. J., "Prediction of the Initiation of Slugs with Linear Stability Theory", Int. J. Multiphase Flow, **12**, 79, (1986).

Maron, D. M., Yacoub, N., and Brauner, N., "New Thoughts on the Mechanisms of Gas-Liquid Slug Flow", Lett. Ht. Mass Transfer, **9**, 333, (1982).

Nicholson, M. K., Aziz, K., and Gregory, G. A., "Intermittent Two Phase Flow in Horizontal Pipes: Predictive Models", Canad. J. Chem. Eng., **56**, 653, (1978).

Nydal, O. J., Pintus, S., and Andreussi, P., "Statistical Characterization of Slug Flow in Horizontal Pipes", Int. J. Multiphase Flow, **18**, 3, pp. 439, (1992).

Scott, S. L., Shoham, O., and Brill, J. P., "Prediction of Slug Length in Horizontal Large-diameter Pipes", SPE Paper 15103, (1986).

Taitel, Y., and Dukler, A. E., "A Model for Predicting Flow Regime Transitions in Horizontal and Near Horizontal Gas-Liquid Flow", AIChE J., **22**, 1, 47, (1976).

Zhou, X., and Jepson, W. P., "Corrosion in Three-Phase Oil/Water/Gas Slug Flow in Horizontal Pipes", Corrosion/94, Paper no. 94026, (Houston, TX: NACE International, 1994).

Modelling

The modelling of transient three-phase flows

M WATSON
Consultant, Frimley Green, UK

A mathematical model is presented for the simulation of transient three phase flows. The governing equations for the conservation of mass and momentum for each phase are solved without making the approximations of homogeneous mixtures. Sophisticated numerical techniques are used to solve the governing equations enabling sharp interfaces such as the fronts of slugs to be tracked along the pipeline. Results from the model are compared with experimental data from the BP Exploration three phase rig. The example considered is that of air flowing over a pool of water and oil in a dip in a pipeline. The model successfully simulates the separation of the water and the oil; the position of the water/oil interface; the maximum excursion of the oil in the upward sloping pipe and the frequency of waves/slugs on the water and oil surfaces.

1 NOTATION

d pipe diameter
g acceleration due to gravity
p pressure
t time
x distance along pipe
A cross-section occupied by fluid
U fluid velocity
θ pipe inclination
ρ fluid density

Subscripts
G gas
L oil
W water

2 INTRODUCTION

The issues relating to three phase flows are of increasing concern to the oil and gas industry. The presence of water in a pipeline can significantly affect the complexity of flow regimes (1,2) and hence, in particular, the properties of slug flow. In addition, water may reside at dips in the pipeline which has important implications for corrosion which costs the industry vast amounts of money every year.

There is thus a need to develop models and tools for the simulation of transient three phase flows. Correlations obtained from experimental or field data are often limited in their usefulness to the range of data from which they have been obtained and may be unreliable when used outside this range. The development of mathematical models and computer programs for the simulation of transient three phase flows would therefore be welcomed by engineers in the industry.

The full governing equations describing transient multi-phase flows in a pipe are complicated and, with present knowledge, very difficult to solve. It is therefore necessary to make some pragmatic approximations in order to make the equations tractable to solve whilst maintaining the key physical mechanisms. This is the approach we take in this paper. Like many other models to simulate transient multi-phase flows the flow properties in the pipe are averaged over the cross-section reducing the number of space dimensions to one - that along the axis of the pipe. However, we do not make assumptions about treating the water phase as a homogeneous mixture in the oil phase, thus approximating the flow as quasi two phase. The water, oil and gas phases are kept distinct and the mass and momentum equations are solved separately for each phase.

Although the restriction to one space dimension may appear drastic the governing equations are still complicated and difficult to solve. Indeed, the mathematical structure of these one dimensional equations is such that sophisticated numerical methods are needed to solve these equations successfully (3). The front of a slug may be treated as a moving discontinuity and appropriate techniques need to be applied in order to capture these discontinuities without smearing or unphysical oscillations. Current well known commercial codes do not use adequate numerical methods to capture discontinuities well.

The model described in this paper solves the governing equations for the conservation of mass and momentum for each of the three phases. However, the mathematical structure of the resulting system of equations is taken account of by the application of sophisticated numerical methods which enable discontinuities at the front of slugs to be captured and tracked without smearing and unphysical oscillations.

3 GOVERNING EQUATIONS

We consider three phase flow in a pipe. The fluids are assumed to be gravitationally separated and flow in a circular pipe of diameter d inclined at an angle θ to the horizontal. If the flow properties are averaged over the cross-section then the mass and momentum conservation equations for the water phase are

$$\frac{\partial}{\partial t}\left(\rho_W A_W \right) + \frac{\partial}{\partial x}\left(\rho_W A_W U_W \right) = 0 \qquad (3.1)$$

$$\frac{\partial}{\partial t}\left(\rho_W A_W U_W\right)+\frac{\partial}{\partial x}\left(\rho_W A_W U_W{}^2\right)=-A_W\frac{\partial p_{WL}}{\partial x}-\rho_W g\cos\theta A_W\frac{\partial h_{WL}}{\partial x}-\rho_W A_W g\sin\theta+F_{WL}-F_W \quad ...(3.2)$$

with similar equations for the oil and gas components (subscripts L and G respectively). The water-oil interface pressure is p_{WL}; the height of the interface is h_{WL} and F_{WL} and F_W are the usual simple algebraic representations of the interfacial and wall frictions respectively (4). The water density is ρ_W; the cross-section area occupied by the water is A_W and the water velocity is U_W.

With the corresponding equations for the oil and gas phases there are six conservation equations. For the purposes of this paper we do not consider the energy equations for the fluids. The emphasis in this work is to present a model and method of solution for the simulation of transient three phase flows. The primary objective is to show how the model and numerical method tackles wave formation, growth and decay whilst maintaining sharp interfaces at the front of waves and slugs. The inclusion of the energy equations would not affect the essence of our objective: it would only have a small effect on the details.

Equations (3.1) and (3.2) above are appropriate for compressible fluids. However, for the purposes of this paper we restrict attention to incompressible liquids (water and oil) and only allow for compressibility in the gas

For the model that we are considering we therefore have six evolution equations. It is possible to reduce this system to a set of four evolution equations along with two algebraic constraints (4). The details will not be repeated here.

4 SOLUTION TECHNIQUE

The governing set of evolution equations can be written in the vector form

$$\frac{\partial \underline{U}}{\partial t}+\frac{\partial \underline{F}}{\partial x}=\underline{G} \qquad (4.1)$$

Such a system is in divergence form and the solutions may exhibit shocks which need to be treated appropriately. Watson (3) has shown how solutions of such a system for two phase flows may grow from infinitesimal sinusoidal disturbances into finite amplitude travelling waves with the fronts as moving shocks. Such behaviour can only be modelled with the appropriate application of a suitable numerical method. The Lax-Wendroff method with Boris & Book anti-diffusion (5) is a suitable choice. The Lax-Wendroff method is a second order accurate explicit finite difference method which is easy to apply to hyperbolic systems such as (4.1) However, this method on its own generates spurious non-physical oscillations in the neighbourhood of shocks *eg* the front of slugs. The anti-diffusion extension of Boris & Book removes these non-physical oscillations so we can be sure that any holdup oscillations that occur during the simulation are due to the physics of the problem and not a feature of the numerical method.

It is all too easy to apply inappropriate numerical methods to a system such as (4.1). Watson (3) has shown that wave and slug formation, growth and decay can indeed be simulated from such a simple looking system as (4.1) provided an appropriate numerical method is applied.

5 EXPERIMENTAL RESULTS

The experimental results are from the BP three phase rig at Sunbury-on-Thames (6). The pipe consists of a 0.1514m (6in) diameter pipe comprising a downward section inclined at 0.983° and an upward section at 1.283°. The pipe is approximately 62m long with the dip at 21m from the inlet.. The fluids have been taken to be water, oil and air with atmospheric pressure and a temperature of 15C. The fluid physical properties are:

	Density (kg/m^3)	Viscosity (Pa s)
Water	1000	0.001
Oil	810.4	0.00237
Air	1.225	1.7E-05

The set of data considered for the comparison between the experiments and the simulation consists of a case with 0.25 m^3 (250 litres) of liquids with 50% water cut settled in the dip in the pipeline. The air superficial velocity was 0.95 m/s.

The experimental results showed that the oil and water separated with the water essentially staying in the dip and the oil moving downstream of the dip into the upward sloping pipe. Slugs were generated in the water phase and passed into the oil phase. The maximum extent oil excursion was found to be about 27.5m from the dip. The frequency of the slugs was about one every 8s.

6 SIMULATION RESULTS

The mathematical model and numerical method described above has been implemented into a code for the simulation of pipeline transient three-phase flows. The program was set up with the initial conditions described in Section 5 above.

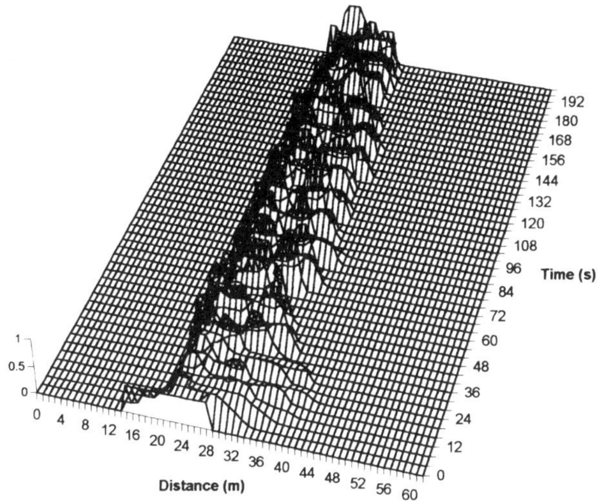

Figure 1 Water holdups

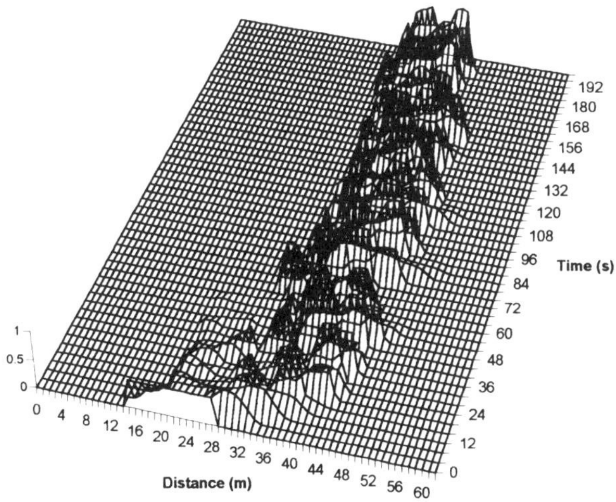

Figure 2 Oil holdups

Figures 1 and 2 give a summary of the whole simulation. The flow is from left to right and the dip in the pipe is at about 21m from the inlet. At time 0s the initial conditions consist of water and oil at rest in the dip. After this time the air flows and the liquids move. It can be seen from these figures that the liquids move downstream of the dip into the upward sloping section of pipe. It can be seen that the layers of water and oil separate with the oil moving further downstream and further into the upward sloping pipe. Large amplitude waves or slugs form in the water layer and progress into the oil layer. The maximum excursion of the oil is about 45 to 50m from the inlet. The experimental data (6) suggest a survival distance of about 48.5m from the inlet. The simulation results are in very good agreement with the experiments. The simulation results also suggest that the frequency of the waves is about on every 10s in reasonable agreement with the experimental results of one every 8s.

More detail can be seen in the following figures which show a sequence of snapshots of the total (water+oil) holdups for simulation times between 142s and 150s.

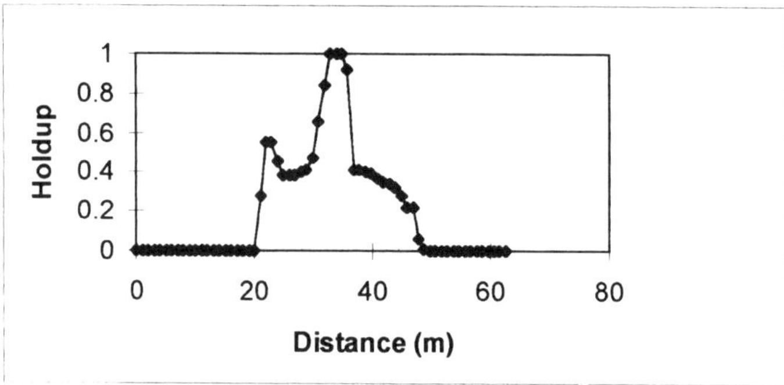

Figure 3 Total holdup at 142s

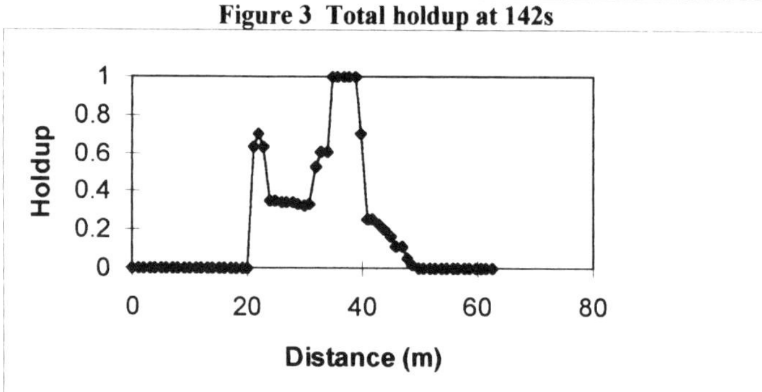

Figure 4 Total holdup at 144s

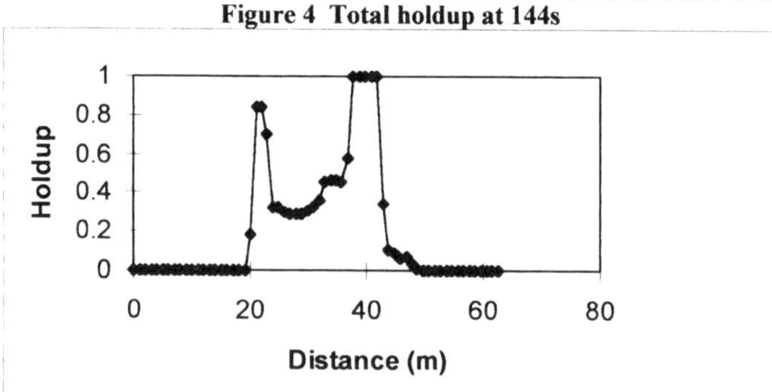

Figure 5 Total holdup at 146s

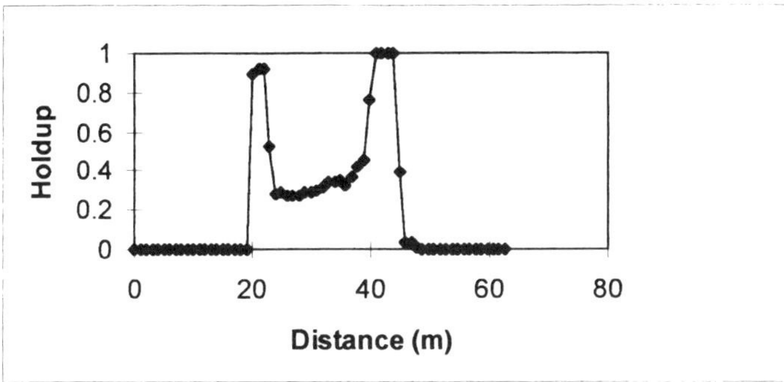

Figure 6 Total holdup at 148s

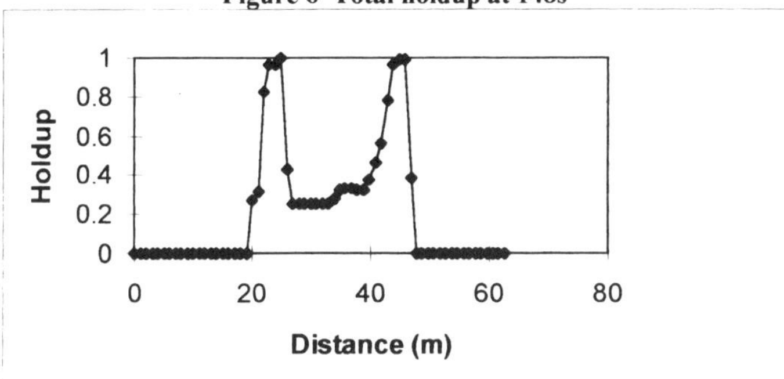

Figure 7 Total holdup at 150s

Figures 3 to 7 above show a sequence of snapshots of the total liquids holdup at 2s intervals between 142 and 150s. The dip in the pipe is at about 21m from the inlet and these figures show that the liquids have been pushed into the upward sloping part of the pie downstream of the dip. Slugs are formed in the liquid nearest the dip (water) and propagate downstream into the oil. The simulations show that the slug holdups are almost one. These figures also show the efficiency of the numerical method in maintaining sharp interfaces. This is vital if slugs are to be modelled properly.

Figures 8 and 9 below show the water and oil holdups at 150s.

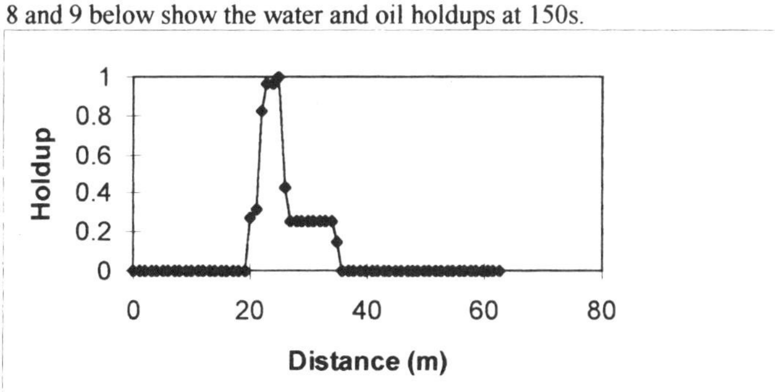

Figure 8 Water holdup at 150s

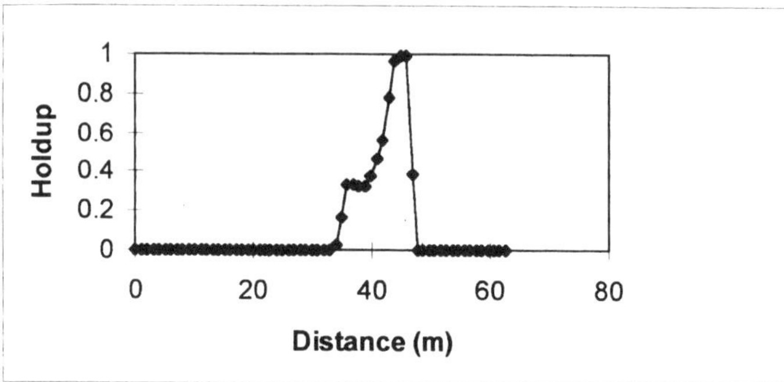

Figure 9 Oil holdup at 150s

These two figures clearly show that the water and oil have separated with the oil downstream of the water.

The simulations were done on a PC with a 486 33 MHz processor with 8MB of memory. They took just a few minutes to set up and run.

Some further simulations have been done at BP on higher gas flowrates. It has been found that although the code works quite well for the case discussed in this paper for the higher flowrates the simulations predict too much liquid pushed out of the pipe (6). For the higher gas flowrates there is more mixing at the water/oil interface and liquid droplets will be entrained in the faster moving gas flow. Neither of these effects are currently modelled in the code.

7 DISCUSSION

A mathematical model and numerical method has been described for the simulation of transient three-phase flows. This has been implemented into a code that runs in just a few minutes on a PC. For systems of equations such as for multi-phase flows it is very important to be able to track the formation of waves and slugs and follow their progress along the pipe. The numerical method described in this paper is able to do this.

The mathematical model described in this paper is based on topologically stratified flows, *ie* gravitationally separated layers of water. oil and gas. Separate conservation eqations for each of the phases are considered without making assumptions about , for example, a quasi two-phase model with a water/oil mixed layer. This model enables interfacial waves to form and grow and the phases to separate in the pipe. As can be seen from the results above this enables us to obtain good qualitative and quantitative agreement with the BP experimental results. A model with a mixed water/oil layer would not be able to reproduce such results.

It has been mentioned above that further simulations have been done at BP with the code for higher gas flowrates (6). It has been found that the code can predict too much liquid to be swept out of the pipe. For the higher flowrates there will be more mixing of the phases. For example at the water/ oil interface the liquids will mix forming an emulsion. The higher gas flowrate will entrain droplets from the liquid layers. These effects are not currently modelled in the code. Indeed, the flow regimes for three-phase flows are much more complicated than for two-phases (1,2). Although, many aspects of three-phase flows have not been considered in this paper, the simulation results described above have shown what can be achieved using a simple mathematical model focussing on the key physics and using an appropriate numerical method. It could be expected that the inclusion of some simple models for mixing and entrainment would increase the range of applicability of the current model.

8 ACKNOWLEDGEMENT

The author wishes to thank BP for permission to use their experimental data and especially Paul Fairhurst.

9 REFERENCES

1 Acikgoz, M., Franca, F & Lahey, RT, 1992, An experimental study of three-phase flow regimes, Int J Multiphase Flow, **18**, 327.
2 Lahey, RT, Acikgoz, M & Franca, F, 1992, Global volumetric phase fractions in horizontal three-phase flows, AIChE Journal, **38**, 1049.
3 Watson, M, 1991, Non-linear waves in pipeline two-phase flows, Proc 3rd Int Conf on Hyperbolic Problems (Uppsala, Sweden, 11-15 June, 1990), (Engquist, B & Gustafsson,B (eds)), Studentlitteratur, Lund, 938.
4 Watson, M, 1993, A model for pipeline three-phase flows, Proc 4th Int Conf on Hyperbolic Problems (Taormina, Italy, 3-8 April, 1992), Notes on Numerical Fluid Mechanics, **43**, (Donato, A & Oliveri, F (eds)), Vieweg, 578.
5 Sod, GA, 1985, Numerical methods in fluid dynamics, Cambridge University Press, Cambridge.
6 Fairhurst, CP & Barrett, N, 1997, Oil/water/gas transport in undulating pipelines - field observations, experimental data and hydraulic model comparisons, SPE Paper 38811.

Friction factor modelling in laminar liquid film in stratified two-phase flow in pipes

F MEKNASSI and **R BENKIRANE**
Faculté des Sciences de Fès, Maroc
A LINÉ
INSAT Complexe Scientifique de Rangueil, Toulouse, France
D LOPEZ
Elf Production, France

Synopsis

Now a days in the oil industry, the production and transportation of oil and gas are mainly operated in two-phase flow. In order to calculate liquid retention as well as pressure drop in pipelines, different models have been proposed and different codes are available. However, new production constraints induce new modelling problems. In particular, the objective of this paper is to improve the modelling of stratified flow when the flow pattern in the liquid film is laminar. Two-phase stratified laminar flow in a pipe has been formally addressed many years ago by Semenov and Tochigin (1), Bentwich (2) who derived analytical solution in laminar-laminar flow. From their derivation Ranger and Davis (3) derived the volumetric flux in each phase. Indeed it is possible to deduce more instructive informations from the analytical expression of the velocity field in the liquid film : as will be shown in the present paper, it is possible to calculate the wall shear stress at the wall in order to deduce the friction factor in the liquid film at the wall. In addition, it is possible to solve the problem in the laminar liquid film with different boundary conditions on the interfacial shear stress ; in this case, the new solution of the velocity field in the laminar liquid film that is proposed in this paper is valid both for laminar or turbulent gas flow. The resulting expression of the friction factor in the liquid film in laminar flow will be compared to experiments and to numerical simulations.

INTRODUCTION

In the transportation of oil and gas in stratified flow, the two-phase flow modelling is based on different assumptions :

- the shape of the gas-liquid interface is imposed ; in this paper, it is assumed to be flat.

- the different friction factors at the wall and at the interface have to be modelled . In the gas phase, the friction factor at the wall is generally expressed by a simple extrapolation of empirical relation developed in single phase flow . Considering the large cross-section of the gas phase, this assumption is shown to be acceptable. Concerning the interfacial friction factor, the number of references is large (see the reviews of Hanratty and McCready (4) and Liné and Fabre (5)). In the liquid phase, there are very few works related to the modelling of the friction factor at the wall (Andritsos (6)) ; however, Lopez (7) has shown in his PhD thesis that the existing modelling of this friction factor lead to inaccurate predictions in the case of thin and/or laminar liquid films.

The objective of the present study is to improve the modelling of the friction factor in laminar liquid film. Instead of developing any new experimental facility ,we first attempted to develop a theoretical approach of the problem. The advantage of any analytical solution lies in the fact that it gives an explicit solution of the problem and consequently an explicit formulation of the friction factor in the liquid film at the wall. This theoretical approach is then validated by comparison to experimental data.

As previously mentioned, the present work is based on the analytical solution of Semenov and Tochigin (1). However, their solution was limited to laminar-laminar two-phase stratified flow with a flat interface. In most of the industrial application, the gas flow is turbulent. Consequently, it has been necessary to modify the previous derivation. In the present study, the analysis has been focused in the laminar film with given boundary conditions at the gas-liquid interface in terms of interfacial shear stress. This boundary condition can account for both laminar and turbulent gas flow. In the case of laminar flow in the gas, the exact analytical solution of the velocity field enabled Semenov and Tochigin (1) to derive explicitly the boundary condition at the gas-liquid interface. Unfortunately, it is not possible to derive an analytical solution in the gas in turbulent flow regime. As long as this local solution of the turbulent gas flow is unknown, a simplified boundary condition at the gas-liquid interface is substituted to the exact interfacial shear stress profile, a constant value is imposed. Consequently, the expression of the friction factor in the liquid film that is proposed in this paper includes an approximation concerning the interfacial shear stress, but it is more realistic that the assumption of laminar flow in the gas. In order to account more precisely for turbulent flow in the gas as well as curved gas- liquid interface in future works, a numerical code is being developed in the PhD of Meknassi (8). The first numerical results are presented here and limited to the comparison and mutual validation of the analytical approach and the numerical one.

Pioneering works of Semenov and Tochigin (1) and Bentwich (2) showed that in two-phase stratified laminar-laminar flow in a pipe, the velocity field in each phase has an exact solution expressed by Fourier integral representation in terms of bipolar coordinates. The choice of bipolar coordinates (figure 1) is the key point of the derivation. Indeed, such bipolar coordinates have been used in different works on stratified flow in circular pipes (Shoham and Taitel, (9), Issa, (10)), the two last references being related to numerical simulations. For sake of simplicity, the details of the theoretical work will not be included in this paper. The resulting expression of the friction factor in the liquid film will be given and discussed.. The basic point lies in the exact solution for the velocity field in each phase proposed by Semenov and Tochigin (1). However, we did not follow Semenov and Tochigin (1) to derive laminar flow in each phase : the derivation is limited to the liquid film with imposed shear stress profile at the interface.

FORMULATION OF THE PROBLEM

The local momentum equation in steady-state fully developed laminar flow writes :

$$-\left(\frac{dp}{dx}\right)_k + \mu_k \Delta U_k + \rho_k g \sin\alpha = 0$$

The first term is the pressure gradient ; it is constant in each phase. The second term is the molecular diffusion of momentum ; μ is the viscosity and U is the local longitudinal velocity. The third term accounts for gravity effects in the longitudinal momentum equation. We consider one directional flow in the x direction. The plane (y,z) corresponds to the cross-section of the flow.

Then, the longitudinal velocity in the plane (y,z) is solution of the resulting equation

$$\Delta U_k = -\frac{J_k}{\mu_k} \qquad \text{with} \qquad J_k = -\left(\frac{dp}{dx}\right)_k + \rho_k g \sin\alpha$$

From the local distribution of the velocity in each phase, it is possible to deduce the superficial velocity by spatial averaging :

$$U_{ks} = \varepsilon_k < U_k >$$

ε_k is the phase fraction.

Since the problem has to be solved in a circular cross section pipe, the use of bipolar co-ordinates is efficient ; they are defined by (ξ,η):

$$y = \frac{\sin\frac{\Omega}{2}\ \sinh\eta}{\cosh\eta - \cos\xi} \qquad\qquad z = \frac{\sin\frac{\Omega}{2}\ \sin\xi}{\cosh\eta - \cos\xi}$$

In the plane (ξ,η), the upper fluid 1 domain and the lower fluid 2 domains correspond to :

fluid 1: $-\infty \le \eta \le +\infty \qquad \frac{\Omega}{2} \le \xi \le \pi$ fluid 2 : $-\infty \le \eta \le +\infty \qquad \pi \le \xi \le \frac{\Omega}{2} + \pi$

The solution of the problem needs the expression of boundary conditions. In particular at the fluid-fluid interface, one can write

- velocity continuity $\qquad U_1\ (\xi,\eta) = U_2\ (\xi,\eta)$

- shear stress continuity : $\qquad \mu_1 \left(\frac{\partial U_1}{\partial \xi}\right)_{\xi=\pi} = \mu_2 \left(\frac{\partial U_2}{\partial \xi}\right)_{\xi=\pi}$

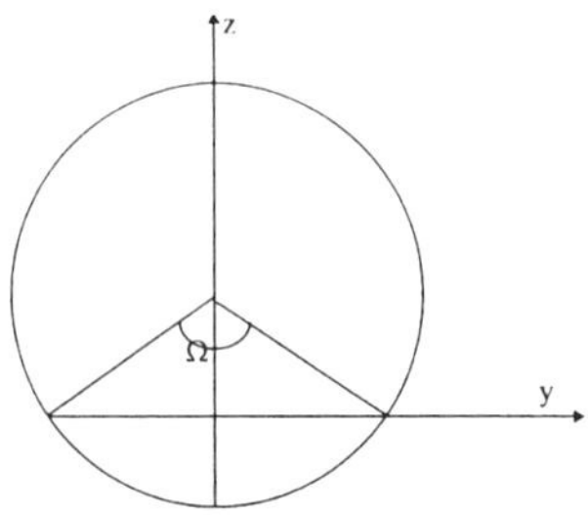

Figure 1 . Cross-section of stratified flow in a circular pipe

FRICTION FACTOR IN LAMINAR LIQUID FILM

Given the analytical profile for the liquid velocity, one can estimate the friction factor at the wall which is classically defined as follows :

$$f^{w}_{L\,(anal)} = \frac{2 < \tau^{w}_{L} >_{(anal)}}{\rho_L \; < U_L >^{2}_{(anal)}}$$

the numerator expresses the mean value of the shear stress at the wall wetted by the liquid, averaged along the wetted perimeter and the denominator expresses the square of the mean velocity averaged over the liquid cross section.
After calculations , one obtains the following expression for the friction factor at the wall :

$$f^{w}_{L} = \frac{16}{\mathrm{Re}_L} f_{corr}$$

with the Reynolds number expressed in the liquid phase : $\mathrm{Re}_L = \dfrac{\rho_L \; < U_L > D_L}{\mu_L}$

where D_L is the hydraulic diameter in the liquid.
One can easily see that the friction factor in the liquid at the wall depart from the single phase flow relation by a corrective term expressed as follows (8) :

$$f_{corr}(\Omega,\rho) = \frac{1}{1-\rho\dfrac{\sin(\Omega/2)}{(\Omega/2)}} \; \frac{\left(1-\dfrac{\sin(\Omega)}{\Omega}\right)^{3}}{\left(1-\dfrac{4}{3\Omega}\sin\Omega+\dfrac{1}{6\Omega}\sin 2\Omega\right)\left(1+R_2(\Omega,\rho)\right)}$$

where: $R_2(\Omega,\rho) = -1+\alpha(\rho)\dfrac{\Omega}{2\pi}$ with $\alpha(\rho) = 2+(2\rho)^2$

The shear stress ratio (shear stress at interface divided by shear stress at the wall) denoted ρ is defined as : $\rho = \dfrac{< \tau^{i}_{L} >}{< \tau^{w}_{L} >}$

then the corrective function depends on 2 parameters, the angle subtended by interface at the centre of the pipe Ω and the shear stress ratio ρ :

$$\boxed{f_{corr} = f_{corr}(\Omega, \rho)}$$

VALIDATION OF THE FRICTION FACTOR LAW IN LAMINAR LIQUID FILM BY EXPERIMENTS

Given the experimental results of Andritsos (6), it is possible to estimate the friction factor at the wall from his experiments in stratified flow in a horizontal pipe of internal diameter 9.5 cm. The gas is air and the liquid is tap water with glycerine concentration up to 80 %. Given the liquid retention, the pressure drop and assuming that the friction factor at the wall in the gas follows a single phase flow trend, it is possible to estimate the experimental friction factor in the liquid at the wall. The shape of the interface which is assumed flat. The resulting experimental values of the corrective function for the friction factor are compared to the prediction of our model and plotted on figure 2 versus the angle subtended by interface at the centre of the pipe Ω for different shear stress ratio ρ .

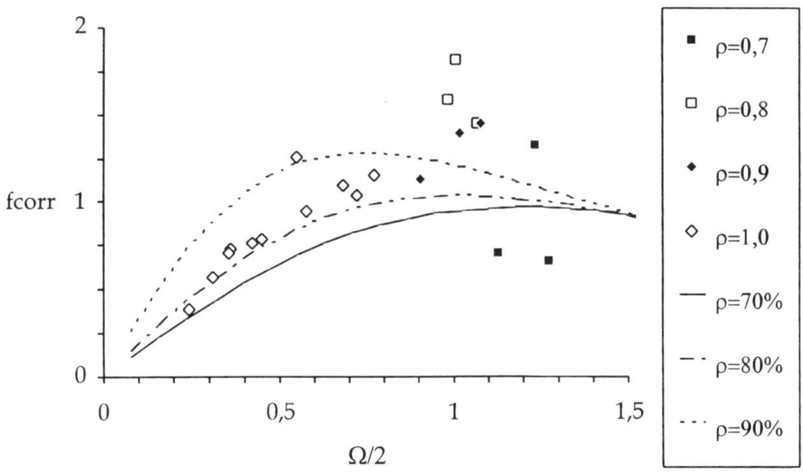

Figure 2. Corrective function of the friction factor in the laminar liquid film versus the angle subtended by interface at the centre of the pipe Ω for different shear stress ratio ρ (symbols after Andritsos data, curves after the present theoretical solution)

The trend predicted by the theoretical relation is shown to be correct. Of course, there is not a complete agreement between the predictions and the experiments. Different explanations can be proposed : in the theoretical approach, the interface is assumed to be flat and the interfacial shear stress is assumed to be uniform ; in the experimental estimation of the friction, the interface is also assumed to be flat leading to erroneous estimation of the interfacial perimeter and wall perimeter wetted by the wall if the interface is curved.

In order to improve the present theoretical approach, different ways can be investigated. It seems reasonable to try to improve the distribution of shear stress at the interface which is not uniform. This can be done via Computational Fluid Dynamics (CFD) by coupling the simulation of laminar or turbulent flow in the gas phase to analytical or numerical solution in the liquid phase. This can be done for any shape of the interface. However, a predictive model of the shape of the interface remains to be developed. Provisionally, it will be considered flat. The last part of this paper is devoted to CFD . The ultimate goal is to solve the stratified two-phase flow for any flow pattern in the gas and in the liquid. The first step presented here is limited to the computation of laminar-laminar stratified two-phase flow. It constitutes a mutual validation of theoretical work and CFD work.

VALIDATION OF THE FRICTION FACTOR LAW IN LAMINAR LIQUID FILM BY CFD

Given the set of equations and boundary conditions previously defined, it is possible to simulate the problem expressed in bi-polar co-ordinates (8). From the numerical solution of the problem, it is possible to derive the simulated friction factor , the wall shear stress being estimated from the normal profiles of longitudinal velocity as follows :

$$f_{L\,(sim)}^{w} = \frac{2 < \tau_L^w >_{(sim)}}{\rho_L \; < U_L >^2_{(sim)}} \qquad \text{and} \qquad < \tau_L^w >_{(sim)} = \frac{2\mu_L}{P_L^p} \int_0^\infty \left(\frac{\partial U_L}{\partial \xi} \right)_{\xi = \pi + \frac{\Omega}{2}} d\eta$$

The comparison between analytical and simulated values of the friction factor are given in table (1), for different runs corresponding to laminar flow in the film. The comparison between analytical approach and numerical approach is excellent.

	Liquid retention	Re_L	$\frac{< \tau_L^i >}{< \tau_L^p >}$	f_{corr}	$f_{L\,(anal)}^p$	$f_{L\,(sim)}^p$
Run (1)	0.059	77.5	0.836	0.791	0.164	0.164
Run(2)	0.072	73	0.817	0.792	0.175	0.175
Run(3)	0.080	53	0.804	0.793	0.240	0.240
Run(4)	0.102	51	0.776	0.796	0.251	0.251
Run(5)	0.227	133	0.647	0.818	0.0983	0.0987
Run(6)	0.420	197	0.499	0.853	0.0693	0.0697
Run(7)	0.521	241	0.428	0.874	0.0582	0.0585
Run(8)	0.810	246	0.211	0.965	0.0628	0.0630
Run(9)	0.943	185	0.0751	1.063	0.0920	0.0918
Run(10)	0.983	144	0.0485	1.120	0.124	0.122

Table 1. Laminar-laminar stratified flow -

CONCLUSION AND PERSPECTIVES

In the present study, a theoretical work has been investigated to express analytically the friction factor at the wall in a laminar liquid film. Starting from previous analytical solution, the present work had two objectives : derive the friction factor at the wall from the knowledge of the local velocity field and extend the previous analytical solution to the case of turbulent flow in the gas. The resulting expression shows that the friction factor value is inferior to the value extrapolated from single phase flow expressions.

This theoretical solution is compared to experiments acquired by Andritsos (1986). The trend is shown to be correct. The proposed relation of the friction factor in a laminar liquid film is valid both for laminar and turbulent gas flow. In order to improve the modelling of the friction factor, two simplified assumptions have to be improved : the interfacial shear stress is assumed uniform in the analytical derivation and the shape of the gas-liquid interface is assumed to be flat. The first assumption seems easier to improve than the second one.

CFD is developed to improve the knowledge of the distribution of interfacial shear stress. In this paper, the very first numerical results are presented. It constitutes a mutual validation of both analytical approach and numerical approach. The objective is to simulate turbulent flow in the gas in order to derive the distribution of interfacial shear stress in the liquid. This distribution will be taken into account to improve the expression of the friction factor in the liquid. The objective remains to propose an efficient analytical relation as simple as possible.

REFERENCES

(1) Semenov, N. I. and Tochigin, A. A., J. Eng. Phys. 4, 29 (1961)
(2) Bentwich, M., Chem. Eng. Sci. 31, 71 (1976)
(3) Ranger, K. B. and Davis, A. M. J. , Steady Pressure Driven Two-Phase Stratified Laminar Flow Through a Pipe, The Can. J. of Chem. Engng, 57, pp 688-691 (1979)
(4) Hanratty, T. J., and McCready, M. J., Workshop on Two-Phase Flow Fundamentals, Imperial College, London,,U.K., April (1992)
(5) Liné, A. and Fabre, J., International Encyclopedia on Heat and Mass Transfer, pp 1097-1101, (1997)
(6) Andritsos N., Ph. D. Thesis, Univ. of Illinois, Urbana, U. S. A., (1986)
(7) Lopez, D ., Thèse de Doctorat, I.N.P. Toulouse, France, (1994)
(8) Meknassi, F., Thèse d'Etat du Gouvernement Marocain, Maroc, (1998)
(9) Shoham, O. & Taitel ,Y. , AIChE J. (30), pp 377-385, (1984)
(10) Issa, R. I., Int. J. Multiphase Flow, (14), pp 141-154, (1988)

Prediction and measurement of wellhead temperatures

A AARVIK and **B STRAND**
Norsk Hydro ASA, Oslo, Norway

1 ABSTRACT

The flowing wellhead temperature of a production well is an important design parameter for an oil field installation. It has impact on the process and pipeline design as well as on the strategies to avoid operational problems such as hydrates, scale, wax and corrosion.

The change of fluid temperature from the reservoir to the wellhead is governed by changes in the thermodynamic state of the well fluid according to the 1st law of thermodynamics. These changes are caused by heat transfer between the well fluid and the well surroundings, including the effects of natural convection of the completion fluid in the well annuli, mechanical energy dissipation in the well fluid and pressure losses due to gravity and frictional losses.

A new model for the prediction of steady state wellhead temperature for offshore wells has been established, including the heat transfer from the well fluid to the sea. Production data from five North Sea wells with different well characteristics have been gathered and analyzed in order to establish reliable wellhead temperature data at various flow rates. Practical problems to be experienced when acquiring production data have been described.

The temperature calculations based on the new model compare favorably with field data, and the model, coupled to a pipeline simulation tool, is capable of predicting the flowing wellhead temperature with an accuracy sufficient for design purposes.

2 INTRODUCTION

A proper design of offshore receiving and processing facilities relies on adequate information on the properties, conditions and rates of the produced well fluids. The temperature of the arriving wellstream is important for the design of the separation and gas processing trains, and to determine strategies to avoid problems related to wax deposition, hydrate formation and pipe corrosion.

In this work, the temperature of the well fluid immediately upstream the topside production choke is referred to as the *wellhead temperature*. Large uncertainties in the predicted wellhead temperature can require an unnecessarily robust and expensive process design, in order to cater for operating conditions that may never occur. Likewise, predicted wellhead temperatures that fail to harmonize with the actual operating conditions could cause large disturbances in the planned production, and lead to significant losses in revenues.

To predict the wellhead temperature of a well with reasonable accuracy, detailed knowledge of production rates, well completion and reservoir conditions must be available. Further, the thermodynamic and transport properties of the well fluid must be known as a function of state (pressure and temperature). Finally, a model must be available for the prediction of the coupled pressure and temperature changes along the well from the reservoir to the wellhead. This model should include effects of frictional, accelerational and gravitational losses, viscous dissipation in the fluid and heat exchange with the surrounding

In the present work, focus is set on the modeling of steady state heat transfer between the well fluid and its surroundings, represented by the completion and the formation. A model for this heat transfer has been established, and is used in connection with Norsk Hydro's in-house process simulation tool, which uses OLGAS (1) for pressure drop prediction. Validation of the model has been done by comparing predicted wellhead temperatures to production data from five North Sea platform wells producing at various flow rates, and important issues regarding collection of reliable production data are covered. A good model not only allows the expected wellhead temperature to be predicted as a function of the well's completion and well fluid properties. Moreover, it can play an important role in well design, as effects of tailoring the well's completion to achieve a desired wellhead temperature can be studied. Typically, effects of insulated production tubing or use of insulating gel can be considered.

A model for prediction of the flowing well temperature has been proposed by Alves, Alhanati and Shoham (2), however without explaining how the heat transfer coefficient between the wellbore and its surroundings should be established. Sagar, Doty and Schmidt (3) present a model for the calculation of wellbore temperature profiles, but their work does not include a description of the heat transfer in the vertical direction (parallel to the wellbore). Beck and Shen (4) present results of a transient analysis of the temperature distribution in a flowing well, concluding that failure to include effects of heat transfer in the vertical direction may lead to errors. The model presented in our paper includes the effect of vertical heat transfer. Durrant and Thambynayagam (5) present a thorough investigation of heat transfer from wells used for injection of steam, including time dependent heat transfer across the formation in both vertical and radial direction. Their calculation model is based on the assumption that the formation may be divided into several vertical layers separated by isothermal boundaries. This assumption is not made in our steady-state model.

3 THEORY AND MODELING

3.1 Heat transfer from the well fluid to its surroundings

If the wellbore is considered to be a pipe with inner diameter D [m], the local heat transfer rate $\dot{q}'(x)$ [W/m] between the flowing fluid and the wellbore's surroundings per unit wellbore length can be expressed as

$$\dot{q}'(x) = \pi \cdot D \cdot U(x) \cdot (T_{wf} - T_a) \qquad [1]$$

where T_{wf} [^{0}C] is the wellbore flowing temperature, T_a [^{0}C] is some ambient temperature and $U(x)$ [W/m^2 ^{0}C] is the *local heat transfer coefficient*, which varies with position x [m] along the wellbore's axis. In the following discussion emphasis will be lent to the description of a method to obtain $U(x)$ for a well.

Figure 1 shows how the heat transfer from a vertical well fluid to its surroundings is assumed to take place for a vertical well.

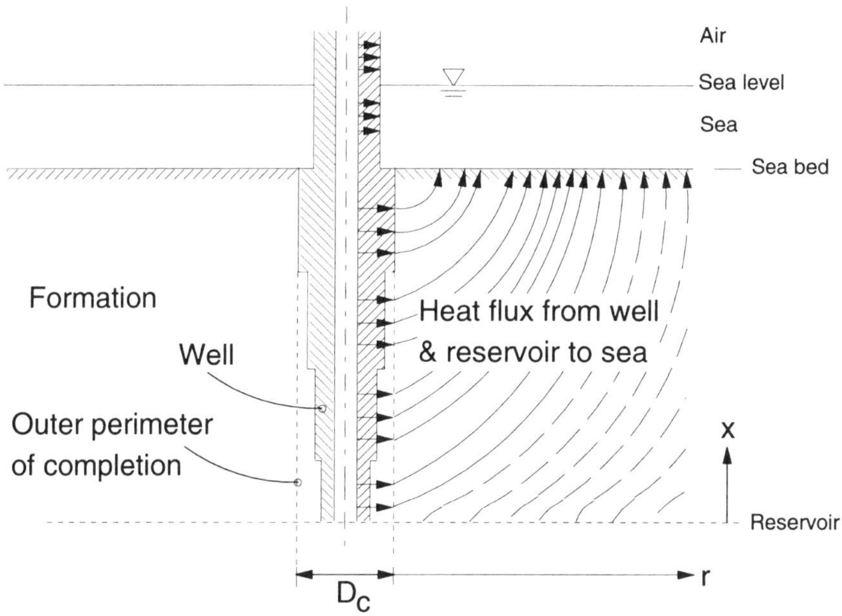

Figure 3.1: Heat transfer from the well to the surroundings

The arrows in Figure 1 illustrate heat flux lines. Because the reservoir temperature is higher than the temperature of the sea, there will always be a natural heat flux from the reservoir to the ocean above. Assuming that the thermal properties of the formation are homogeneous and isotropic, the direction of heat transfer will be parallel to the vertical coordinate. When a vertical well is producing, it acts as a cylindrical heat source with a temperature at or below the reservoir temperature. Assuming that the temperature of the well fluid is always higher than that of its immediate surroundings, a heat flux from the well through the formation to the ocean is established. The solid lines in Figure 1 indicate how the heat flux from the well to the ocean takes place, displacing the heat flux lines from the reservoir, indicated with dashed lines in Figure 1.

The configuration of the different layers of steel, cement, completion fluid, etc. constituting the well's completion will vary along the wellbore, as indicated in Figure 1. Defining now the outer perimeter of the completion to be the surface of a cylinder with a diameter D_c [m] equal to the maximum diameter of the completion, the local heat transfer rate between the flowing fluid and the wellbore's surroundings per unit wellbore length can be written as

$$\dot{q}'(x) = \pi \cdot D \cdot U_c(x) \cdot (T_{wf}(x) - T_c(x)) = \pi \cdot D \cdot U_f(x) \cdot (T_c(x) - T_s) \qquad [2]$$

where

- $U_c(x)$ [W/m^2 ^{0}C] is the local coefficient for heat transfer between the well fluid and the outer perimeter of the completion
- $T_c(x)$ [^{0}C] is the local surface temperature of the outer perimeter of the completion
- $U_f(x)$ [W/m^2 ^{0}C] is the local coefficient for heat transfer between the outer perimeter of the completion and the ocean
- T_s [^{0}C] is the sea bed temperature

If U_c and U_f are known, Equation 1 may be written as

$$\dot{q}' = \pi \cdot D \cdot U_w(x) \cdot (T_{wf}(x) - T_s) \qquad [3a]$$

where the total coefficient of heat transfer between the well fluid and the ocean is

$$U_w(x) = \left[\frac{1}{U_c(x)} + \frac{1}{U_f(x)}\right]^{-1} \qquad [3b]$$

3.2 Heat transfer across the well's completion

The steady state heat transfer across the well's completion is expressed in terms of the parameter U_c. If the heat transfer across the completion is assumed to be strictly radial, one may write

$$U_c(x) = \left[\frac{1}{h_i} + \left(\sum_{j=1}^{n-1} \frac{r_{j-1}}{k_j} \cdot ln\frac{r_j}{r_{j-1}}\right) + \frac{D_c}{2 \cdot r_n} \cdot \frac{1}{h_o}\right]^{-1} \qquad [4]$$

in a standard manner. In Equation 4,

- h_i [W/m^2 ^{0}C] is the convection coefficient for heat transfer between the well fluid and the internal wall of the wellbore
- n [-] is the number of concentric layers constituting the well's completion
- r_j [m] is the internal radius of layer j
- k_j [W/m ^{0}C]] is the thermal conductivity of layer j
- h_o [W/m^2 ^{0}C] is the convection coefficient for heat transfer between outer perimeter of the completion and its surroundings
- D_c [m] is the diameter of the outer perimeter of the completion.

For the riser, which is surrounded by sea or air, h_o will have a finite value, (1000 W/m^2 ^{0}C for sea and 30 W/m^2 ^{0}C for air is used in the present work) whereas h_o will assume a value of infinity for the part of the well surrounded by formation, assuming the heat transfer between the completion and the formation takes place purely by conduction. A value of 2000 W/m^2 ^{0}C is used throughout for h_i.

It is not uncommon for several of the concentric, vertical annuli of the completion to be filled with a completion fluid, which can be a salt-water mixture. The difference in temperature between the outside of the completion and the well fluid represents a driving force for natural convection flow in an annulus, moving warmer fluid upwards along the inside wall of the annulus, and colder fluid downwards along the outside wall of the annulus. This convection will improve the heat transfer across these layers. As a result, heat transfer coefficients U_c calculated assuming conduction only across the completion fluid layers is likely to be too low. Correlations for free convection heat transfer across vertical annuli has been established by Nagendra, Tirunarayanan and Ramachandran (6) based on experiments with wires and long cylinders in annuli. The work of Nagendra et al. covers laminar flow only, and it is assumed that the annular gap is sufficiently wide to avoid the boundary layers on the walls to meet. Their work is based on small scale experiments (typically 10" long cylinders with 0.72 - 0.315" gaps). Laminar free convection from the outer surface of a vertical cylinder has been analyzed both by Le Fevre and Ede (7) and by Minkowycz and Sparrow (8), whereas the free convection heat transfer across vertical plane layers has been studied experimentally and analytically by Emery and Chu (9). However, the literature does, to our knowledge, not offer any description of free convection in vertical annuli that covers geometries and Rayleigh and Grashof number ranges that apply to the vertical well annulus problem. Further, it would have to be justified that the boundary layer thickness assumption made by Nagendra et al. is valid for our problem. In order to not disregard this important effect completely, the thermal conductivity of the completion fluid has throughout been taken as approximately three times that of water, and has been given a value of 2 W/m^{0}C. The annulus natural convection problem is left for later investigation.

3.3 Heat transfer across the formation

The steady state heat transfer across the formation may be defined as a problem in the coordinates x and r in the cylindrical coordinate system, where r is the radius measured from the center of a vertical well, and x is the vertical coordinate measured from the bottom hole. The following boundary conditions are defined:

- $r = \frac{D_c}{2}$ $\quad T = T_c(x)$ [5a]
- $r \rightarrow \infty$ $\quad T = T_g(x)$ [5b]
- $x = 0$ $\quad T = T_r$ [5c]
- $x = H$ $\quad T = T_s$ [5d]

In Equation 5c, T_r is the reservoir temperature and H is the elevation difference between the reservoir and the sea bed. Simplifying the problem by assuming one stand-alone well only, the temperature at some distance away from the well - infinitely far away in theory - will approach the geothermal temperature $T_g(x)$ which can be written

$$T_g(x) = T_r + \frac{T_r - T_s}{H} \cdot x \qquad [6]$$

Equation 6 is correct if the reservoir and sea bed temperatures are constants in r, the formation is homogeneous and isotropic, and if the heat transfer between the reservoir and the sea takes place purely by conduction. These assumptions are not likely to be valid for all formations, but are however necessary to obtain an analytic solution to the problem of heat transfer problem across the formation. One of the most important parameters for the model presented, is the conductivity of the formation. Available measurements indicate that the formation conductivity varies from 2 to 4 W/m ^{0}C. Wells can be surrounded by a porous, liquid filled formation, and heat transfer across this formation can set up natural convection currents in this liquid, which in turn increases the rate of heat transfer. The high value of 4 W/m ^{0}C has been chosen as an attempt to quantitatively account for this effect. A more thorough analysis of this problem would not only have to include more detailed information on the properties of the formation, but would probably also have to include a thorough analysis of convection in porous media.

The formation's steady state temperature field $T(r,x)$ may then be determined by solving the Laplace equation

$$\nabla^2 T = 0 \Rightarrow \frac{\partial^2 T}{\partial r^2} + \frac{1}{r} \cdot \frac{\partial T}{\partial r} + \frac{\partial^2 T}{\partial x^2} = 0 \qquad [7]$$

for a cylindrical coordinate system subject to the boundary conditions 5a - 5d. A solution $T(r,x)$, $r \geq \frac{D_c}{2}$, to Equation 7 is given in Appendix A.

Assuming at this point that $T_c(x)$ is known, one may write

$$U_f(x) = -k_f \cdot \frac{\partial T(r,x)}{\partial r} \cdot \frac{1}{T_c - T_s} \qquad [8]$$

where $\frac{\partial T}{\partial r}$ is evaluated at $r = \frac{D_c}{2}$. Equation 2 may be rewritten as

$$T_c = \frac{U_f \cdot T_s + U_c \cdot T_{wf}}{U_f + U_c} \qquad [9]$$

Equations 3b, 4, 8 and 9 form the basis of solution of the temperature field of the formation. As an example, Equation 7 is solved for a 2700 m long vertical well where T_c varies linearly from a bottom hole value of 105 ^{0}C to a sea bed value of 5 ^{0}C. The reservoir temperature T_r is 105 ^{0}C, and the sea bed temperature T_s is 5 ^{0}C. Figure 3.2 shows calculated formation isotherms for this example. Radial distance from the completion's perimeter in meters is given on the abscissa, whereas depth in meters relative to the sea bed is given on the ordinate.

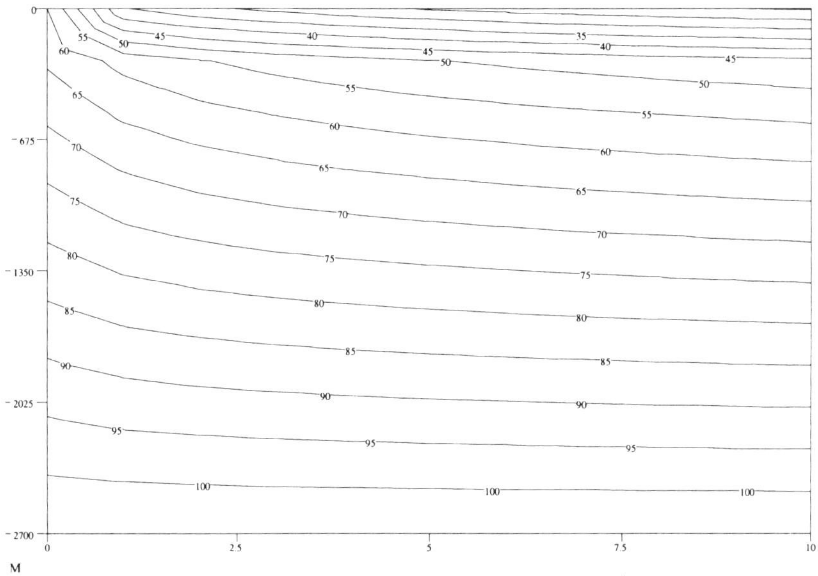

Figure 3.2: Calculated isotherms

Remembering that heat flux lines will be parallel to the isotherms, it can be seen from the Figure that the heat flux from the well must be directed towards the sea bed.

3.4 Solution

Based on the described model, the temperature profile of a vertical well can be calculated, provided the following is available:

- Data on completion geometry, materials and lithology
- Bottom hole temperature and pressure
- Production rate
- Environmental data (the sea temperature T_s, which is the boundary condition for the calculation of heat transfer between the production tubing and the sea)
- A steady state pipeline simulation tool capable of predicting temperature profiles. This tool should be able to calculate the coupled pressure and temperature changes along the well from the reservoir to the wellhead, subject to frictional, accelerational and gravitational losses, viscous dissipation in the fluid and heat exchange with the surroundings. Clearly, the thermodynamic and transport properties of the well fluid must be well known as a function of pressure and temperature.

A solution algorithm with reference to the relevant equations is given in Appendix B. Our experience indicates that only two to three iterations are needed to achieve sufficient convergence in the calculated temperature profiles and heat transfer coefficients when using this algorithm. The results presented in this paper are obtained using Norsk Hydro's in-house process simulation tool.

Many wells are drilled with a deviation relative to a vertical line through the wellhead. Due to this deviation, there is a difference between the True Vertical Depth (TVD) of the well, and the Measured Depth (MD) which is the distance along the well's axis. Typically, a well can have MD = 3800 m and TVD = 2800 m.

Due to the deviation, the temperature field surrounding the well can not be expected to be symmetrical with respect to a vertical axis through the wellhead, as will be the case for a purely vertical well. We simplify the problem and apply our model to a deviating well by the following procedure:

- The overall heat transfer coefficient U_w is calculated for a purely vertical well with a TVD corresponding to the real well.
- The heat transfer coefficients calculated based on the vertical well are given as a function of TVD, and are transferred to the corresponding TVD-values of the actual well. For simulation purposes, the real well deviation is used, implying both a correct TVD and MD.

This approach is of course a shortcut compared to a complicated model of a true, deviating well. However, it is assumed to be a both reasonable and sensible approach, due to the following:

- A detailed model of heat transfer from a deviating well satisfying the Laplace equation would not lend itself easily to analytical methods, and would probably require a finite element computational method or similar.
- For typical North Sea wells, the first 400 - 600 m below sea bed are vertical or near vertical. Since the heat transfer rate per unit wellbore length between the well and the sea is believed to be most significant close to the sea bed (due to limited "insulating" formation material between well and sea), the chosen method is believed to be sufficiently accurate.

4 FIELD DATA ACQUISITION

It is a challenge to acquire good field data for a production well, suitable for validation of a simulation model. This chapter illustrates the problems to be faced when searching for reliable test data.

One of the objectives when gathering field data in order to verify the new temperature calculation model was to have high confidence data. Prior to selection, the wells were submitted to the following selection criteria:

- Accurate temperature and pressure measurements upstream- and downstream of the well choke had to be available
- All test wells should produce towards a test separator
- The test separator had to be equipped with accurate flow measurements for gas, oil and water
- Accurate information was required for either reservoir or bottom hole pressure and temperature
- A representative fluid characterization of the wellstream was required
- A production test of the selected wells including data for at least three different flow rates had to be available
- The test period at each flow rate had to be sufficiently long to reach steady state conditions before any adjustment of the production choke
- The well fluid composition was required to remain constant during the test period.

It was also considered necessary to select wells with different parameters, such as:

- Gas to oil-ratio (GOR)
- Water cut at standard conditions
- Water depth
- Reservoir conditions (pressure and temperature)
- Well length and completion

Although data from a very large number of wells was available with a variety of well characteristics, it was extremely difficult to find well candidates satisfying all the above criteria. The main reasons were:

- Many wells include subsea wellheads with poor temperature data measurements
- Some installations are producing off plateau, meaning that an extended well test to reach steady state condition at less than maximum rate would have an economic impact
- Some installations had poor flow data from the test separator for one of the three outgoing phases
- For some installations, the temperature indicators were situated within the pipe walls or even outside the piping and did not represent the actual bulk wellstream temperature
- When collecting data from wells with high GOR and watercut due to gas or water breakthrough, the compositions changed during the well test when adjusting the choke, making it difficult or impossible to reach steady state conditions.

Figure 4.1 below summarizes the production data required to get adequate test data:

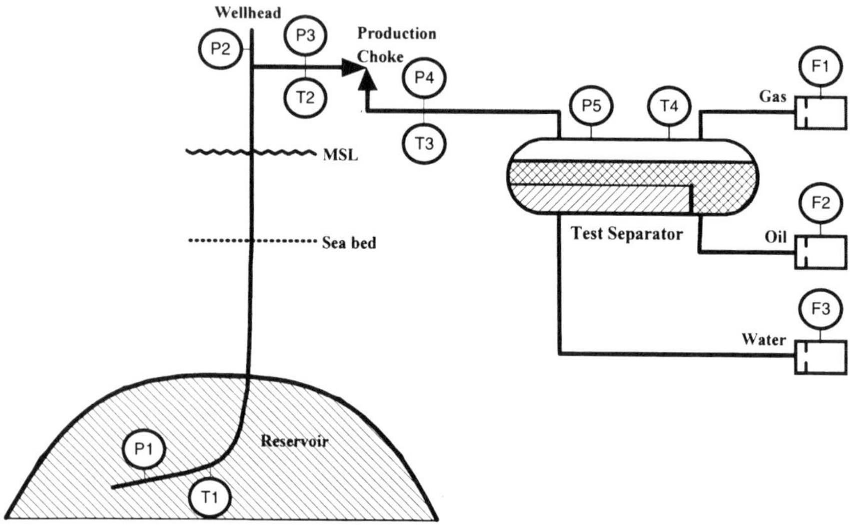

Figure 4.1: Well production data to be recorded during a well test (if available)

For a well with stable composition (i.e. no water or gas breakthrough), the upstream choke pressure together with an allocation curve is used to estimate the well flow rate. However this flow rate is very often related to the volumes from a well passing through the export oil or gas metering station and is not related to a single stage flash of the wellstream.

One of the most important features when collecting field data from a well is to have a good determination of the wellstream flow rates. Most of the commercial simulation tools used for such flow calculations require flow rates in volume per time (hour or day) as input. However the gas, oil or water volumes have to be coupled to actual flow conditions (i. e. temperature and pressure) to be uniquely defined. Since the test separator is the only place where actual volumes are coupled to pressure and temperature conditions, all test wells have been simulated including the test separator. Hence a realistic wellstream composition is obtained and the total wellstream production rate is given in kg/s, which is a less confusing unit than m^3/hour or m^3/day.

Among the available wells, test data from five wells were collected, four oil producers and one gas producer. Main wellstream characteristics and flow rates are given in Table 4.1. Test data were averaged over ten minute intervals were recorded automatically for well 1, 2, 3 and 4. However, the upstream choke temperature was only recorded manually by reading a local transmitter for wells 3 and 4.

Well 5 was selected since it is a gas producer. All data for this well were manually recorded. The upstream choke temperature was recorded by a temporary digital temperature element taped to the piping and insulated. Hence, the test data from this well are somewhat less reliable than data from the other wells.

Table 4.1: Overview of test well characteristics

Well no.	GOR [Sm^3/Sm^3]	Water cut [Vol%]	Production Rates [kg/s]	[Sm3/d] [1)]
1	155	1	69.3	5850 [1)]
			52.1	4400 [1)]
			17.8	1505 [1)]
2	162	1	63.8	5356 [1)]
			45.5	3816 [1)]
			14.3	1200 [1)]
3	312	10	27.5	1890 [1)]
4	186	6	63.2	4946 [1)]
			31.1	2430 [1)]
5	4723	1	41.5	3040000 [2)]
			29.2	2141000 [2)]
			22.8	1670000 [2)]

Notes: *1) Oil flow rate in a single stage flash of wellstream to standard conditions*
2) Gas flow rate in a single stage flash of wellstream to standard conditions

As explained in the beginning of this chapter, one of the objectives of this data acquisition was to get test data from each well including at least three different choke openings. However, the compositions changed during the test for wells 3 and 4. For this reason the wells did not reach steady state, and for some choke openings test data have been rejected for these two wells.

It should be noted that the five selected wells are located on three different offshore platforms each having between 20 and 40 production wells. Hence, thermal effects of neighboring wells probably affect the collected data and this, although complicated to model, should be kept in mind when viewing the results.

5 RESULTS

The five wells presented in Chapter 4 were modeled as described in Chapter 3, based on available information on completion, fluid properties and production rates, and wellbore flowing temperatures were calculated accordingly.

5.1 Well 1

Figure 5.1 shows the calculated values of the completion heat transfer coefficient U_c, the formation heat transfer coefficient U_f and the total heat transfer coefficient U_w as a function of elevation relative to the wellhead (zero elevation), using a sea water temperature of 5 °C as boundary condition T_s. All heat transfer coefficients refer to the inner diameter of the production tubing, and results are included up to an elevation of -151 m, which represents the sea bed. U_w is provided all the way up to the wellhead, and it is noted that $U_c = U_w$ for the riser.

Observing the differences in magnitude between U_c and U_f, it is clear that the resistance to heat transfer through the formation is dominating the total heat transfer coefficient U_w, by virtue of Equation 3b. U_f increases almost linearly with elevation at first, whereas the increase is more progressive for the upper 500 m of the well, close to the sea bed.

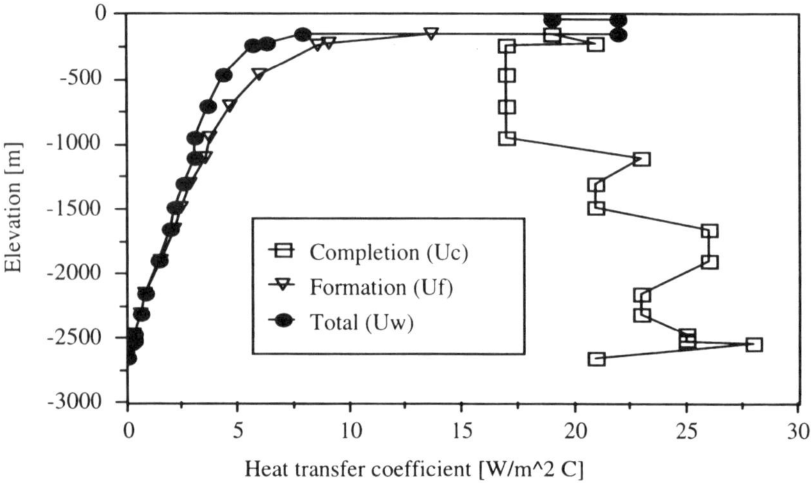

Figure 5.1: Calculated heat transfer coefficients for well 1

The calculated wellbore flowing temperatures for well 1 are given in Figure 5.2 for three production rates, together with the well trajectory. It can be observed from this Figure how much of the temperature drop that seems to take place in the upper part of the well, where the heat transfer coefficient is at is highest, both in the riser and in the first 500 m below the sea bed. This justifies the assumption made previously that the heat transfer rate per unit wellbore length between the well and the sea is most significant close to the sea bed.

Table 5.1 compares measured and predicted wellhead temperatures, and it is seen that the wellhead temperature is overpredicted by 4 - 5 ^{0}C for this well.

Table 5.1: Calculated and measured wellhead temperatures for well 1

Production Rates		Temperatures [°C]			
			Wellhead		
	[Sm^3/d]	Reservoir	Calculated	Measured	Difference
Low	**1505**		90	86	+ 4
Medium	**4400**	107	96	91	+ 5
High	**5850**		96	91	+ 5

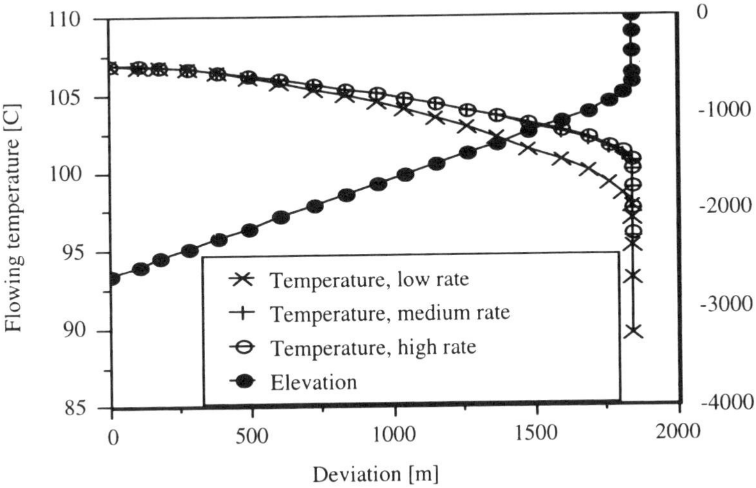

Figure 5.2: Calculated wellbore flowing temperatures for well 1

5.2 Well 2

Figure 5.3 shows calculated heat transfer coefficients for well 2. As for well 1, the characteristic progressive increase in heat transfer coefficient in the upper part of the well can be seen, combined with the observation that the total heat transfer coefficient is dominated by the value for the formation, U_f . The calculated wellbore flowing temperatures for this well are shown in Figure 5.4. Measured and predicted wellhead temperatures are given in Table 5.2. The wellhead temperatures are overpredicted by 2 ^{0}C for the two highest rates, whereas there is an underprediction of 4 ^{0}C for the lower rate.

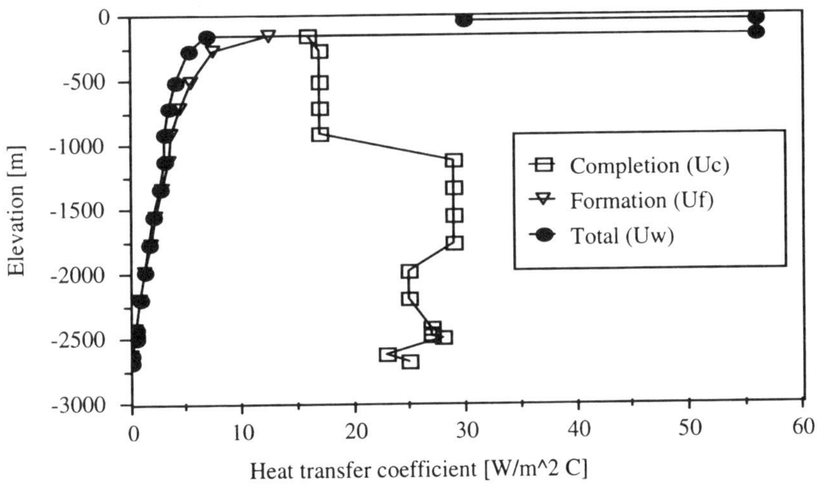

Figure 5.3: Calculated heat transfer coefficients for well 2

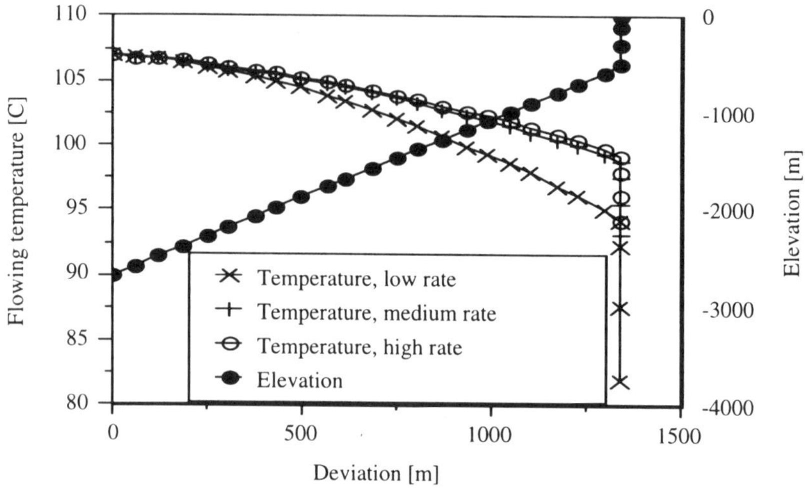

Figure 5.4: Calculated wellbore flowing temperatures for well 2

Table 5.2: Calculated and measured wellhead temperatures for well 2

Production Rates		Temperatures [°C]			
			Wellhead		
	[Sm³/d]	Reservoir	Calculated	Measured	Difference
Low	**1200**		83	86	- 3
Medium	**3816**	107	93	91	+ 2
High	**5356**		94	92	+ 2

5.3 Well 3

The Figures 5.5 and 5.6 show calculated heat transfer coefficients and wellbore flowing temperature, respectively, for well 3. Again, the same observations can be done regarding heat transfer coefficients and temperature drop along the wellbore. Only one temperature measurement was available for this well, and the wellhead temperature is overpredicted by 3 ^{0}C compared to the measurements.

Table 5.3 Calculated and measured wellhead temperatures for well 3

Production Rate		Temperatures [°C]			
			Wellhead		
	[Sm³/d]	Reservoir	Calculated	Measured	Difference
Medium	**1890**	107	90	87	+ 3

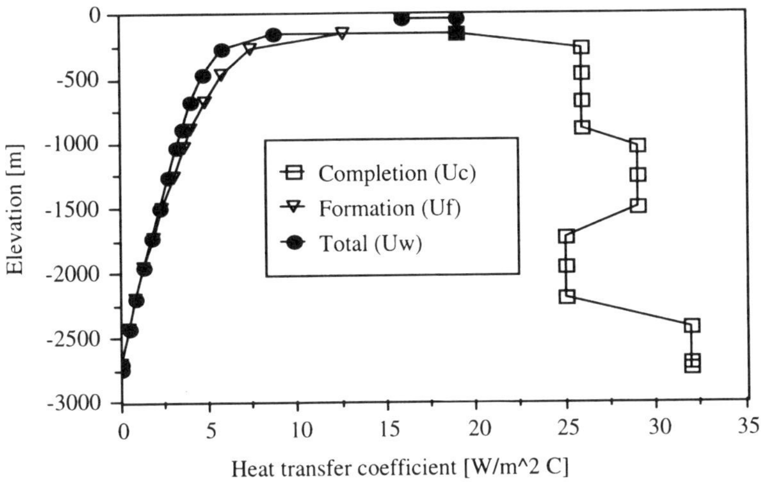

Figure 5.5: Calculated heat transfer coefficients for well 3

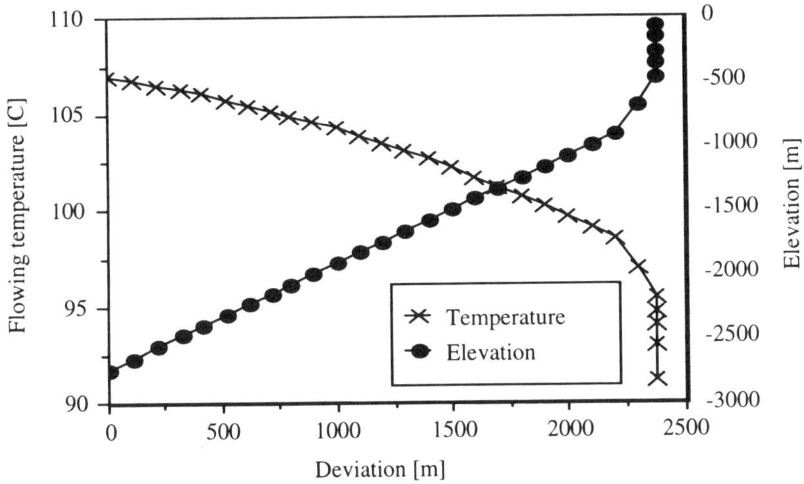

Figure 5.6: Calculated wellbore flowing temperature for well 3

5.4 Well 4

Heat transfer coefficients and calculated wellbore flowing temperatures for well 4 are shown in Figures 5.7 and 5.8, and Table 5.4 compares measured and calculated wellhead temperatures.

Table 5.4 Calculated and measured wellhead temperatures for well 4

Production Rates		Temperatures [°C]			
			Wellhead		
	[Sm³/d]	Reservoir	Calculated	Measured	Difference
Low	**1200**	107	93	91	+ 2
High	**5356**		96	93	+ 3

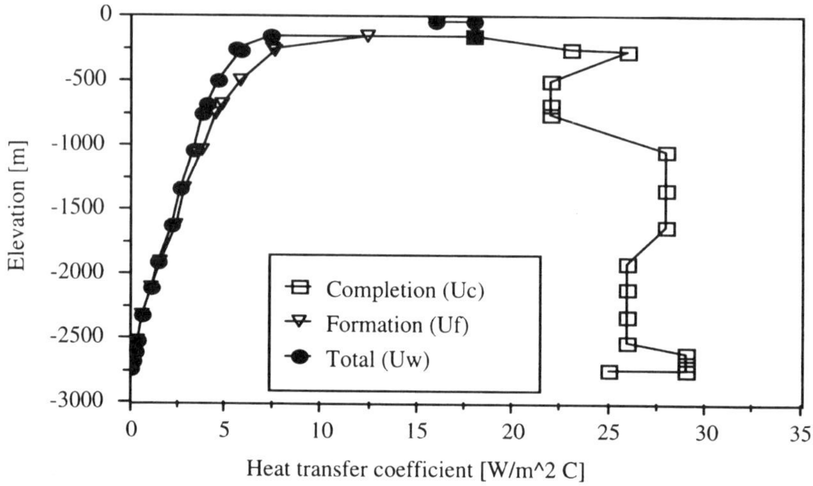

Figure 5.7: Calculated heat transfer coefficients for well 4

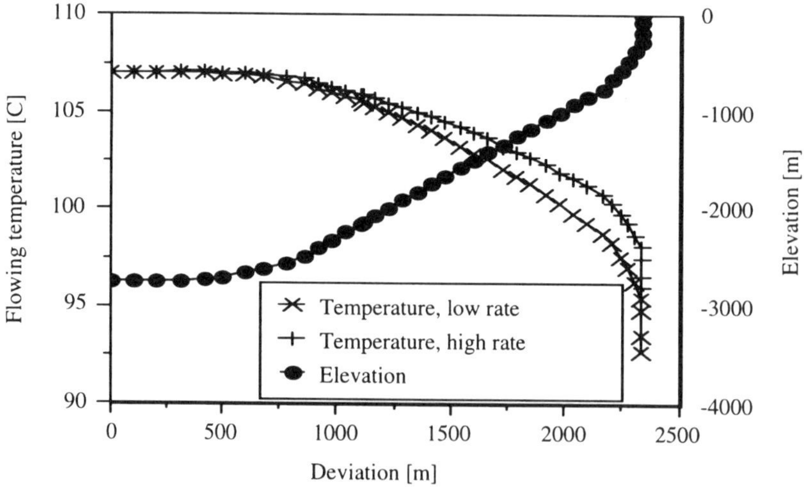

Figure 5.8: Calculated wellbore flowing temperatures for well 4

5.5 Well 5

Heat transfer coefficients and calculated wellbore flowing temperatures for well 5 are given in Figures 5.9 and 5.10, and Table 5.5 compares measured and calculated wellhead temperatures.

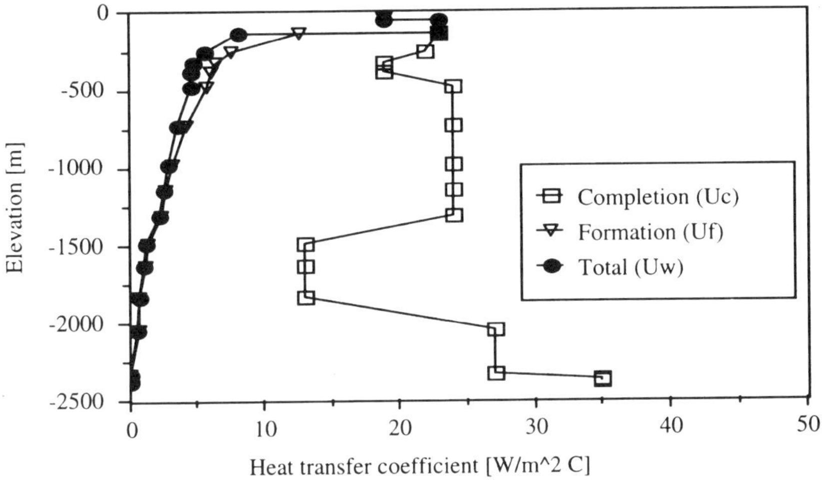

Figure 5.9: Calculated heat transfer coefficients for well 5

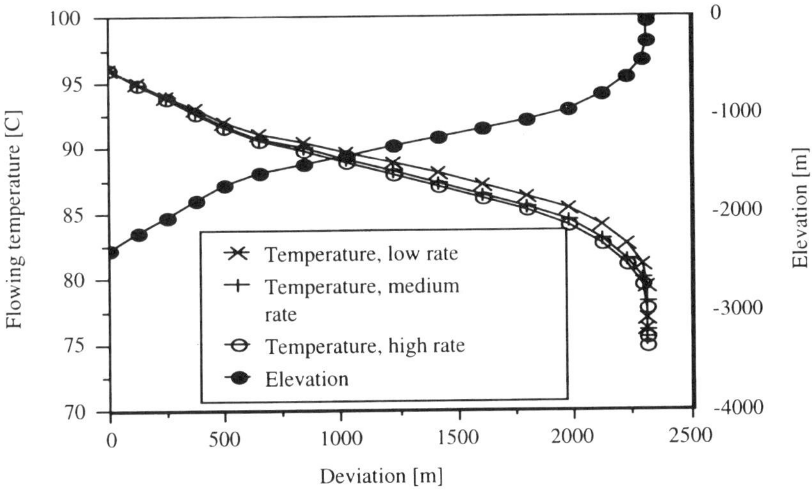

Figure 5.10: Calculated wellbore flowing temperatures for well 5

Table 5.5 Calculated and measured wellhead temperatures for well 5

Production Rates		Temperatures [°C]			
			Wellhead		
	[10^3 Sm^3/d]	Reservoir	Calculated	Measured	Difference
Low	**1670**		76	71	+ 5
Medium	**2837**	96	75	73	+ 3
High	**3040**		75	73	+ 2

6 DISCUSSION

When a well starts to produce, the heat transfer is likely to be mainly radial, with flux lines parallel to the horizontal plane for a vertical well. However, once steady state is reached, the vertical component of the heat transfer becomes important. Steady state is approached when the formation is warmed up by the well, and when this condition is reached, heat transfer will be from the well via the formation to the ocean, which acts as a heat sink. Differing from traditional well temperature calculation models, this effect is included in the current model.

The temperature model has been compared to production data for a gas well, although the test data for this well are less reliable than for the others. The calculations show that for this well, the most important cause of temperature reduction from the reservoir to the wellhead is the Joule-Thomson effect and change in potential energy. A pressure reduction from 225 bara at reservoir conditions (pressure and depth) to 135 bara upstream the production choke gave a calculated adiabatic (calculated with zero heat transfer coefficient) temperature reduction at maximum flow rate of 18 ^{0}C, compared to a total temperature reduction of ~20 ^{0}C (21 ^{0}C calculated and 23 ^{0}C measured). It is interesting to notice that for this well both the measured and calculated temperature falls at high production rates when increasing the flow rate, since the Joule-Thomson effect dominates over the effect of the reduced heat loss to the surroundings.

Comparison between the presented temperature calculation model and measurements from all five wells show that the calculated temperatures are within - 3 to +5 ^{0}C of the measured data. It is difficult to explain whether these discrepancies are due to shortcomings in the presented model or errors in the measured data.

If focus is put on the measured data, the most likely sources of error summarize as:

- Errors in temperature (especially) or pressure measurements
- Wrong fluid characterization and PVT data
- Errors in determination of the wellstream composition/rate either due to poor instrumentation or due to liquid entrainment in test separator gas, water entrainment in test separator oil or oil entrainment in test separator water
- A constant tubing/casing/conductor conductivity of 50 W/m ^{0}C has been selected. The conductivity could very well be in the range 20 - 60 W/m ^{0}C

- Inaccuracies in the well completion diagram with respect to steel, completion fluid, and cement layers
- Uncertainties in values selected for formation, completion fluid and cement thermal conductivity
- For each production rate the choke opening remained constant until steady state conditions were reached. However, it is questionable whether 4 - 5 hours at each choke opening is sufficient for the temperature to stabilize. It is more probable that small temperature variations were suppressed by wrong calibration of the temperature transmitters and their signals

The model itself has room for improvement due to shortcomings such as:

- Natural convection in the annuli has not been accurately modeled.
- The extent of the annulus convection could possibly be influenced by neighboring wells transporting hot well fluid.
- The model is based on a purely vertical well, whereas a real well will have a deviation, leading to a much more complicated solution of the Laplace equation.

The proposed model matches the acquired data sufficiently to be used for process design or to establish operating philosophies to avoid undesired problems related to fluid effects. Due to the complexity of such calculations and the many sources for possible errors in the measured data, it will require a substantial effort to calculate the wellhead temperature of a random production well within an uncertainty of 3 - 4 °C for a large span of rates.

7 CONCLUSIONS

A new well temperature calculation model has been established, based on a solution of the Laplace equation, including the effect of heat transfer from a well and it's completion towards the sea.

The model has been validated by comparison with production data of well tests from five different North Sea wells, all having different well characteristics. The calculation model compares quite well with the production data, even though no adjustments have been done to fit the data. However, it seems to predict somewhat high wellhead temperatures, up to 5 °C above measured data. It is assumed that the sources of discrepancy between the measured data and the model are equally distributed among measurement errors, input data uncertainties and a shortcomings of the presented model.

The new temperature calculation model could be improved if the phenomenon of natural convection in the well's annuli was fully understood. This effect has only been estimated in the present work. Effects of neighboring wells transporting hot well fluid have been neglected.

If the effect of heat transfer from neighbor wells was included, it is believed that this would have an effect on the calculated wellhead temperature, probably improving the match between measured and calculated data.

If the mechanisms affecting the flowing wellhead temperature are understood, this temperature could be manipulated when designing a well. However, due to all uncertainties regarding measurements and physical parameters, it seems difficult to predict the flowing wellhead temperature of a random well at a given rate with perfect match to field data.

Acknowledgment: The work presented was carried out with funding from the field data program of the SSH (Statoil, Saga Petroleum and Norsk Hydro) cooperation.

8 REFERENCES

1 OLGAS 89, released by Scandpower A/S, Kjeller, Norway 1989

2 Alves, I.N., Alhanati, F.J.S. & Shoham, O., A Unified Model for Predicting Flowing Temperature Distribution in Wellbores and Pipelines, *SPE Production Engineering,* November 1992, pp. 363 - 367 (also available as SPE paper 20632)

3 Sagar, R., Doty, D.R. & Schmidt, Z., Predicting Temperature Profiles in a Flowing Well, *SPE Production Engineering,* November 1991, pp.441 - 448 (also available as SPE paper 19702)

4 Beck, A.E. & Shen, P.Y., Temperature distribution in flowing liquid wells, *Geophysics,* Vol. 50, No. 7, July 1985, pp. 1113 - 1118

5 Durrant A.J, Thambynayagam R.K.M, Wellbore Heat Transmission and Pressure Drop for Steam/Water Injection and Geothermal Production: A simple Solution Technique, *SPE Reservoir Engineering,* March 1986, pp.148 - 162

6 Nagendra, H.R., Tirunarayanan, M.A. & Ramachandran, A., *Chemical Engineering Science*, Vol. 25, 1970, pp. 605 - 610

7 Le Fevre, E.J. & Ede, A.J., Laminar free convection from the outer surface of a vertical circular cylinder, *Proceedings, Ninth Int. Congr. Appl. Mech., Brussels, Vol. 4, 1956, pp. 175 - 183*

8 Minkowycz, W.J. & Sparrow, E.M., Local Nonsimilar Solutions for Natural Convection on a Vertical Cylinder, *Journal of Heat Transfer*, May 1974, pp. 178 - 183

9 Emery, A. & Chu, N.C., Heat Transfer Across Vertical Layers, *Journal of Heat Transfer*, February 1965, pp. 110 - 116

10 Özisik, M.N., *Heat Conduction*, New York: John Wiley and Sons, Inc., 1980

11 Kreyszig, E., *Advanced Engineering Mathematics*, 6th ed., New York: John Wiley and Sons, Inc., 1988

12 Abramowitz, A. & Stegun, I.A. (ed.), *Handbook of Mathematical Functions*, New York: Dover Publications, Inc., 1965

APPENDIX A SOLUTION OF THE LAPLACE EQUATION

A.1 Simplification of the problem

Problem:

$$\nabla^2 T = 0 \Rightarrow \frac{\partial T}{\partial r^2} + \frac{1}{r} \cdot \frac{\partial T}{\partial r} + \frac{\partial^2 T}{\partial x^2} = 0 \qquad [A.1]$$

Boundary conditions:

- $r = \frac{D_c}{2}$ $\quad T = T_c(x)$ $\quad$ [A.2a]
- $r \rightarrow \infty$ $\quad T = T_g(x)$ $\quad$ [A.2b]
- $x = 0$ $\quad T = T_r$ $\quad$ [A.2c]
- $x = H$ $\quad T = T_s$ $\quad$ [A.2d]

A partial differential equation such as A.1 can be solved by the method of separating variables, (10, 11), provided the problem has only one inhomogeneous boundary condition (10). The above problem has inhomogeneous boundary conditions at $r = \frac{D_c}{2}$ and as $r \rightarrow \infty$. Consider the temperature fields defined as

$$\Delta T(x) = \left(1 - \frac{x}{H}\right) \cdot (T_r - T_s) - T_s = T_s \cdot \frac{x}{H} + T_r \cdot \frac{H-x}{H} \qquad [A.3]$$

and

$$\Theta(r,x) = T(r,x) - \Delta T(x) \qquad [A.4]$$

The Laplace equation A.1 is both linear and homogeneous, and may be written

$$\nabla^2 T = \nabla^2(\Theta + \Delta T) = \nabla^2\Theta + \nabla^2\Delta T = 0 \qquad [A.5]$$

The linearity means that solutions (temperature fields) found by solving

$$\nabla^2\Theta = 0 \qquad [A.6]$$

and

$$\nabla^2 T = 0 \qquad [A.7]$$

may be added to yield a solution to A.1. The solution A.3 to A.7 is already known. By subtracting A.3 from the boundary conditions A.2, the problem of solving A.1 is then reduced to solving A.7 with the boundary conditions:

- $r = \frac{D_c}{2}$ $\quad \Theta = \Theta_w(x) = T_c(x) - T_s \ldots$
  $\ldots - (1 - \frac{x}{H}) \cdot (T_r - T_s)$ $\quad$ [A.8a]

- $r \rightarrow \infty$ $\Theta = 0$ [A.8b]
- $x = 0$ $\Theta = 0$ [A.8c]
- $x = H$ $\Theta = 0$ [A.8d]

In this new problem, the only inhomogeneous boundary condition is A.8a.

A.2 Solution to Equation A.6

By the method of separating variables, we are looking for a solution of the form (11)

$$\Theta(r,x) = R(r) \cdot X(x) \qquad [A.9]$$

By substituting A.9 into A.6 we get

$$\frac{1}{R} \cdot \frac{d^2R}{dr^2} + \frac{1}{r \cdot R} \cdot \frac{dR}{dr} + \frac{1}{x} \cdot \frac{d^2X}{dx} = 0 \qquad [A.10]$$

and can write

$$\frac{d^2R}{dr^2} + \frac{1}{r} \cdot \frac{dR}{dr} - \eta^2 \cdot R = 0 \qquad [A.11]$$

$$\frac{d^2X}{dx^2} + \eta^2 \cdot x = 0 \qquad [A.12]$$

where η^2 is a non-negative constant (choosing $-\eta^2$ as a constant will not satisfy the boundary conditions). Introducing the substitution $\alpha = r \cdot \eta$, A.11 can be written as

$$\frac{d^2R}{d\alpha^2} + \frac{1}{\alpha} \cdot \frac{dR}{d\alpha} - \alpha^2 \cdot R = 0 \qquad [A.13]$$

Equation A.13 is the *modified Bessel equation.* Solutions to A.13 and A.12 are, (11)

$$R(r \cdot \eta) = C_1 \cdot I_0(r \cdot \eta) + C_2 \cdot K_0(r \cdot \eta) \qquad [A.14]$$

$$X(x \cdot \eta) = C_3 \cdot \sin(x \cdot \eta) + C_4 \cdot \cos(x \cdot \eta) \qquad [A.15]$$

I_0 is the *modified Bessel function of the first kind of order 0*, whereas K_0 is the *modified Bessel function of the second kind of order 0*. C_1, C_2, C_3 and C_4 are constants. Tabulated values for I_0 and K_0 are readily available (12). With the boundary conditions A.8b - A8.d, a solution to A.9 is

$$\Theta(r,x) = \sum_{\lambda=1}^{\infty} A_\lambda \cdot \sin\left(\frac{\lambda \cdot \pi \cdot x}{H}\right) \cdot K_0\left(\frac{\lambda \cdot \pi \cdot r}{H}\right) \qquad [A.16]$$

The only unknown in Equation A.16 is the constant $A_\lambda = C_2 \cdot C_3$, and the only boundary condition not used thus far is A.8a. At $r = \frac{D_c}{2}$ the solution to A.16 must be

$$\Theta_c(x) = T_c(x) - T_s - (1 - \tfrac{x}{H}) \cdot (T - T_s) = \sum_{\lambda=1}^{\infty} A_\eta \cdot \sin(\tfrac{\lambda \cdot \pi \cdot x}{H}) \cdot K_0(\tfrac{\lambda \cdot \pi \cdot D_c}{2 \cdot H}) \qquad \text{[A.17]}$$

The *orthogonality property* of the sine function implies that

$$\int_0^H \sin(\lambda \cdot \tfrac{\pi \cdot x}{H}) \cdot \cos(m \cdot \tfrac{\pi \cdot x}{H})\, dx = \left\{ \begin{array}{ll} 0 & if\, m \neq \lambda \\ \frac{H}{2} & if\, m = \lambda \end{array} \right\} \qquad \text{[A.18]}$$

Combining A.17 and A.18 and integrating both sides then gives

$$A_m = \tfrac{2}{m \cdot \pi} \cdot \left[K_0(m \cdot \tfrac{\pi \cdot D_c}{2 \cdot H}) \right]^{-1} \cdot ((T_r - T_s) \cdot (\tfrac{\sin(m \cdot \pi)}{m \cdot \pi} - 1) \ldots$$

$$\ldots + T_s \cdot (\cos(m \cdot \pi) - 1) + \int_0^H T_c(x) \cdot \sin(m \cdot \tfrac{\pi \cdot x}{H})\, dx) \qquad \text{[A.19]}$$

The x axis is partitioned into n segments, not necessarily of equal length, and the boundary condition A.8a is discretisized by writing

$$T_c(x) \approx T_c^j + \tfrac{T_c^{j+1} - T_c^j}{H^{j+1} - H^j} \cdot (x - H^j) \qquad H^j \leq x \leq H^{j+1} \qquad \text{[A.20]}$$

where $H^1 = 0$, $H^{n+1} = \text{H}$, and $1 \leq j \leq n+1$. At the position of a node with height H^j above H^1, the corresponding temperature is T_c^j. With A.20 substituted into A.19, combine A.19 with A.16 writing $\lambda = m$ to obtain $\Theta(r, x)$. This solution together with A.3 and A.4 gives the solution to the temperature field outside the completion for $r \geq \frac{D_c}{2}$ as

$$T(r, x) = \left[\sum_{m=1}^{\infty} \tfrac{2}{m \cdot \pi} \cdot \left[K_0(\tfrac{m \cdot \pi \cdot D_c}{2 \cdot H}) \right]^{-1} \cdot C_m \cdot K_0(\tfrac{m \cdot \pi \cdot r}{H}) \cdot \sin(\tfrac{m \cdot \pi \cdot x}{H}) \right] \ldots$$

$$\ldots + T_s \cdot \tfrac{x}{H} + T_r \cdot (\tfrac{H - x}{H}) \qquad \text{[A.21]}$$

where

$$C_m = (T_r - T_s) \cdot (\tfrac{\sin(m \cdot \pi)}{m \cdot \pi} - 1) + T_s \cdot \cos((m \cdot \pi) - 1) + \ldots$$

$$\ldots \left[\tfrac{H}{m \cdot \pi} \cdot \tfrac{T_c^{j+1} - T_c^j}{H^{j+1} - H^j} \cdot (\sin(\tfrac{m \cdot \pi \cdot H^{j+1}}{H}) - \sin(\tfrac{m \cdot \pi \cdot H^j}{H})) + T_c^j \cdot \cos(\tfrac{m \cdot \pi \cdot H^j}{H}) - T_c^{j+1} \cdot \cos(\tfrac{m \cdot \pi \cdot H^{j+1}}{H}) \right]$$

This concludes the solution of the Laplace equation for the temperature field outside the well.

APPENDIX B SOLUTION ALGORITHM

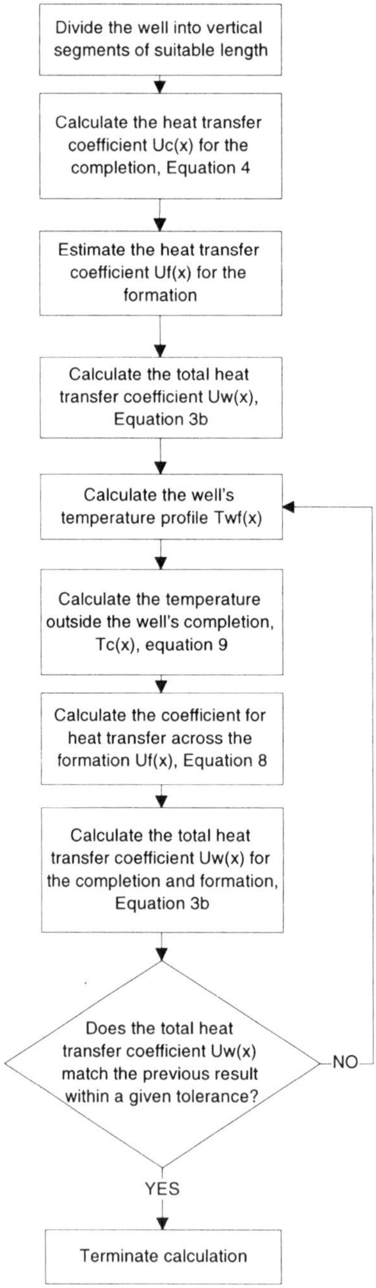

Authors' Index